Advanced Sciences and Technologies for Security Applications

The series Advanced Sciences and Technologies for Security Applications comprises interdisciplinary research covering the theory, foundations and domain-specific topics pertaining to security. Publications within the series are peer-reviewed monographs and edited works in the areas of:

- biological and chemical threat recognition and detection (e.g., biosensors, aerosols, forensics)
- crisis and disaster management
- terrorism
- cyber security and secure information systems (e.g., encryption, optical and photonic systems)
- traditional and non-traditional security
- energy, food and resource security
- economic security and securitization (including associated infrastructures)
- transnational crime
- human security and health security
- social, political and psychological aspects of security
- recognition and identification (e.g., optical imaging, biometrics, authentication and verification)
- smart surveillance systems
- applications of theoretical frameworks and methodologies (e.g., grounded theory, complexity, network sciences, modelling and simulation)

Together, the high-quality contributions to this series provide a cross-disciplinary overview of forefront research endeavours aiming to make the world a safer place.

The editors encourage prospective authors to correspond with them in advance of submitting a manuscript. Submission of manuscripts should be made to the Editor-in-Chief or one of the Editors.

More information about this series at http://www.springer.com/series/5540

Cristina Alcaraz
Editor

Security and Privacy Trends in the Industrial Internet of Things

Editor
Cristina Alcaraz
Computer Science Department
University of Malaga
Malaga, Spain

ISSN 1613-5113 ISSN 2363-9466 (electronic)
Advanced Sciences and Technologies for Security Applications
ISBN 978-3-030-12332-1 ISBN 978-3-030-12330-7 (eBook)
https://doi.org/10.1007/978-3-030-12330-7

This Springer imprint is published by the registered company Springer Nature Switzerland AG.
The registered company address is: Gewerbestrasse 11, 6330 Cham, Switzerland

Preface

We are increasingly witnessing how the industry in general is modernizing its underlying critical systems to move toward the fourth industrial revolution, commonly known as Industry 4.0. This new industrial paradigm encompasses, among others, the Industrial Internet of Things (IIoT) as one of the most relevant technologies of the today's industry. Through IIoT, it is possible to open the industrial connections to address effective and more extensive controls, allowing monitoring from anywhere, at any time, and in anyhow, and in this way to increase the effectiveness and reliability of production states, reduce operational costs, and improve the overall market economy.

Although there exist already consortiums and bodies working on the deployment of this technology, there are also diverse entities (academy, governments, international organizations, and industries) working on many aspects related to security and privacy. Particularly, certain interest issues deserve to be considered in their own right. For example, the hardware and software limitations of the vast majority of the IIoT devices do not help provide complex and robust security approaches; and the current predominance to lead advanced persistent attacks in the diverse industrial sectors brings about numerous security risks. There exists a special attraction to track and exploit zero-day vulnerabilities in order to proceed with potential attacks related to information exfiltration, data manipulation, false data injection, or end users' privacy violation. In addition to this, the incorporation of IIoT-related technologies in Industry 4.0 does not help avoid these types of risks. Cyber-physical systems, cloud/fog computing, big data, digital twins, and the diverse emergent technologies that need to collaborate each other for the convergence IT (information technologies) – OT (operational technologies) certainly add new security and privacy risks that should widely be considered from the security point of view.

Therefore, the present volume highlights all these issues from the beginning, showing the current research challenges and ongoing work lines, with an eye toward keeping the operability of the underlying critical systems and their monitoring infrastructures. Diverse standpoints are addressed, capturing a theoretical analysis of the current situation and the benefits and drawbacks that the IIoT technology itself can bring to the operational processes. Part of these analyses likewise involves the provision of lightweight approaches based on cryptographic algorithms, access control, anomaly detection, intrusion detection methodologies, or remote attestation algorithms. But beyond this, privacy techniques are also addressed in this book to evaluate the impact of the problem and its occurrence in determined critical environments such as smart health ecosystems. In counterpart to the theoretical procedures, practical researches in the IIoT security field are equally keys to demonstrate the validity of the approaches and their applications in critical scenarios. In this case, the design of IIoT-based testbeds and their influence on research procedures undoubtedly constitute a fundamental part to consolidate the new security and privacy trends on IIoT and its real application.

This book can therefore serve as a timely introduction to the state of the art of the technology of IIoT, trying to aid researchers to gain an overview of a field that is still largely unexplored, industries interested in modernizing their infrastructures from a secure perspective, and lecturers wishing to prepare future Industry 4.0 experts with solid criterion and contents.

Malaga, Spain
December 2018

Cristina Alcaraz

Contents

List of Abbreviations

2PAKE	Two-party password authenticated key exchange
6LoWPAN	IPv6 over low-power wireless personal area networks
ABAC	Attribute-based access control
ACC	Authenticated control center
ACI	Access control information
ACL	Access control list
ADF	Access control decision function
ADI	Access control decision information
ADS	Anomaly detection system
ADU	Application data unit
AE	Authenticated encryption
AEAD	Authenticated encryption with associated data
AEF	Access control enforcement function
AI	Artificial intelligence
AJAX	Asynchronous JavaScript and XML
ALKE	Authenticated lightweight key exchange
AM	Attribute manager
AMI	Advanced metering infrastructure
AMQP	Advanced Message Queuing Protocol
API	Application programming interface
APT	Advanced persistent threat
ASC	Authenticated Stream-Cipher
ASLR	Address space layout randomization
BIBD	Balanced incomplete block design
BLE	Bluetooth low energy
C&C	Command-and-control
CAD	Computer-aided design
CAM	Computer-aided manufacturing
CapBAC	Capability-based access control
CDM	Central detection module
CH	Context handler

CI	Critical infrastructure
CIM	Computer-integrated manufacturing
CIR	Channel impulse response
CNN	Convolutional neural networks
COM	Component object model
CPPS	Cyber-physical production system
CPS	Cyber-physical system
CTF	Capture-the-flag
D-IDS	Distributed-IDS
DAG	Directed acyclic graph
DAO	Destination advertisement object
DAO-ACK	Destination advertisement object acknowledgment
DCOM	Distributed component object mode
DCS	Distributed control system
DDH	Decisional Diff ie-Hellman
DDoS	Distributed denial of service
DHT	Distributed hash table
DIO	DODAG information object
DIS	DODAG information solicitation
DNP3	Distributed network protocol 3
DODAG	Destination-oriented DAG
DoS	Denial-of-service
DPI	Deep packet inspection
DTMC	Discrete-time Markov models
EHR	Electronic health record
EKF	Extended Kalman filter
EMS	Energy management system
ENIP	EtherNet/IP
EPC	Electronic product code
EPIC	Electric power and intelligent control
EPP	Event processing point
ETSI	European Telecommunications Standards Institute
ETX	Expected transmission time
GDPR	General data protection regulation
GE	Gate equivalent
GPAKE	Password authenticated group key exchange
GQ	Generalized quadrangles
H4.0	Hospital 4.0
HIPAA	Health Insurance Portability and Accountability Act
HIS	Hospital information systems
HITECH	Health Information Technology for Economic and Clinical Health
HMI	Human machine interfaces
I4.0	Industry 4.0
IAC	Industrial automation and control
ICS	Industrial control system

ICT	Information and communication technology
IDS	Intrusion detection system
IED	Intelligent electronic device
IEEE	Institute of Electrical and Electronics Engineers (IEEE)
IETF	Internet Engineering Task Force
IIC	Industrial Internet Consortium
IIoT	Industrial Internet of Things
IMIoT	Industrial Medical Internet of Thing
IoT	Internet of Thing
IoTEED	IoT Trusted Execution Environment for Edge Devices
IPFIX	Internet Protocol Flow Information eXport
IPS	Intrusion prevention system
IR	Industrial Revolution
ISA	International Society for Automation
IT	Information technology
ITS	Intelligent transportation system
ITU	International Telecommunications Union
LAN	Local area network
LC	Lightweight cryptography
LLN	Routing protocol for lossy and low power network
LS	Local stage
LSFA	Leaked-state-forgery attack
LTI	Linear time-invariant
LUT	Look-up table
LWAE	Lightweight AE scheme
M2M	Machine to machine
MAC	Message authentication code
MACRA	Medicare Access and CHIP Reauthorization Act
MI	Mutual information
MILS	Multiple independent levels of security
MitM	Man-in-the-middle
ML	Machine learning
MMS	Manufacturing message specification
MQTT	Message queuing telemetry transport
MR	Medical record
MSPC	Multivariate statistical process control
MTU	Master terminal unit
NFC	Near field communication
NGAC	Next generation access control
NIDS	Network intrusion detection systems
NIST	National Institute of Standards and Technology
OCM	Online compression model
OF	Objective function
OLE	Object linking and embedding
OPC	Process control

ORNL	Oak Ridge National Laboratories
OS	Operating system
OSI	Open system interconnection
OT	Operational technology
P2P	Peer-to-peer
PAP	Policy authorization point
PC	Principal component
PCA	Principal component analysis
PDP	Policy decision point
PDU	Protocol data unit
PEP	Policy enforcements point
PID	Protocol identifier
PIP	Policy information point
PKI	Public key infrastructure
PLC	Programmable logic controller
PRBG	Pseudorandom bit generator
PS	Policy store
PUB	Public utility board
PV	Photovoltaic
RAM	Random access memory
RAP	Resource access point
RAT	Remote Access Trojan
RBAC	Role-based access control
RF	Radio frequency
RFID	Radio frequency identification
RIO	Remote input/output
RN	Reference number
RO	Reverse osmosis
ROP	Return-oriented programming
RPL	Routing protocol for low power and lossy network
RSS	Received signal strength
RTS	Real-time digital power systems
RTU	Remote terminal unit
SBIBD	Symmetric Balanced Incomplete Block Design
SCADA	Supervisory control and data acquisition
SCVAE	Squeezed Convolutional Variational AutoEncoder
SDN	Software-defined networking
SM	Session manager
SMAC	Sequential missing attribute collection
SPC	Statistical process control
SPKI	Simple public key infrastructure
SPL	Smart production logistic
SQL	Structured Query Language
SSL	Secure Sockets Layer
SUTD	Singapore University of Technology and Design

SWaT	Secure water treatment
TAE	Tweakable authenticated encryption
TC	Traff ic control
TID	Transaction identifier
TIHM	Technology Integrated Health Management
TLS	Transport Layer Security
TPM	Trusted Platform Module
UCIoT	Usage control in the Internet of Things
UCON	Usage control
UCS	Usage control system
UDP	User Datagram Protocol
UID	Unit identifier
UPS	Uninterruptible power supply
VLAN	Virtual local area network
VM	Virtual machine
VPN	Virtual private network
W3C	World Wide Web Consortium
WADI	Water distribution
WAN	Wide area network
WPT	Water plant testbed
WSN	Wireless sensor network
XACML	Extensible Access Control Markup Language
ZDI	Zero-day initiative

Part I
Security Analysis and Advanced Threats

Securing Industrial Control Systems

Marina Krotofil, Klaus Kursawe, and Dieter Gollmann

Abstract We propose controllability, observability, and operability as the core security objectives of a control system, whilst the much-used triad of confidentiality, integrity, and availability captures the security requirements on IT infrastructures. We discuss how the deployment of IT in industrial control systems has changed the attack surface, how this invalidates assumptions about independent failure modes crucial in safety design, and explain why stronger IT infrastructure security does not necessarily imply better ICS security. We show how process physics can be used to carry attack payloads and thus become an instrument for the attacker, and argue that ICS security standards should expand their scope to the physical processes layer.

Keywords ICS security · IIoT · Controllability · Observability · Operability · Integrity · Veracity · Safety

1 Introduction

We will not tell the reader why security is important for industrial control systems. We will discuss how operational changes have brought about new *threat vectors*, which options there exist for addressing those threats, and give guidance for

M. Krotofil (✉)
Security in Distributed Applications, Hamburg University of Technology, Hamburg, Germany
e-mail: marina.krotofil@tuhh.de

K. Kursawe
GridSEC, The Hague, The Netherlands

D. Gollmann
Security in Distributed Applications, Hamburg University of Technology, Hamburg, Germany

SCSE, Nanyang Technological University, Singapore, Singapore
e-mail: diego@tuhh.de; dgollmann@ntu.edu.sg

C. Alcaraz (ed.), *Security and Privacy Trends in the Industrial Internet of Things*,
Advanced Sciences and Technologies for Security Applications,
https://doi.org/10.1007/978-3-030-12330-7_1

assessing which *security mechanisms* are fit for purpose in an Industrial Control System (ICS).

Security professionals habitually decry that clients are only stung into action by actual security incidents. Potential problems, acknowledged as important in principle, are not treated with any sense of urgency before. In turn, the security community has become prone to resorting to scare stories, known in the trade as FUD (fear, uncertainty, and dread), developing narratives that freely mix highly sophisticated technical exploits and easy-to-launch spear phishing attacks targeting a human actor.

The media and the IT security sector keep cultivating the image of the hooded hacker, an uber-geek always a step ahead of a defender faced with the *unknown unknowns*. Little actionable advice on handling this situation can be given to parties running an IT infrastructure. Tracing this attacker is the domain of specialist security companies. Focusing on this attacker, however, distracts the view from well organized criminal communities that have acquired their own toolsets to exploit new opportunities as they arise in a changing world. Once their set of attack methods gets known, defenses can be developed by security specialists and then be deployed in the field. At some point, it becomes *unforgivable*[1] not to implement known defenses against known attacks. These classes of known defenses will be our topic.

To sketch how *digitization* impacts ICS, we start at the archetypal *control room*, a large room with a control panel covering an entire wall. From the control panel, wired connections link to *sensors* and *actuators*. The traffic flowing on those wires follows industry specific communication protocols, sometimes proprietary to a device manufacturer. Control units are programmed in languages not covered in standard Computer Science degree programs. Programs running on process controllers execute in infinite loops, unlike programs in many other applications. This is a very brief introduction to the IT aspects in ICS. More details follow in Sect. 3.1.

The control room belongs to an organization enforcing a *name tags must always be worn* policy, and *visitors must always be accompanied*. Only engineers with business in the control room are granted permission to enter, not arbitrary staff from other business units or from administration. Overall, there is a high degree of *physical security*; to access the control system, an attacker already needs to be on premise. Security *perimeters* abound. Defenses against unauthorized access to critical areas would be commonplace in such organizations, but once you are inside, you are *trusted*.

There has since been change in the IT domain. When wires gave way to wireless connections, physical access control to communication links got abandoned with a stroke. Consider a fenced railway shunting yard where drivers can leave their engine to steer it from a belly-pack. An observer beyond the fence can eavesdrop on the traffic, eventually reverse engineer the communication protocol employed, and potentially take control of the engine.

---

[1]Inspired by [10].

Proprietary communication protocols are migrating towards standards based on internet protocols. Software for control units is written in more conventional programming languages. This reduces the learning curve for would-be attackers, and raises the learning curve w.r.t. software security for developers. Otherwise, they may repeat mistakes already familiar from other software domains. Buffer overrun bugs are a prime example. Mobile devices take the endpoints of control channels out of the control room. They are multi-purpose devices running operating systems similar to the operating system one may find on a PC. Determined attackers will find ways of infiltrating those devices. Hence, quoting a speaker from the US DHS from memory:

> We are less concerned when you query the status of your plant from your mobile device; we get worried when you use your device to control the plant.

On the other end of a control channel, *IoT devices* like light bulbs have become *web servers*; they can be controlled via the *http* protocol. Again, there is a considerable body of knowledge about potential vulnerabilities in web based solutions. At the core, intentionally malformed user inputs have to be handled securely.

Previously data collected in an industrial control system were retained on servers within a company's network. Now, data are sent into the cloud for collaborative usage and for processing by external specialists. Finally, more flexible company structures and management processes promising efficiency gains give more people a reason to access data from the industrial control system. Their devices then also become entry points to the control system. This modern ICS has been fashionably relabelled as the *Industrial Internet of Things* (IIoT), although the internetworking of devices is one of the older features of these architectures.

2 Urban Myths

We have sketched the new attack surface that has emerged in Industrial Control Systems. It is tempting, not the least from a marketing perspective, to claim that ICS security is substantially different from what had been known before. For that purpose, a bogeyman called *traditional security* is constructed, without specifying what is 'traditional', that then serves as a fictive counterpoint to new ICS security.

There are three specific points in this argument that get perpetuated from paper to paper on ICS security. The first states that contrary to traditional IT security, availability and integrity are more important than confidentiality in ICS security. The seminal work by Clark and Wilson [11] on *commercial security* states:

> In particular, for that core of commercial data processing that relates to management and accounting for assets, preventing fraud and error is the primary goal. This goal is addressed by enforcing the integrity rather than the privacy of the information [11].

Equally, the importance of availability in commercial settings has been emphasized for a long time. The goal of security is to protect the business at hand. The business of *military* security, dominating work on IT security in the 1970s and 1980s, had been keeping secret information secret, hence the emphasis on confidentiality then.

Hard real-time requirements are the second feature that supposedly sets ICS security apart from traditional security. Such requirements may be relevant in industrial control settings, but they can also be found in other areas. Prominent examples are financial trading systems, where delays or outages can cause considerable financial losses and where regulatory demand on fraud control is high. It may then be wiser to point to lessons that can be learned on reconciling security and real-time requirements than to stipulate a novel challenge.

Third, it is observed that industrial installations have lifetimes spanning decades. Hence, security devices would have to be designed in a way that supports later security upgrades in the field. The base premise may be correct, but it applies to the installation overall. Parts are subject to wear and tear and will be regularly replaced. There are thus opportunities to integrate the update of security devices in existing maintenance processes. The observation on longevity has a dual side. Complex systems may grow 'organically' over time so that eventually old parts remain because the effects of change are too difficult to fathom. Once more, this may well be a real problem but *legacy software* no one dares to touch is equally known in the world of software engineering.

There is, however, one aspect that is characteristic for industrial control systems, viz. the co-existence of safety and security requirements. *Safety* deals with the mitigation of damages in the physical world, *security* with intentional unauthorized attempts to manipulate a system.

3 Information Technology

Information technology makes the world *smart and connected*. The device logic is implemented in software and devices are networked. When a device is connected, it will listen on *input ports* for incoming traffic, process the input, and optionally send a response. (We ignore the case of devices that only send outputs without even offering support for remote maintenance.) In this generic sense, information technology is to be found in many places in an organization. Software and the communication protocols may differ though depending on where one looks.

In ICS, a convention has been adopted whereby the information technology in the office (management domain) is known as IT and the information technology in the actual industrial control system as OT (operational technology). There may be differences in the instantiation of information technology with respect to software, communication protocols, and devices. E.g., in modern IT a few megabyte of memory come cheaply. On the other hand, process control devices may operate in environments normal chips are not designed for. A smart meter in Scotland must survive several harsh winters, requiring specialized and thus expensive hardware.

The fundamental security questions on the IT components, however, are the same. Is the software running on a device the software intended? Can that software handle intentionally malformed inputs? Can the device distinguish between authorized and unauthorized sources of input, be they users or devices? Who is authorized to update the software? Who is authorized to receive the output from the device? Research on these questions had started around 1970. Today, a comprehensive set of security solutions is available, often with theoretical foundations to back up the security properties promised.

3.1 Operational Technology

In industrial automation, we are dealing with systems that often have grown over several decades, and that had started digitizing before even a rudimentary form of the Internet existed. During most of this time, components and protocols were highly specialized for industrial applications, and were designed with a priority on safety, robustness, and longevity.

On the *Field Level*, there are sensors and actuators, i.e., measurement devices and devices that control physical aspects of a process. Those devices do not carry much intelligence, but mainly provide data for and react to commands from a higher level.

On the *Control Level*, the data from sensors is evaluated and commands are sent to actuators. The devices on this level usually have comparatively simple code. The main programming paradigm for a Programmable Logic Controller (PLC), for example, is ladder logic, which closely simulates physically hardwired control structures. The Control Level is often physically close to the field level, i.e., near the physical process it controls. An important subset of devices on the Control Level are protection devices. Those devices get measurements of critical process data, e.g., temperature, voltage levels, and immediately apply countermeasures if the process approaches an unsafe state.

The *Supervising & Production Control Level* provides the high level control, and is already close to the IT systems one finds in an office environment. This level has the Human Machine Interfaces (HMI), which is used by the operators to steer the process, historians to keep track of what is happening, and supervisory process control, which regulates the higher level aspects of the process that require the collaboration of many control devices on the lower level. Increasingly, this level implements a digital representation of the entire process, a.k.a. *digital twin* [16], allowing easier representation to the operators as well as the simulation of various actions. For example, a grid operator can simulate what will happen if a specific line is deactivated for repairs, or simulate accidents and test how the system reacts to them.

The *Information/Enterprise level* is a classical IT system that interfaces with the outside world, providing planning, contact to customers, order management, etc.

Note that in some settings – such as the electrical grid or a water supply system – the field level, and to some extent the higher levels, might be in remote locations that are difficult to secure physically. Security experts may be familiar with perimeter-based security architectures that assume attackers coming from the outside, e.g., from the Internet, or residing in an outer ring, e.g., on an office PC, trying to work their way towards the critical systems in an inner ring. For some Industrial Control Systems, this may be turned inside out, as the field devices are the easiest to physically access. This is known as *hacking upstream* in pentesting jargon. In rare settings, such as smart metering, the field devices as well as some of the Control Level might even be on the physical premises of a potential attacker. The move towards Industry 4.0 will only increase this trend. In this setting, industrial processes will take into account sensor data from suppliers and other external parties, thus drastically increasing the connectivity to the outside world on the lower levels.

4 ICS Security Properties

Section 2 had argued that the bogeyman 'traditional' security does not help in identifying the essence of ICS security. Debates about which ordering of the CIA security properties best fits industrial control systems have also attempted to extend these concepts to physical processes and to the control infrastructure. The CIA triad was conceived for communication systems (ISO 7498-2:1989) in the 1980s, i.e., for a *digital infrastructure*. ICS are engineered physical systems. CIA in the original sense certainly applies to the supporting IT infrastructure, i.e., to protecting data in transit and storage. An ICS is, however, more than its IT infrastructure. It is a control system, *nomen est omen*. Even if one would try to adapt the CIA triad to control systems and find satisfactory explanations, terms would be overloaded with different interpretations in different parts of a system. For example, should an up-and-running but uncontrollable process be termed 'available'?

The alternative is to approach control system security from first principles. To do so, one needs a *threat model*, just as it is the case in the IT domain. The threat model has to capture the fundamental possibilities of the attacker. What can she attacker do to a control system?

4.1 Controllability, Observability and Operability

Control is the adjustment of available degrees of freedom (manipulated variables) to assist in achieving an acceptable operation of the process. In order to control a process, one needs to measure (observe) it first. If the state of the process cannot be determined due to incorrect or incomplete process measurements, one is effectively controlling a fiction, i.e., an arbitrary process in an arbitrary state, but not the actual process. This applies both to the attacker and the defender. To ensure 'secure control' one thus needs process-oriented security metrics and properties.

An attacker must be able to *observe* what is going on inside a control system in order to understand it; she has to be able to derive its *state*. To do "whatever she wants with a given system", she must make sure that she controls the system. Her goal is to acquire controls that are sufficient for bringing the process into a state she desires, and at the same time to take control away from the control system installed and from the operator. To illustrate these concepts, we refer to a security incident at Kemuri Water Company (KWC) reported in 2016 [31], where hacktivists had manipulated the valves controlling the flow of chemicals, though to no particular effect.

> In at least two instances, they [attackers] managed to manipulate the system to alter the amount of chemicals that went into the water supply and thus handicap water treatment and production capabilities so that the recovery time to replenish water supplies increased. Fortunately, based on alert functionality, KWC was able to quickly identify and reverse the chemical and flow changes, largely minimizing the impact on customers [31].

While the attackers were potentially in a position "to do whatever they wanted", they did not obtain observability of the system to measure the effect of their attack. Also, the attackers did not obtain full controllability of their target and could not prevent human intervention which thwarted attack execution

Observability and controllability are thus crucial for control systems, and are therefore core ICS security objectives. These terms were formally defined for linear dynamic systems by Kalman in the 1960s. They can be informally defined as follows.

- *Controllability*: A system is *controllable* if and only if for any two (valid) states A and B, there exists an input and a time t, so that the input takes the system from initial state A to state B by time t.
- *Observability*: A system is *observable* if and only if the value of its initial state can be determined from the system output observed up to some given time t.

It is particularly desirable that a system always remains controllable at least to a degree where it can be brought to a *safe state*.

The third relevant property is *operability*. Operability stands for the ability of a plant to achieve acceptable operations [27]. What is acceptable depends on the situation and includes the ability of a plant to perform satisfactorily under conditions different from its nominal design conditions. E.g., during a process upset or an attack, a plant may operate inefficiently from a cost standpoint, but it would be acceptable operability for the given situation.

Operability can be further refined into resilience, survivability, and graceful degradation. For a process to be resilient, the physical design and the control strategies have to be designed so that they can withstand malicious process manipulations. *Resilience* is thus the ability to absorb process disturbance caused by attacks. *Survivability* is the ability to maintain plant operations albeit at suboptimal level, e.g., with higher production costs or reduced product output, but the plant continues delivering its services. *Graceful degradation* means that a system is capable of maintaining limited but safe functionality and can be safely shutdown.

The three security properties for process control then abbreviate to COO or CO_2 (Fig. 1). Note that the effects of CIA type of attacks on ICS are popularly described

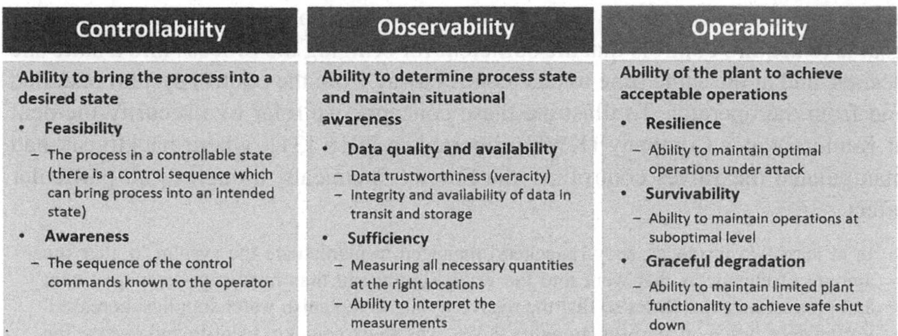

Fig. 1 Security properties for control systems

as 'loss of view' and 'loss of control', which may exaggerate their effect. Even if the operator is blinded due to a modified view at the HMI, a process may still be observable through data in the historian or any database that stores process data, or through the input data in controller memory and physical gauges. Similarly, if the attacker manages to compromise one or several sensor readings, the system may still remain observable through proxy sensors, i.e., related or correlated sensors, or through derived process variables. From the attacker's angle, in order to make the operator lose view, she has to determine an attack strategy that ensures loss of observability by the operator and the control system, i.e., their inability to determine the true state of the process.

4.2 Ensuring ICS Security

Ensuring observability is related to sensor placement methods and derived measurement algorithms from process and control theory [27]. Currently, processes are not designed to support observability for systems under attack, and no approaches are used to determine spoofed process views. Process data may be false at source (manipulated process, faulty or manipulated sensor) or in transit or storage (tampered data). The first case refers to data veracity and is traditionally discussed in terms of *data quality*. The second case refers to data integrity and is mostly discussed in terms of *data security*. With the extension of the attacker model to physical processes and sensors, veracity becomes a critical security property for process data. Research questions related to observability as a security objective therefore include:

- Do we measure the right process variables at the right places to be able to detect possible cyber-physical attacks? An attack could be observed by measured or by derived process variables.

- Ability to interpret the measurements by the operator and identify a possible ongoing attack, type of attack, its possible root cause, etc.
- Ensuring veracity of process data and detection of spoofed sensor readings to ensure trustworthy observation of the process state.

When considering the latter security question, ICS specific security controls can take the form of data consistency or plausibility checks directly through correlated sensors, or indirectly through proxy and related sensors or additional calculations, see e.g., [20]. Section 7.1 gives a further example.

Similar considerations apply to controllability. An attacker managing to break a valve will not necessarily cause loss of control for the operator or the control system; processes are built to be robust and recoverable. Control algorithms or the operator will adjust the process control to account for a lost degree of freedom (valve). From the offensive security standpoint, on one hand, the attacker needs to:

1. Identify controls which would allow her to bring the process into a desired state. Can the process even be moved into a state wanted by the attacker?
2. Ensure that both control system and operator lose controllability of the process, i.e., prevent their access to controls that would allow them to intervene in the attack or to bring the process into a normal state. Alternatively, an attacker can design an attack which cannot be mitigated with the available controls.

From the defensive security standpoint (process design), ensuring controllability implies, e.g., adding additional degrees of freedom (actuators) and overriding controllers to preserve controllability during process upsets caused by an attack. Research questions where the defender first identifies inputs which should be compromised by the attacker in order to manipulate more states with less effort and then invests more in protecting the information flow of the actuators depending on the ranking of each actuator are formulated, e.g., in [4]. In another work, control capacity and process restoration possibilities are studied from the structural controllability point of view [1]. Structural controllability is an evolution of traditional control theory that focuses on providing the concept of "controllability" (properties of control and their relationships) through graph-based structures. Additionally, the authors suggest a solution for a reliable propagation of alarms when a situation threatening structural controllability is detected.

5 IT Defenses

We will now give a survey of some well-established security measures from the IT domain and point to issues that need to be addressed when deploying those measures in an ICS setting.

5.1 Cryptography

Cryptography does not solve security problems, it transforms security problems into *key management problems*. In symmetric key cryptography, the peers securing their communication need to share secret keys. With such shared keys, the integrity of data can be protected, be it stored data or data in transit. With shared keys, the confidentiality of data can be protected through *encryption*. Encryption does not necessarily protect the integrity of data.

Consider now a scenario where communication to multiple devices needs to be protected. The same shared secret key (or password) could be installed on every device. The controller handling the devices needs to store just a single key, but compromising the key on one device will compromise all other devices too. Security argues for one key per device, which demands disciplined key management. Errors in key management will disrupt other services impacting on availability. Experience from other areas[2] shows that operational reasons often militate against more intricate key management regimes. The security community knows this as the trade-off between convenience and security, but this is a biased security view; there is a trade-off between security and other operational requirements.

There exist further trade-offs between encryption and other safety and security requirements. Consider a safety-critical incident in an installation where external emergency services are being called in. Sensor readings from devices in the installation may give valuable cues to the nature of the incident. If those readings are encrypted, they can only be displayed on devices that hold the relevant keys and encryption algorithms. An emergency service called in that had no prior relationship with the organization affected will, by definition, not have the keys required and be hampered in its work. Observability has been reduced. An emergency service with a prior relationship that has been furnished with the keys would be a security risk; there is yet another place where keys can be compromised.

During operations, an Intrusion Detection System (IDS) may monitor internal network traffic for suspicious patterns. When network traffic is encrypted, the IDS cannot inspect the content of the traffic and would be reduced in its efficiency. There may thus be valid reasons why a dynamic and safety critical application such as air traffic control is adopting protocols such as *Automatic Dependent Surveillance-Broadcast* (ADS-B) that do not offer encryption by default.

Asymmetric key cryptography builds on public and private keys. *Digital signatures* generated with private signing keys provide integrity protection and authentication. Successful verification of a signed document establishes a strong mathematical link between the document and the public verification key. For confidentiality protection, public encryption keys serve for encryption, private decryption keys for decryption. In both cases, public keys are cryptic strings devoid of meaning. The main key management tasks are installing public keys on all devices that need

[2]https://www.cyberark.com/blog/four-ssh-vulnerabilities-you-should-not-ignore/

them, and protecting private keys on the devices that use them. To link public keys to meaningful entities (persons, devices, etc.) a *Public Key Infrastructure* (PKI) is needed. Together with such a PKI, digital signatures can link documents to entities. Claims such as *this document is digitally signed so you can trust it* reflect that the source of the document can be authenticated, but do not mean that the content of the document is factually correct; the entity may have signed wrong information. Cryptography provides means for ensuring confidentiality, integrity, and authenticity but not for establishing the veracity of data.

Cryptographic algorithms keep ageing, in particular asymmetric algorithms. More powerful computers and advances in analysis methods keep pushing key lengths up. Official guidance on approved algorithms and key lengths is therefore typically issued for periods of about 6 years. For systems intended to be used for very long time, being able to update the cryptography is thus vital. In the past decades, several algorithms once considered secure had to be phased out either due to the discovery of attacks, e.g., RC4, or insufficient key length, e.g., DES. Even though those algorithms had long known to be insecure, it has proven hard to eliminate them in practice. The main reason is that once a device changes its cryptography and with it its communication protocol, all devices it communicates with need to be able to understand the new protocol as well. If protocols are not designed for algorithm agility, this is near impossible to do in a large system without shutting down the entire system, and even then it is easy to miss something.

It is, however, very rare that a well-established cryptographic algorithm fails unexpectedly. Rather, algorithms can show small weaknesses indicating that they are getting old before they can be broken, giving the community enough time to develop a successor and phase out the old algorithms. This is also the reason why the argument "there is no practical attack yet, only a theoretical one" is nonsensical. This is like claiming a bridge does not need to be repaired because there are only cracks, but it has not collapsed yet. Well-designed protocol such as TLS therefore use a protocol negotiation step when two parties communicate. They first agree on which cryptography to use. This allows for devices to support several versions at the same time, and thus to slowly phase out an outdated protocol.

We conclude with a word of warning. While it might seem easy to create or modify cryptographic algorithms, do-it-yourself cryptography must be strictly avoided. Inspecting the results of encryption to check whether the ciphertexts look random is useless as an indicator of the actual strength of an algorithm. The challenges in designing cryptographic algorithms, are illustrated by the NIST competition for the SHA-3 hash function standard; of the 51 accepted submissions, 10 were withdrawn by their authors on the grounds of being broken, while 16 were eliminated due to substantial weaknesses. Another 11 were rejected as they were not clearly broken, but either had design weaknesses or performance issues. Note that these were participants in a cryptographic standardization effort, so only organizations with established cryptographic expertise attempted it in the first place.[3]

There exist standards for encryption algorithms that capture the state of the art; it is strongly recommended to pick a suitable standard together with a competent implementation. Insecure implementations of secure algorithms are not unheard of. Going with the stream and deploying 'industry standard' products trusts that someone else has done the necessary analysis. The Heartbleed attack [9] on *OpenSSL* has shown that this approach can be too optimistic. Hence, due diligence suggests that one should check that proper security checks had been performed on the crypto library of choice. If, for some reason, the standardized technologies do not work, experts on cryptography must be consulted. The ICS protocol domain is a graveyard of broken protocols, providing little more than the illusion of security [18, 21].

5.2 Intrusion Detection

There are currently great expectations that one only has to learn deeply enough to construct mechanisms that automatically detect when a system is under attack. This hope also extends to Industrial Control Systems. An excellent introduction to intrusion detection in ICS that is based on practical experience is given in [14] and is summarized in the following. *Network intrusion detection systems* (NIDS) monitor network traffic to detect when something goes wrong. Detecting when something goes wrong can be done in two ways. Recognize behavior known to be wrong, i.e., *know your enemy*, or recognize behavior known to be correct and react to behaviors not classified as correct, i.e., *know thyself* – ΓΝΩΘΙ ΣΑΥΤΟΝ. Hence, either a model of malign behavior or a model of legitimate behavior is needed.

The intrusion detection community refers to the first type as *knowledge based* or *misuse based* intrusion detection. *Behavior based* intrusion detection stands for the second type, and can be further divided into *anomaly based* and *specification based* intrusion detection. Anomaly based NIDS pursue a *domain agnostic* approach, building a model of the behavior of the target system from observed clean network traffic using machine-learning methods. This process can be automated to a large extent. Specification-based NIDS pursue an *attack agnostic* approach, developing a specification of legitimate system behavior from the specification of the system, expected use cases, and security policies. This is usually an expensive manual process requiring domain knowledge.

Knowledge-based NIDS can handle known attacks. They are thus a useful first line of defense but can be circumvented by skillful attackers. Anomaly-based NIDS do not recognize attacks, they detect anomalies. When the learning methods used

[3]https://en.wikipedia.org/wiki/NIST_hash_function_competition#Entrants_with_substantial_weaknesses

to construct the model of legitimate behavior had abstracted away all information about the semantics of the target system, then alerts cannot refer to the semantics of the target system. Such an IDS can only make limited contributions to observability or to actionability. *Actionability*, i.e., the quality of advice on how to react to an aler, is one of the IDS benchmark criteria in [14].

Anomaly-based NIDS where the model, and hence the alerts, capture some high-level semantics of the target system are called *whitebox systems* in [14]. Whitebox detection can tell something about the semantics of an anomaly and about the state of the target system at the time of an alert. Hence, insight in the alert can be more detailed, but domain expertise is required to build the model of legitimate behavior.

The observations in [14] conclude with a call for a *discipline of writing software that is supervisable (and privacy-preserving) by design.* For intrusion detection to be effective, it needs the right *sensors* (as the term is used in the IDS community) in the right places. This call carries over from IDS to ICS. Systems should be constructed so that they are observable (supervisable) by design, even when under attack.

5.3 Trusted Computing

Observability requires that the operator of a plant is informed about its state. This includes information about the state of the physical process in the form of sensor readings but also information about the current state of software and configuration on the various devices. Once programmable devices get deployed in a plant, there is a new attack vector to consider. The attacker, when compromising a device, may record its current valid state, store it in compressed form, and return the saved valid state when queried. Techniques of this kind had been used in *stealth virus* design in the 1990s. The attacker could do the same with sensor readings, record and store readings from a process before starting to interfere with it, and return stored readings during the actual attack.

A device could thus construct a *digest* of its current state and provide the digitally signed digest as evidence. This service is known as *attestation*. The recipient needs a public verification key of the device, and has to know which digests correspond to legitimate states. The management overhead required for coping with regular software changes is considerable and attestation has hardly been deployed in practice. Trusted Platform Modules [3] are an example for devices that provide an attestation service; software based solutions are under active research.

While a TPM can help in verifying the integrity of a device at boot time, other mechanisms are needed to check integrity during execution to prevent runtime attacks. In modern PCs, tasks like this are performed within trusted execution blocks, pieces of code that are protected from manipulation by the processor and can thus carry security critical code. On embedded systems, several new technologies are now emerging that can provide such isolation, though they still need to be integrated into the existing codebase, which, given the long device lifetimes in control systems, can be a daunting task.

One should also note that remote attestation is only one building block in a larger infrastructure. Especially in industrial control systems, one cannot simply exclude devices once they fail attestation. Any detection mechanism must be embedded in a larger crisis response infrastructure and policy.

As an alternative to attestation, a device can be constructed in such a way that it only accepts authorized software. Authorized software could be digitally signed by the provider. The device needs to store the public verification key of the provider in a tamper-resistant location. It could then install only digitally signed software from valid sources. UEFI secure boot implements such a service at the level of boot loaders with the DRAM as the hardware root of trust [29].

6 Safety vs. Security

While security is a relatively new issue for most process control systems, safety is an aspect in which the ICS community has a wealth of experience. From a safety point of view, any device that can fail is an additional risk. A firewall might physically catch fire, or it might be misconfigured and suppress important messages. As emergency cases are rare, uncommon messages will be sent especially often in cases when they are important. Message encryption makes it harder to monitor the communication channels. Message authentication failures might refuse valid messages. Key management may introduce further potential points of failure.

Safety and security are related and require similar mindsets, e.g., investing resources for something not to happen, but there are also some fundamental differences in the way of thinking. In a safety system, failures are usually assumed to be accidental. Failure probabilities of individual components are thus by and large independent. If the probability of an engine failure in an airplane is 1 in 100,000 flights, the probability of two engines failing is 10^{-12}; this is then the failure probability for a plane that can land safely with one engine. In security, failures are not independent. An attacker attempting to manipulate one device is likely to also attack the fallback. An attacker's next step may *depend* on what has been observed so far. Events deemed so unlikely that they are simply not accounted for, such as a name containing valid SQL commands, may be triggered intentionally by an attacker.

Finally, security and safety have opposing update regimes. IT security employs *frequent updates* such as installing patches, upgrading firmware or adding new firewall rules to react to new *security advisories*. On the other hand, an engineering system once approved is expected to *remain safe if left untouched*. Any change in software or operational practices must be followed by an extensive safety revision. Failing to do so can result in casualties. For example, after an update of the SAP-based maintenance software at DuPont without a subsequent safety review, an alarm notifying on a due date for a hose change disappeared. As a result, a hose used to transfer phosgene from a cylinder to a process wore out and catastrophically failed spraying a worker in the face resulting in his death [30].

6.1 Information Technology in an OT Environment

Many lessons have been learned in IT security over the last decades, and sometimes at a high price. Good practices have been established that are well applicable to the OT sector, but which have not been applied to many OT systems yet. Due to the criticality of those systems, we cannot afford a similar learning curve again, especially since potential attackers have refined their techniques in the last decade. On the other hand, many good practices from IT do not translate directly into a process control setting, and – as should be common sense in all areas – one should always verify that the approach taken makes sense in the environment it is taken in.

6.1.1 Security Updates

The IT sector has made the experience that poorly documented systems that have grown over time in an ad-hoc fashion, with legacy software no one dares to touch, become brittle under stress. 'Organically grown' is the euphemism for systems in such a state. OT often finds itself in this situation today.

In IT, updates are often scheduled on a weekly basis, or faster when a critical vulnerability is reported. Periods of unavailability due to software updates are accepted within limits, but there are processes in place to keep critical components up and running. OT components may have long lifecycles, both the hardware, with devices in use for decades, and the software. Many constraints are hard to test and, due to the organic nature of many OT systems, often not even known. Hence, software updates on an ICS system need substantially more care than updates on most other IT systems. Even when patching in the IT and not in the OT domain, stability problems may occur if patches have not been sufficiently tested. A recent example are patches for Spectre and Meltdown vulnerabilities that disrupted normal functionality of ICS applications across multiple product lines of multiple vendors. There was public pressure to patch against these vulnerabilities as soon as possible. As a result, ICS vendors had issued patches without comprehensive testing and later warned asset owners not to apply patches [28].

6.1.2 System Shutdown

In a process control system, information itself is usually not the primary security concern. The main goal is to protect the process, which implies ensuring the controllability of the process to maintain its operability. This can put different requirements on a process control system than are needed for most information systems. In a large industrial process, for example, the emergency shutdown procedure may take several days or longer. Thus, if an attack is detected, one cannot simply shut down the system and start recovery; some components must be kept running and under control no matter what.

6.1.3 Software Security

Dealing with intentionally malformed digital inputs has been the domain of software security for the past 20 years. This line of investigation needs now to be extended to physical inputs. At first sight, sensor readings from the physical world fall into reasonably predictable value ranges, so the security problem would be less pronounced. However, even planned changes in an application may change value ranges with unexpected consequences. One of the reasons for the Ariane 5 crash was the reuse of a software module for the Ariane 4, but the forces observed during a launch of Ariane 5 exceeded the values that module could handle [24]. An ICS security analysis would have to consider intentionally malformed sensor readings, either caused by manipulating the environment around a sensor or by manipulating a sensor. Protecting sensor readings in transit is the familiar task of communication security.

6.2 Independence Assumptions

In safety engineering, the analysis of a design is often built on the assumption of *independent failure modes*. To meet this assumption, devices may be sourced from different vendors. With smart devices, independence has to extend to the software running on the devices. However, many Linux based embedded systems, and thus a large fraction of IoT devices, use the *BusyBox* software for a more compact implementation of Linux commands. Hence, devices that appear to be completely independent may in the end all use the same software library. Any issue with this library immediately scales up to millions of devices. This dependency is nearly invisible, as most application vendors or end-users do no analyze their software to that level once a device has been bought. Thus, devices that should run independent software for redundancy reasons might end up having exactly the same software. A further example is an authentication bypass in the CoDeSys library that had been used in more than 200 models of PLCs from a variety of vendors [23].

This argument can be, and has been taken a step further. Software written in the same programming language may exhibit dependent vulnerabilities, see e.g., the discussion about a C++ monoculture in software security [13].

A second dangerous dependency relates to the data sources. With the increasing interconnectivity, it becomes harder to keep an overview where data comes from, and how critical it is. A power grid provider may integrate the weather forecast into its planning; many systems already use GPS time for time synchronization. In addition, the rise of machine learning algorithms is creating an ecosystem of small service providers offering data analytics services. Once data is available inside an organization, there is no telling where it will be used next. Metadata on data quality such as origin or reliability are rarely documented and integrated into a risk assessment. Poisoning external data sources will also bypass any traditional firewall or zoning defense. Even if a subsystem cannot be directly attacked, it can still be fed misleading data that can result in the system behaving in the wrong way. This can get especially bad if external service companies such as data analytics services or the weather forecast provide input for several operators. This would allows an attacker to distribute bad data on a large scale with comparatively modest effort.

This is also a point where risk assessment for 'normal' IoT devices gets next to impossible. A smart lightbulb might be in a living room where an incident is at most an inconvenience, or in the operating theatre of a hospital. Similarly, a sensor might be used in contexts from watering garden plants to monitoring a nuclear power plant. In an ideal world, the latter would adhere to substantially stricter security requirements. In practice, we have often seen devices or code repurposed for use cases that were never intended in the initial risk assessment, and by far not satisfying the requirements the new environment would impose.

7 Information Channels Beyond IT

In an IT system, components are designed to interact with each other only via information channels that are part of the IT system, the notable exceptions being the human user. In IT security research it is well-known that this is an abstraction that can be violated by using *covert channels* [22]. Covert channels use measurable effects such as memory usage or response time to convey information from one process to another, thus bypassing the process isolation implemented in the IT system. Covert channels had been a concern in military IT systems already in the 1970s. *Side channels* based, e.g., on execution time or power consumption can leak information about the internal state of a system. Timing channels have played a part in the recent Meltdown [25] and Spectre [19] attacks on modern CPUs.

In a process control system, the process itself can become both a communication channel and an attack vector. *Transduction attacks* exploit the physics of a sensor to manipulate its output or induce intentional errors. For example, acoustic interference can use the resonant frequency of a sensor to create fake output [15].

With access to a large number of power consumers, electric heaters or vehicles, or laser printers, for example, an attacker can bring and keep down an electricity grid without touching any of the grid's information systems. The attacker modulates the power consumption, of these devices over the internet at a rate too fast for the grid's protection mechanisms to react, pushing the grid into unstable states that trigger automated load shedding or tie-line tripping [12].

7.1 Zones and Conduits

The ISA 99/IEC 62443 set of standards[4] introduces the concept of zones and conduits as a way to segment and isolate the various sub-systems in a control system. A zone defines a grouping of logical or physical assets that share common security

[4]https://www.isa.org/isa99/

requirements based on factors such as criticality and consequence. Equipment in a zone has a security level capability. If the capability level is not equal to or higher than the requirement level, then extra security measures, such as implementing additional technology or policies, must be taken [8]. Any communication between zones must be via a defined conduit. A *conduit* is a path for the flow of information between two zones. It provides the security functions that allow different zones to communicate without introducing security threats to each other.

Consider now a piping infrastructure at an industrial facility. Once two devices are inserted into a process, they can become related to each other by the physics of that process. The physical process then becomes a communication medium and may be used for delivering malicious payloads even if the devices are segregated electronically. With that, the IEC 62443 approach to defining zones and conduits is violated when an attacker can use equipment in a lower security zone to deliver an attack payload to an equipment in a more secure zone via process physics, rather than in electronic form via a configured conduit (Fig. 2). The security zoning analysis must thus be extended to an analysis of the physical process and its component interactions. Although a physical process is a conduit that cannot be eliminated, it can be monitored and attack propagation can be detected and mitigated.

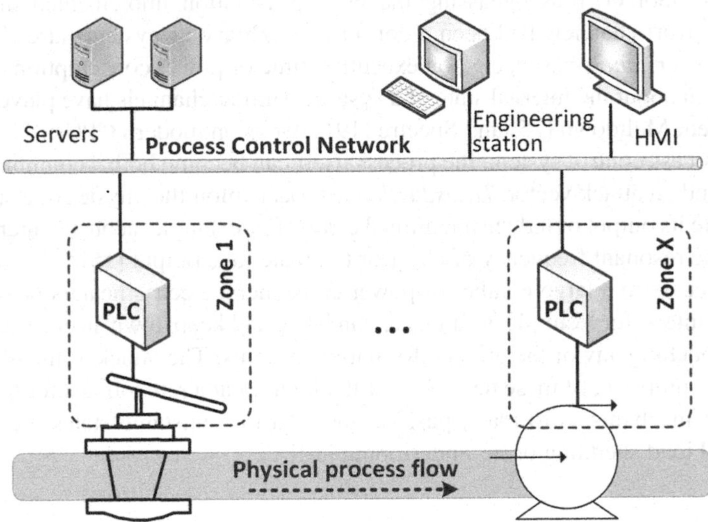

Fig. 2 Delivery of attack payload via process physics and violation of the network security zone architecture

7.2 Delivering Attack Payloads via Process Physics

Results from a physical experiment were presented at Black Hat 2017 using a setup where a valve and a pump did not have any electronic communication between them.[5] The goal of the attack was to damage the pump. The pump was analogue and did not include exploitable IT components. Instead, the attacker exploited the positioner of an electronic valve to manipulate the valve position to cause cavitation in the pump. The damaging payload took the form of cavitation bubbles which were successfully delivered to the pump via process physics (fluid in the pipe).

Although the cavitation process harms the pump, breaking a pump in this way takes some time. To prevent operator response, the attacker may want to implement routines to spoof the reading of the flow sensor and the status of the valve positioner to make the process state appear normal. Detecting an ongoing cavitation attack is related to the research question on ensuring observability outlined in Sect. 4.

In this experiment, even though the attacker was able to cause loss of view with respect to the operator, she could not achieve loss of process observability. The pump was instrumented with a few additional analogue sensors (e.g., vibration, temperature) for pump diagnostic. By manually collecting pump diagnostic data and using pump documentation, the operator could establish that the process state was not consistent with the pump state. Furthermore, using pump documentation and correlating pump measurements, it was possible to derive the true flow meter reading and the state of the valve positioner. Since analogue sensors were only accessible locally, their measurements could be trusted. In a geographically distributed control systems, it was previously suggested to protect the security of few known physical measurements with cryptographic means and use these measurements for detecting falsified states of the system [26].

Also, aggregated diagnostic data could be sent to the cloud via independent one-directional communication links and become accessible for the analysis in real time.[6] It was previously argued that IIoT technology introduces additional attack surfaces to OT. While this is true, IIoT can also become part of the solution when addressing OT security.

7.3 Information Flow via Process Physics

Moreover, a physical process may constitute an *information flow* channel. Such effects have been demonstrated in a Matlab model of a process for producing vinyl acetate [17], where the attacker attempted to deactivate a catalyst by raising the temperature in the reactor in a plant, without being detected. According to the

[5]https://www.blackhat.com/docs/us-17/wednesday/us-17-Krotofil-Evil-Bubbles-Or-How-To-Deliver-Attack-Payload-Via-The-Physics-Of-The-Process.pdf

[6]https://www.flowcontrolnetwork.com/how-iiot-monitoring-improves-pump-maintenance/

process design, the reactor outflow passed through a heat exchanger to cool down before being further fed into a piping infrastructure. Due to increased temperature in the reactor, reactor outflow was unusually hot, yielding a condensation problem in the heat exchanger and causing a 'condensation' alarm.

To conceal the ongoing attack, the attacker needed to implement a routine to suppress the digital alarm in the heat exchanger control loop. Suppressing the digital alarm resulted in hiding information about the deteriorated state of the physical process from the operator and from the controller, but did not change the physical state of the process. The 'unhealthy' process state propagated into a neighboring plant section causing a low liquid level in an absorber vessel. Even though the absorber alarm was also suppressed to prevent compensating actions from the operator, the degraded state of the process kept propagating downstream, eventually reaching an unsafe limit in the distillation column causing a safety shutdown of the plant. Even though the flow of the digital alarms over the communication link was interrupted, the plant subsystems continued to communicate with each other via the physics of the process. This is a natural information flow in physical processes that cannot be prevented. However, it is important to be aware of such information flows as they may be used by attackers for coordinating attack activities. By implementing routines for detecting specific process states, the attacker can effectively coordinate attack activities between multiple malicious implants throughout the plant without sending messages via digital communication links.

8 Security Architecture

Given the long lifetime on many ICS devices, the generally low security maturity in the field, and the difficulty of software updates and patching, it appears unfortunate but unavoidable that many field devices will remain insecure for quite some time to come. This requires a solution that is orthogonal to trusted computing. While trusted computing helps to establish trust in remote devices, we need to be able to handle devices that are untrusted.

One approach to this end goes back to the first multi-user computers and to military security, when the focus was on controlling communication between users and devices at different security levels. The Bell-LaPadula [5] and Biba [6] models respectively, assign all components to security levels. The Bell-LaPadula model focuses on confidentiality of information; devices on a higher level can never send information to devices on a lower level. Thus, highly confidential information cannot leak to a lower security level. The Biba integrity model works in a dual fashion. Information may only flow from higher integrity levels to lower levels, thus protecting the more secure devices and data from being influenced in any way from a lower level. For example, the engine control of a car can tell the entertainment system to adjust the radio volume to the engine noise, but the entertainment system has no means of communicating back.

8.1 Multiple Independent Levels of Security Architecture

To capture the complexity of more advanced systems, these concepts got generalized towards the Multiple Independent Levels of Security (MILS) architecture [2]. To fully go into the details of MILS would exceed the scope of this Chapter, and we only highlight how some of the ideas that are helpful for securing control systems.

The first step is to decompose a system into small units that are reasonable easy to understand, for which use cases and interfaces can be clearly defined, and for which a security/safety policy or a separate risk assessment can be made easily. Decomposition is done along functional lines, and can go into quite some detail. A sensor, for example, could be decomposed into measurement unit, authentication, and communication. On a process level, decomposition goes both by functionality, e.g., the protection system. Process control systems add to the original MILS structure the need to consider process criticality. In an airplane, it is tolerable if one engine fails, but not if all fail. Thus, it is vital in this setting that all engines are in separated, isolated compartments. More generally the whole process usually fails only if a chain of individual failures occur. The more independent the links in the chain are from each other, the easier it is to secure the process.

At the component level, once the decomposition has happened and, ideally, every device has a clean definition of its functionality, technical requirements, and assurance methods to guarantee it satisfies its requirements, the architecture defines the communication flow between the modules. This can be done on a network level, e.g., by a data diode, or also within an individual device, e.g., through a microkernel.

At the control system level, decomposition and definition of the required communication paths are getting more complicated. There may be devices with exactly the same level of criticality that require to be separated strongly. E.g., it is desirable to stop an attacker who has physically breached one electric substation to connect to all other substations from there. Other devices are supposed to provide redundancy, which makes it even more important that they are strongly separated, e.g., different sensors measuring similar data. Finally, some form of graceful degradation should be built into the architecture. While it may help in normal operation that all devices talk to the same network time server, in crisis mode it is necessary to separate the components, let them run on their own time, and later reconnect them again.

8.2 Role of the Vendor

A further issue with implementing a secure architecture is that equipment vendors need to be on board at an early stage. From a security point of view it is an unfortunate development that devices become more and more integrated, and it thus gets harder to separate functionality according to risk levels. While there are plenty of ways to implement strong separation inside a device, this cannot be done by the system integrator but has to be provided by the vendor. Moreover, in some cases vendors have only tested their devices within a specific architecture and refuse to provide product warranty if the architecture is changed. If this issue has not been addressed during procurement, there is little chance to customize the architecture as no sane operator will voluntarily waive product warranty.

8.3 Example: Smart Meters

As a concrete example for this approach on the device level, consider a remote disconnect switch for smart meters. The communication protocols used in many smart meter architectures are quite complex, and in many cases penetration tests have shown serious implementation issues. If the worst-case attack would be some level of fraud, the problem would be manageable. Grid operators have a lot of experience in expected usage patterns and detecting *non-technical losses*. As the meter is linked to the consumer's home address, it is sufficient to detect fraud in the long run and provide sufficient evidence to handle hacked meters. A different issue, however, is a meter with a remote disconnect switch. If an attacker can compromise a meter from a distance and disconnect the household, the potential damage goes well beyond simple fraud. The meter could be a component for a large-scale attack on the grid. Further usage scenarios also envisage smart meters to actively control household devices to help with the grid's load management, which points to a similar threat scenario.

Following the basic ideas of independent security levels, we would separate the code into the parts of communication, authentication and key management, metrology (which needs to be separated anyhow, as this is certified by the government authorities and thus cannot easily be updated), and remote disconnect, as well as load balancing for more advanced scenarios. This could be done on a software level, e.g., through a microkernel, through secure computing compartments on a chip level, or by providing completely independent hardware. In the latter solution, the remote off-switch might have a small processor to do its own key validation, so that the main processor has no means to activate it without knowing the key, and values coming from the metrological unit are authenticated by a smart-card like device before being handed over to the communications module.

While this seems to imply additional effort and thus cost, such an implementation will save expenses on the long run. The security requirements for the bulk of the implementation code are no longer those of critical infrastructure, but of a communications relay for billing, which implies substantially less government regulation, certification, and maintenance effort. In this setting, the most complex software part – the communication module – cannot do much more damage than a (temporary) denial of service attack, which is easy to detect and usually has no impact on the reliability of grid operations.

9 Conclusion

The buzzword “IIoT” is misleading as it highlights the internetworking of devices at the expense of control, the very reason why devices are networked in industrial installations. ICS security is thus not IT infrastructure security with priorities re-arranged. ICS security is control systems security. The important security goals are controllability – being able to bring a process into a desired state – and observability – knowing the state of the process and its components – and operability – keeping a process always within its safety capabilities, even without human interaction. These are general requirements on a control system. They become security requirements when they have to be met in the presence of intentional interference by an adversary. Ensuring these security properties must also be addresses from the point of process and control systems design theory.

IT security solutions developed over the past decades in other sectors have a role in ICS security, but have to be adapted to these new domains. Security requirements on the IT infrastructure may conflict with control systems security requirements, such as a conflict between confidentiality and observability. Cryptographic key management has to consider restrictions such as devices that are not physically accessible. Software update schemes have to take into account that the IT system must never be taken off-line and that safety-related applications may require re-validation after each update. Intrusion detection systems have to learn domain specific rules.

Safety engineers have to appreciate the fundamental difference between reliability and safety on one side, and security on the other. In safety, one starts from assumptions about failure probabilities and about the independence of certain failures. Then a safety system is designed and its effectiveness analyzed based on the given probabilities. In security, the defender moves first and may already field a set of security mechanisms. Once the system is deployed, the attacker starts probing for vulnerabilities. The attacker is not bound to adhere to assumptions made by the defender about possible adversarial behavior. In fact, it is an effective attack strategy to search for assumptions that can be violated. For the attacker, the control system is a *weird machine* [7] commandeered by unconventional inputs used in ways not intended by the designer. To constrain the attacker, systems ultimately have to be designed and operated in a way so that as little as possible can be done with them beyond their intended use. This may include replacing critical digital components with immutable alternatives.

References

1. Alcaraz C, Lopez J (2017) A cyber-physical systems-based checkpoint model for structural controllability. IEEE Syst J 12:3543–3554
2. Alves-Foss J, Oman PW, Taylor C, Harrison WS (2006) The mils architecture for high-assurance embedded systems. Int J Embed Syst 2(3–4):239–247

3. Arthur W, Challener D (2015) A practical guide to TPM 2.0: using the Trusted Platform Module in the new age of security. Apress, Berkeley
4. Barreto C, Cárdenas AA, Quijano N (2013) Controllability of dynamical systems: threat models and reactive security. In: International Conference on Decision and Game Theory for Security. Springer, pp 45–64
5. Bell DE, LaPadula LJ (1973) Secure computer systems: mathematical foundations. Technical report, MITRE CORP BEDFORD MA
6. Biba KJ (1977) Integrity considerations for secure computer systems. Technical report, MITRE CORP BEDFORD MA
7. Bratus S, Locasto M, Patterson M, Sassaman L, Shubina A (2011) Exploit programming: from buffer overflows to weird machines and theory of computation. {USENIX; login:}
8. Byres E (2012) Using ANSI/ISA-99 standards to improve control system security. White paper, Tofino Security
9. Carvalho M, DeMott J, Ford R, Wheeler DA (2014) Heartbleed 101. IEEE Secur Priv 12(4):63–67
10. Christey S (2007) Unforgivable vulnerabilities. Black Hat Brief 13:17
11. Clark DD, Wilson DR (1987) A comparison of commercial and military computer security policies. In: Proceedings of the 1987 IEEE Symposium on Security and Privacy, pp 184–194
12. Dabrowski A, Ullrich J, Weippl ER (2017) Grid shock: coordinated load-changing attacks on power grids: the non-smart power grid is vulnerable to cyber attacks as well. In: Proceedings of the 33rd Annual Computer Security Applications Conference. ACM, pp 303–314
13. Duntemann J (2004) The lessons of software monoculture. SD Times, p 28, 1 Nov 2004
14. Etalle S (2017) From intrusion detection to software design. In: European Symposium on Research in Computer Security. Springer, pp 1–10
15. Fu K, Xu W (2018) Risks of trusting the physics of sensors. Commun ACM 61(2):20–23
16. Glaessgen E, Stargel D (2012) The digital twin paradigm for future nasa and us air force vehicles. In: 53rd AIAA/ASME/ASCE/AHS/ASC Structures, Structural Dynamics and Materials Conference 20th AIAA/ASME/AHS Adaptive Structures Conference 14th AIAA, p 1818
17. Gollmann D, Gurikov P, Isakov A, Krotofil M, Larsen J, Winnicki A (2015) Cyber-physical systems security: experimental analysis of a vinyl acetate monomer plant. In: Proceedings of the 1st ACM Workshop on Cyber-Physical System Security. ACM, pp 1–12
18. Jovanovic P, Neves S (2015) Practical cryptanalysis of the open smart grid protocol. In: International Workshop on Fast Software Encryption. Springer, pp 297–316
19. Kocher P, Genkin D, Gruss D, Haas W, Hamburg M, Lipp M, Mangard S, Prescher T, Schwarz M, Yarom Y (2018) Spectre attacks: exploiting speculative execution. arXiv preprint arXiv:1801.01203
20. Krotofil M, Larsen J, Gollmann D (2015) The process matters: ensuring data veracity in cyber-physical systems. In: Proceedings of the 10th ACM Symposium on Information, Computer and Communications Security. ACM, pp 133–144
21. Kursawe K, Peters C (2015) Structural weaknesses in the open smart grid protocol. In: 2015 10th International Conference on Availability, Reliability and Security (ARES). IEEE, pp 1–10
22. Lampson BW (1973) A note on the confinement problem. Commun ACM 16(10):613–615
23. Leverett E, Wightman R (2013) Vulnerability inheritance programmable logic controllers. In: Proceedings of the Second International Symposium on Research in Grey-Hat Hacking
24. Lions J-L, Lübeck L, Fauquembergue J-L, Kahn G, Kubbat W, Levedag S, Mazzini L, Merle D, O'Halloran C (1996) Ariane 5 flight 501 failure report by the inquiry board
25. Lipp M, Schwarz M, Gruss D, Prescher T, Haas W, Mangard S, Kocher P, Genkin D, Yarom Y, Hamburg M (2018) Meltdown. arXiv preprint arXiv:1801.01207
26. McQueen M, Giani A (2011) 'Known secure sensor measurements' for critical infrastructure systems: detecting falsification of system state. In: International Workshop on Software Engineering for Resilient Systems. Springer, pp 156–163
27. Sharifzadeh M (2013) Integration of process design and control: a review. Chem Eng Res Des 91(12):2515–2549

28. Tung L (2018) Meltdown-spectre: more businesses warned off patching over stability issues. https://www.zdnet.com/, 15 Jan 2018
29. Unified Extensible Firmware Interface specification, Version 2.5, April 2015
30. U.S. Chemical Safety Board (2011) Dupont corporation toxic chemical releases: investigation report, July 2011
31. Verizon (2016) Data breach digest. Scenarios from the field

Towards a Secure Industrial Internet of Things

Georgios Spathoulas and Sokratis Katsikas

Abstract The Industrial Internet of Things (IIoT), being one of the underlying and enabling technologies of the Industry 4.0 initiative, brings about expectations for unprecedented value creation opportunities in industry. Unfortunately, these do not come without a price; in this case the price to pay is the increased vulnerabilities, the increased threats and the increased attack surface that result when industrial systems originally designed with little or no cybersecurity in mind connect to the Internet. Consequently, the cybersecurity of the IIoT becomes of paramount importance. Research has started focusing on this area, as well as on the related areas of cyber-physical systems security and industrial network security, but a multitude of issues still remain to be addressed. In this chapter, we review recent research results in the area of IIoT security, with an eye towards identifying trends on one hand and areas where research seems to lag behind on the other, by classifying research results using the security lifecycle model of the National Institute of Standards and Technology (NIST) framework for improving the cybersecurity of critical infrastructures.

Keywords IoT · Industrial IoT · Security · NIST framework · Industry 4.0

G. Spathoulas
Department of Computer Science and Biomedical Informatics, University of Thessaly, Volos, Greece
e-mail: gspathoulas@dib.uth.gr

S. Katsikas (✉)
Department of Information Security and Communication Technology, Norwegian University of Science and Technology, Gjøvik, Norway

School of Pure and Applied Sciences, Open University of Cyprus, Nicosia, Cyprus
e-mail: sokratis.katsikas@ntnu.no; sokratis.katsikas@ouc.ac.cy

C. Alcaraz (ed.), *Security and Privacy Trends in the Industrial Internet of Things*, Advanced Sciences and Technologies for Security Applications,
https://doi.org/10.1007/978-3-030-12330-7_2

1 Introduction

The use of computers in manufacturing processes started as early as the late 1960s, with software used to control machine tools that would be able to build 2-D or 3-D objects designed by Computer Aided Design (CAD) software [38]. The term used to describe this type of software is Computer Aided Manufacturing (CAM). Even though this technology managed to facilitate a faster production process and produce components and tooling with more precise dimensions and material consistency, it cannot reason as a skilled machinist can. Therefore, the use of CAM requires heavy interaction with and involvement of the user.

The next evolutionary step in the use of information and communication technologies (not only computers) in manufacturing came in the early 1970s, with the concept of Computer Integrated Manufacturing (CIM) i.e., the manufacturing approach of using computers to control the entire production process, by allowing individual processes to exchange information with each other and initiate actions, thus creating automated manufacturing processes [15]. The concept was not to be turned into reality before the mid 1980s. Notwithstanding its many advantages over CAM, CIM was still missing the ability to reason and connectivity was limited to the inside of the CIM installation.

In the beginning of the twenty-first century, the term "cyber-physical systems" was coined. According to the original definition [11], "Cyber-physical systems are physical, biological, and engineered systems whose operations are integrated, monitored, and/or controlled by a computational core. Components are networked at every scale. Computing is deeply embedded into every physical component, possibly even into materials. The computational core is an embedded system, usually demands real-time response, and is most often distributed. The behavior of a cyber-physical system is a fully-integrated hybridization of computational (logical) and physical action". A little earlier, in 1999, Kevin Ashton coined the term "Internet of Things (IoT)", to describe the extension of the Internet to cover networking with and among everyday objects.

In 2011, the term "Industry 4.0" was first used [14]. It is a collective term for technologies and concepts of value chain organizations which draws together cyber-physical systems and the Internet of Things, and is commonly referred to as the fourth industrial revolution. The term "Industrial Internet of Things (IIoT)" is used to describe the application of the IoT concept to industrial environments and comprises the technologies underlying the implementation of the Industry 4.0 initiative. Despite the semantic difference between the IIoT and Industry 4.0, it is more often than not that the two terms are being used interchangeably.

The promising value creation opportunities expected to derive from the realization of Industry 4.0, through the exploitation of underlying technologies, including the IIoT and artificial intelligence -that aims, inter alia, at providing the reasoning ability that CAM and CIM missed- do not come alone. With them comes also the realization that connecting industrial processes, that were originally designed under the assumption of no connectivity outside the controlled access industrial environment, to the Internet raises serious cybersecurity risks [34]. Hence, the need to understand and mitigate such risks, which has recently given rise to relevant cybersecurity research.

The objective of this chapter is to provide an overview of recent research results in the domain of IIoT cyber security. In order to do this in a systematic and organized manner, a classification scheme for such results is needed. To this end, we have selected to use the NIST Framework for Improving Critical Infrastructure Cybersecurity [7], that provides a common organizing structure for multiple approaches to cybersecurity and offers a flexible way to address cybersecurity. Even though developed to assist US critical infrastructure owners and operators in identifying and mitigating cyber risk, the framework can also be used -and has been used- in other sectors or communities. The framework describes five core functions, namely Identify, Protect, Detect, Respond, and Recover that together provide a high-level, strategic view of the lifecycle of an organization's management of cybersecurity risk. The remaining of the chapter discusses recent research results classified in each of the five framework functions in corresponding sections, followed by our conclusions, in which areas where further research is needed are identified.

2 Identify

Security for the IIoT has been recently identified as an important field and a lot of research has focused on identifying what are the major relevant security issues and on evaluating the accordant risk.

One of the early attempts to define the major risks in such networks [5] discusses the transformation from traditional industrial control systems to more flexible and versatile schemes based on wireless IoT sensors. The authors present the historical evolution of industrial control systems; and they discuss the commonly used architecture of Remote Terminal Units (RTUs) which enables the remote monitoring of data produced by sensors or systems in remote locations of an installation. They then describe recent advancements in internet and wireless sensors and how these technologies can be employed in order to create more efficient industrial systems. They also identify some of the major security risks such technologies may introduce; these are mostly related to access control, nonexistent or poor use of cryptography and insufficient network or application security countermeasures.

Industrial systems are often related to critical infrastructures, so it is essential to consider the main non-functional requirements of such systems before discussing security. Specifically, high quality maintenance of the infrastructure is required in order for it to be in the best possible state. The reliability of such systems is one of the most important factors, as erroneous functioning or limited availability may induce deviations in the work flow and consequently significant financial losses. Finally, the overhead in processing and communications shall be monitored and limited, in order to ensure the unobstructed functioning of the system.

2.1 Security Requirements

Regarding security, the main identified issues are as follows:

- **Attack impact** In industrial systems the impact of successful cyber attacks is normally high, due to the critical nature of such systems. This makes industrial systems very popular targets for attackers.
- **Secure communications** It is essential to maintain secure communication channels between all nodes of the system. Bearing in mind the diversity of nodes along with their limited flexibility, in terms of software customization, significant effort is required in order to apply traditional encryption approaches for communication among them.
- **Authentication/Authorization** Another important issue that has to be carefully addressed is the authentication and authorization mechanisms used. Machine to machine authentication techniques shall be employed, to provide the required layer of trust between nodes that function without human intervention.
- **Accountability** The size of possible damages from security related events implies that it is crucial to be able to detect the source of the problem. It is important to record the interactions between users and nodes of the system.
- **Trust Management** It is common in such implementations for multiple nodes to provide the same service, for redundancy purposes. In a heterogeneous environment, where not all nodes operate in the same way, it is crucial to be able to select the nodes to collaborate with. Trust management is required to measure and share the reputation of different nodes to allow them to selectively collaborate with the most reliable counterparts.

2.2 Integration Approaches

An important factor that defines most of the security parameters, is the integration approach selected for integrating wireless sensor networks into existing industrial systems. Some of the most common setups are as follows:

Front-End The nodes of the wireless sensor network are not directly connected to the internet. There is a concentrator device that manages the interaction with the outside world and at the same time communicates with the nodes of the network, implementing specific IoT communication protocols. The concentrator, a major component of the installation, manages all data flows and is a critical part of the whole system.

Gateway In this scenario there is a device that acts as a gateway. In practice it translates all IoT specific protocols used by the devices in the installation to TCP/IP and it forwards connections from inside of the installation to outside and vice versa. In comparison to the front-end scheme the gateway device is still related to all communications, but in this case it only forwards traffic instantiated by other devices or hosts.

TCP/IP In this approach all devices implement the TCP/IP protocol (or protocols compatible to TCP/IP), so that they are able to communicate to the Internet without requiring any middleware device. Any host can open a direct connection with them, and vice versa, if the access control mechanisms allow it. Even with such mechanisms in place, the direct connection of devices to the internet enlarges the attack surface for the installation and makes its protection a very difficult task.

The main advantages of the TCP/IP approach are full integration with the Internet, support for gradual updates, and resilience to node failures. On the other hand, more complex security mechanisms are required, it is significantly vulnerable, and specific industrial WSN protocol optimizations cannot be used. The Front-End approach can utilize standards for security mechanisms, it can use specific industrial WSN protocols, and produces networks that are simpler to maintain. The main disadvantages are the complexity of the upgrade procedure and the high risk of the Concentrator becoming a single point of failure. Finally, the Gateway approach has the same advantages with the Frond-end approach, with the addition of support for gradual updates. On the other hand, it requires more complex security mechanisms, while both maintenance and upgrade procedures are harder [5].

According to the analysis in [3], designing a strategy for interconnecting CPSs to the Internet is a complex procedure, as multiple factors have to be taken into account. Real-time performance shall not be affected by complex access management mechanisms. The dependability of industrial systems is crucial, as in most cases they constitute critical systems; therefore the requirements for any interconnection mechanism to be employed are the same. Scalability and extensibility have also to be taken into account, as the integration mechanism should have the required capacity to serve the interconnected CPSs in the case of expansion or upgrade of the latter.

The selected approach has an important effect on the fulfillment of the security requirements mentioned above. The existence of a middle-ware device provides various alternatives with regards to authentication, authorization, accountability, and trust management. On the other hand, such a device may hinder the implementation of end to end encryption or it may act as a single point of failure for the installation.

2.3 Industrial IoT Security Parameters

In order to identify the main security issues that come up in the context of the IIoT, the security related parameters of the domain have to be highlighted. The transition from cyber-physical systems to the internet of things in industrial systems has altered the security parameters for industry 4.0 [30].

Firstly, it has to be taken into account that designing security and prevention solutions for the IIoT is a **trade-off between security and availability**. In the event of a detected cyber attack the system should probably go offline; this may be impossible in cases when availability is fundamental.

Another important issue is related to **the integrity verification of devices** in industrial installations. It is common to find in a production environment devices with modified firmware or even devices in which counterfeit hardware has been used. When such devices are used in industrial environments then the risk of failure and significant damage is high, while the scenario of industrial espionage is also possible. Mechanisms to verify integrity in such systems must be used as a matter of high priority.

One of the most important problems in industrial systems occurs when upgrading from an older system, which did not account for global connectivity, to a modern internet of things system [35]. Such upgrades end up with **legacy devices connected to the internet**, either directly on indirectly. The usual case is that existing installations are transformed through the use of new technology gradually and this hides an important risk. The legacy system may have networking capabilities, but these are built in such devices in order to occasionally connect to local area networks mainly for maintenance reasons. When legacy systems are partly replaced or extended by the use of modern IoT systems, then the older systems that remain in use are also connected to a network that is directly connected to the internet. Legacy devices are not protected by any security mechanisms, and connecting these to the internet practically exposes multiple vulnerabilities and provides attackers with the opportunity to attack the installation.

Encryption is very important, in order to protect data privacy and to limit the possible damage in the case of a cyber attack [23]. Encryption can be employed either on the network layer or on the application layer. The former approach creates a fully encrypted connection (for example Transport Layer Security – TLS), while the latter approach only encrypts the data payload. Network layer encryption is a better option from the security point of view, as apart from protecting privacy it also defends the system from replay attacks. Application level encryption is more flexible in terms of collaboration between light devices and concentrator nodes. Gateway devices may be required to forward or inspect data sent from IoT devices; this is impossible when network level encryption is used. In IIoT communications the diversity of installed devices and firmware makes **applying encryption a relatively complex procedure**.

Even for installations that are not directly connected to the internet, the risk of being targeted by attacks is relatively high [22]. As it has become evident by recent attack attempts [19, 39], **attackers may use alternative ways to gain access to restricted industrial units**. As there are many humans working inside such installations, attackers usually take advantage of multiple techniques that are based on the human factor. Social engineering is a common way to extract information about a victim user and use it to gain access to their accounts. Additionally, a more recent issue has arisen through the proliferation of the Bring your Own Device paradigm, as it is common for users to connect their mobile devices, such as smartphones or laptops, to corporate networks. No matter how well the corporate network is secured, if such mobile devices are infected or vulnerable, then they provide an alternative option for the attacker to infiltrate the corporate network.

One of the most common root causes for cybersecurity issues in the IoT domain is that **IoT devices are not manufactured with security as a main design goal** [40]. Usually such devices are as cheap as possible, in order to survive the competition. The relatively low cost of such devices will increase significantly if manufacturers start implementing complex security mechanisms. Consequently, the general rule is that devices themselves are not going to be protected, so additional third-party security mechanisms must be employed in each installation. This scheme is not easy to be applied as many compatibility issues may arise and a performance overhead may be induced into the work-flow of the industrial system.

The development cycle of devices used in the IIoT does not allow resolving issues discovered in the long run [29]. These devices are being designed and produced in one cycle, with limited feedback from the end users that install them in real world scenarios. When a vulnerability is disclosed, then it is not certain that manufacturers will resolve the issue. Even if that happens, updating devices is not easy, as firmware upgrade may not be trivial or a hardware replacement may be required for the device. This means that security problems disclosed throughout the lifetime of the devices are rarely fixed in deployed installations. Practically, **security vulnerabilities of devices are rarely fixed in deployed installations**.

It is usual in industrial systems to use the Radio Frequency Identification (RFID) technology [41]. RFID is a communication technology that allows data transfer without establishing a mechanical or optical contact between a system and a specific target. This can be useful in an industrial workflow, especially if it is combined with IoT devices. On the other hand, **the use of RFID technology additionally enlarges the attack surface** of the system. There are multiple attacks that can be launched because of the use of such technology [24], such as jamming of the communication aiming to interfere with the normal functioning of the system; relay attacks aiming to change the content of the communication; cloning or spoofing of the RFID tags to achieve impersonation or eavesdropping, aiming to violate data confidentiality in the installation.

Protecting IIoT systems against zero day attacks is very hard [9]. Due to the nature of such installations, it is not easy to make changes in the configuration, in the hardware, or the firmware at short notice. For traditional computer systems, when new vulnerabilities or possible attacks are identified and disclosed, the most efficient approach is to render the system inactive until an update that fixes the problem is published. Then the system shall be updated and booted up again. This procedure is not possible to be applied in IIoT systems. Due to the critical nature of the installation, **finding a time slot to install an update is very hard**, while deactivating the system until an update is issued is completely out of the question.

Industrial IoT systems are very complex. They are made up of multiple different devices that are usually built without taking into account cyber-security [37]. Applying traditional systematic security approaches in such systems is ineffective due to various collaborating subsystems that practically adhere to completely different rules. Even if each component is sufficiently protected, the collaboration and the data exchange between dissimilar subsystems may trigger additional security risks that are hard to cope with. **Maintaining oversight of security within such systems is a hard task** and should not be underestimated.

By default, devices used in the IIoT are limited to hardware resources, while in some cases they operate on battery power. These characteristics make such installations **vulnerable to resources exhaustion attacks** [6]. This is a kind of vulnerability that is exploited by depleting the devices' hardware resources, such as processing power, RAM, storage or communication bandwidth. Specifically for battery operated devices, if a malfunctioning state of the device that keeps it busy is invoked, then the battery can be quickly depleted. Such attacks can make the device temporarily unavailable and cause significant problems to the installation.

Security by design is always the most efficient approach for protecting a computer system or network [25]. It is always easier to act in a proactive manner instead of trying to resolve security issues in an installation after it has already been attacked. In the IIoT, such a proactive approach is not as common as it should be. Either because such systems are designed by people with relatively low security awareness or because the existing tools and methods are not applicable in this context, **security is not taken into account as seriously as it should be in the design phase**; this almost always creates significant problems in the long run.

3 Protect

As the impact of cyber security attacks against IIoT systems is very high, it is crucial to protect such systems in the best possible way. Recent research efforts propose the use of appropriate security protection mechanisms when designing such systems. The main objective of applying such mechanisms is to minimize the impact of cybersecurity attacks in the IIoT context. The relevant literature reveals two main directions for such mechanisms, namely hardware security measures, and communication protection measures.

3.1 Hardware Security Measures

One of the most vulnerable components of industrial systems is the actual IoT devices. Due to causes analyzed in Sect. 2, it is hard to protect IoT devices throughout their lifetime. An interesting approach to enhancing the security of such devices is to use a secured execution environment [21], a secure area of the main processor that guarantees code and data loaded inside it to be protected with respect to confidentiality and integrity. The use of isolation is proposed, which is based either on special processor extension or external hardware module solutions. In practice, such approaches provide the device with an isolated environment of enhanced security, in order to cryptographically protect the execution integrity for critical parts of the device's software.

While the isolation approach has been proven successful in the general computing context, it remains to be proved that it can efficiently offer the same or similar features when it comes to critical real-time applications, such as IIoT scenarios. In this context, an enhanced implementation of the trusted execution environment, called IoT Trusted Execution Environment for Edge Devices (IoTEED), is proposed [26]. Security is assured while the systems real-time properties remain nearly intact. The authors have designed a new Trust-Zone-based architecture that implements a trusted execution environment, while their implementation is able to meet critical real time requirements.

Configuration of smart sensors used in industrial environments is needed throughout their complete life-cycle. The use of Near Field Communication (NFC) is proposed in order to secure the configuration or update process of sensors [36]. The architecture discussed therein includes a portable device that is used for committing the configuration of the smart sensors, and a back-end server. The portable device communicates through NFC with the smart sensors, to minimize the risk of a man-in-the-middle attack. There are two levels of confidentiality to which a configuration action may adhere. In the case of lower requirements in terms of integrity, the attestation of integrity is carried out between portable device and sensor, while in the case of more strict requirements the back-end server is also involved in the procedure.

3.2 Communication Channels Security

Another proposed protection approach is to strengthen the security features of communication protocols between devices. Due to the fragmentation of communication protocols in the IoT landscape, there are many cases in which the chosen protocols do not provide enough security countermeasures. From this point of view, it is crucial to make the most secure options available under the constraints imposed by the need for compatibility among installed devices. Additionally, when possible, the security of selected communication protocols has to be enhanced. In a recent research effort [33], the authors propose the use of a 5G radio access network for the industrial and tactile Internet of Things. They use a centralized Software Defined Networking (SDN) control plane, deterministic low-jitter scheduling, and lightweight encryption in Layer 2, to design a new approach for providing wireless security and enhanced performance at the same time.

In a similar context, the use of the Message Queuing Telemetry Transport (MQTT) protocol is proposed in order to support secure and reliable communication between IoT devices in the industrial sector [16]. MQTT is an IoT connectivity protocol that runs on top of the TCP protocol. It is a message-oriented protocol, that implements a publish/subscribe interaction model. There is a main broker entity that is interfering with all clients and enables them to exchange messages. The protocol supports the authentication of clients, while authorization can be achieved through the use of Access Control Lists (ACLs) in the broker side. The confidentiality of messages can also be protected by either using application level encryption (payload encryption) or network level encryption (client to client or client to broker encryption).

Securing communication between lightweight devices is not straightforward because of the limited resources available in such devices. A lightweight authentication mechanism [10] based only on hash and XOR operations has been proposed to be used in machine to machine communication. The proposed algorithm requires a router with a Trusted Platform Module (TPM), and a security chip (cryptocontroller) that performs cryptographic functions. On the device side the mechanism creates minimal overhead in terms of computation, storage and communication, so it is easy to be applied in already deployed installations. It achieves mutual authentication, session key agreement, and device identity confidentiality, while it offers resistance against multiple attacks, such as replay attacks, man-in-the-middle attacks, impersonation attacks, and modification attacks.

Communication security is of high importance for industrial installations, that are segmented in different geographical areas or even in different networks. Using network tunnelling (Virtual Private Network – VPN) is the most common approach, in order to enable devices to communicate securely through less secure networks such as the Internet. When the communicating nodes are limited in terms of resources, the choice of a tunnelling architecture and of encryption algorithms becomes a complex task. Most used VPN solutions have been tested in this context [27] and IPsec seems to perform better in terms of throughput or key exchange rate requirements. On the other hand, openVPN is much easier to be installed in various architectures and platforms. IPsec setup is relatively complicated and employing it in the IoT domain requires installing it in various devices of different vendors. Depending on the flexibility that these devices offer, this procedure may guide to insecure compromises or even be completely impossible to carry out. While traditional VPN technologies seem to provide an initial solution to setting up tunnel connections between IIoT nodes, it is obvious that novel methods, which will be more lightweight and compatible to IoT device firmware, have to be proposed.

3.3 General Protection Approaches

The diversity of application domains in which the IIoT is used makes it harder to protect installations by adopting a common, unified approach. While in traditional computing environments most of the requirements and parameters of different installations are similar, in the IIoT the functional and non-functional requirements are strongly coupled with the domain of each different use case. Practically, a one-size-fits-all approach is usually not efficient, and there is no unique methodology that can protect all different IIoT installations [12]. When designing security systems in such scenarios, a thorough analysis is required in each different industrial application domain. Security risks and possible countermeasures significantly differ with regards to the installed devices or with the general workflow of the system. Security analysts need to conduct attack vector analysis in each different scenario and provide custom solutions, based on available tools and mechanisms for each different case.

Usually data monitored by IoT sensors is sent and stored in cloud servers, so an important issue arises regarding both privacy and integrity of information. Data can be encrypted before being uploaded to the server, but if traditional encryption techniques are used, then the whole dataset must be downloaded and decrypted, in order to be usable. Taking into account the volume of data collected over time, the overhead added by traditional encryption may render it an inapplicable solution to the privacy and integrity problem mentioned above. Recently, more flexible encryption algorithms, that enable more options than just encrypting and decrypting data, have been studied. Specifically for the IIoT, a public key searchable encryption scheme has been proposed [13]. This public key scheme makes the distribution of keys a straightforward procedure and at the same time enables searching for keywords through the ciphertext. Practically, it is an important step towards combining the protection of privacy and integrity of data being collected with the usability of being able to access part of this data, according to specified criteria, while the data remain encrypted.

Blockchain technology has also been proposed as a means of increasing security in the IIoT sector [32]. Blockchain can provide trust in a decentralized network of nodes and protect the integrity of data being exchanged between those. In practice, if an IIoT system is viewed as a peer to peer system of multiple nodes, then blockchain can add interesting features, such as data integrity or time stamping; these may be very useful in critical real-time applications. Additionally, blockchain can provide fault tolerance in systems based on more than one similar sensing nodes, as one of its main characteristics is Byzantine fault tolerance. Employing blockchain can ensure data consensus among nodes, even in the case when a significant percentage of those fails or becomes compromised.

4 Detect

Due to the impact of cybersecurity attacks in IIoT installations, a novel research area that has recently seen more activity is the detection of such attacks.

Recently, Convolutional Neural Networks (CNNs) have been proven to be one of the most important advances in machine learning. There are a lot of successful applications of such systems, but their use requires a lot of processing power. This characteristic hinders the use of CNNs in scenarios with limited hardware resources. In the IIoT case there are a lot of devices producing significant volumes of data and CNNs would be suitable for processing this data to detect anomalies in the functioning of industrial systems. Squeezed Convolutional Variational AutoEncoder (SCVAE) is a model for unsupervised anomaly detection in time series sensor data.

Its use has been proposed [18] in edge computing environments, where processing power is restricted. It has been proven that the proposed system is able to detect abnormal states for industrial systems with relatively low processing resources requirements. Consequently, it can be employed to detect cyber attacks, with the procedure being carried out locally in the IoT system/network, without using cloud based solutions that induce additional privacy risks.

In industrial systems the detection of an irregular state of the system can be based upon the actual data that the devices are sensing. SysDetect [17] is a system that monitors the behavior of industrial processes (I/O values) in real time and detects the occurrence of critical states. An a-priori algorithm is employed, in order to detect frequent (normal) states of the system. Expert knowledge is used in order to certify that the functioning of the system is truly legitimate and, after some iterations, critical states of the system can be detected. The authors have also tested the scalability of the system by checking the produced results by varying the number of sensors and actuators and the percentage of critical states against all possible ones.

Another interesting approach is tackling a special kind of attacks called sequence attacks, which are based on the misplacement of normal events in order to cause damage, while going undetected [8]. Most detection systems detect on a per event (traffic packet, system call etc.) basis, by classifying each event as malicious or not. The proposed system identifies patterns of Industrial Control Systems (ICS) network events, extracts their semantic meaning and models known behaviors over time. In practice, the normal behavior of the system is modeled with the use of discrete time Markov chains and then the current activity of the system is compared to the normal one by computing a weighted distance between Markov chain states.

Network intrusion detection has been excessively studied for computer networks and similar approaches were also recently proposed for industrial systems. Network telemetry intrusion detection [28], a method that takes into account meta-data of network traffic has been proposed for protecting industrial systems from cyber attacks. Specifically, the data monitored are the time it takes the client to respond to messages, the amount of client-side dropped packets, the time between the repeated packet transmissions and the number of server-side dropped packets. By feeding such data to various classifiers it was shown that it is possible to detect known attacks by observing meta-data on network connections.

A common attack related with IoT in general is Distributed Denial of Service (DDoS) and a multi layer DDoS mitigation framework has been proposed to protect IIoT networks from such attacks [42]. The three layers of the proposed methodology are the perception layer, the network layer and the application layer, which correspond to edge, fog and cloud computing levels. The system is compatible with SDN network environments and it is hierarchically structured in order to efficiently process all incoming data and accurately make detection decisions. The edge level is responsible for protecting the devices with various techniques, such as firmware security checks, access control, malicious firmware/software detection or vulnerability scanning. The fog level of the system actually protects the IoT network from the DDoS attacks as it collects traffic data, it detects DDoS attacks based on this data, it restrains attacks based on detection and it perceives the network state by using honeypots. Finally, the cloud level offers the required processing power to commit the required data analytics to detect the attacks. In fact, its resources are occasionally used by the fog level components to execute the algorithms required for the detection of attacks.

Industrial systems are complex in general and usually include mobile devices to enable users to control such installations. Such devices operate on well known mobile operating systems and are vulnerable to malware attacks such as viruses, worms, trojan horses, and rootkits. Thus, it is crucial for mobile control devices used in industrial environments, to be adequately protected from such attacks. In a recent paper [31], malware mobile app detection is discussed and the effects it can have on IIoT systems security are highlighted. The detection methodology can either be static, so apps are characterized according to features such as the permissions required for their functioning, or dynamic, so apps are characterized according to features such as the system calls used when executed. Additionally hybrid approaches, that combine both previous methodologies are analyzed. The authors also analyze the method's shortcomings, such as inefficient detection of zero day attacks; this may limit the positive effect for IIoT systems.

5 Respond and Recover

There in no significant research activity in the IIoT domain regarding the two remaining functions of the NIST framework, namely respond and recover. There are some research attempts [20], but those are in initial stages and no concrete results have been presented yet. The related subjects of resilience and structural controllability for cyber-physical systems have been addressed in [1, 2] and in [4] respectively. While the respond and recover functions are related to the final stages of security planning and designing, they are equally important to the rest of the functions. Regardless of whatever actions and measures are taken in order to identify risks, protect the system and detect attacks, we have to assume that it is impossible to completely avoid a security related event. Therefore, both respond and recover actions are required in order to mitigate the damage caused by such events.

6 Conclusions

The research activity regarding security in IIoT systems has been reviewed and analyzed according to the five security functions identified in the NIST Framework. These functions are identify, protect, detect, respond and recover. In general the literature review has come up with significant findings for only the three first functions, while few research efforts have been identified for the respond and recover functions.

In the scope of the identify function there is a lot of activity, as many different researchers have tried to enumerate the main parameters that make IIoT security a difficult problem to solve. It has become evident that a lot of effort will be needed in the near future, in order to provide mature security solutions for this domain. Security is a difficult problem to solve even in traditional computing environments. IoT installations add a lot of additional issues to cope with. Specifically, installed firmware in IoT devices is usually out of date, required security mechanisms are not embedded into the devices and the quality of the parts of such devices is questionable. If someone adds to these issues the special conditions that hold in the industrial domain, then more problems arise. Industrial systems are infrastructures whose any interruption of availability or any malfunctioning may cause significant problems. From this perspective the risk from possible cyber attacks is high, while applying various security countermeasures may be a much more complicated procedure, compared to other domains.

Regarding the protection of IIoT systems, there have been two main findings. Recent research has proposed either the use of hardware security measures or the use of encryption techniques in order to proactively protect IIoT installations. The trusted execution environment has emerged as a novel approach to enhance the security protection of electronic devices, by utilizing a special functionality offered by modern processors. Device manufacturers can use an isolated part of the processor and ensure the integrity and the confidentiality of the data or the critical execution logic of the system. By this concept IoT devices can be built to be more resilient to any kind of attack. The other main trend in protecting IIoT installations is the use of encrypted communication schemes. While this is not a huge innovation, overcoming issues imposed by the characteristics of devices to be protected makes it difficult to employ. Applying encryption for communication in such environments requires creating encryption schemes that can function under limited hardware resources of the involved devices. Additionally, proposed solutions have to be easy to install, as the firmware on devices does not provide flexibility in terms of software installation.

The use of techniques and approaches from the traditional computer security domain has been proposed for detecting attacks in IIoT systems. Monitoring network traffic or relevant metadata can reveal significant patterns related to security attacks. Additionally, the abnormal functioning of such systems may be concluded by observing the actual, monitored application data and stores by such systems. It is common for a malfunctioning sensor to send inaccurate data, possibly also in different than normal rates or for a malfunctioning actuator to strangely interact with the installation in an observable way. Most of the methods that detect attacks require that significant processing of data has to be carried out in real time. In order to apply such techniques, researchers had to either use additional hardware for the implementation of the detection procedure or customize the algorithms by limiting their detection scope to make them lighter and appropriate for running on low power devices.

A lot of effort is required to create both respond and recover schemes for IIoT. As it has been mentioned in Sect. 5, the number of relevant references in the literature is limited. Bearing in mind the potential damage that can be caused by security attacks in IIoT environments, there is an urgent need to come up with ways to either prevent detected events from happening or to be able to recover quickly and in a cost-effective manner, after the occurrence of an attack.

References

1. Alcaraz C (2017) Resilient industrial control systems based on multiple redundancy. Int J Crit Infrastruct (IJCIS) 13(2/3):278–2951
2. Alcaraz C (2018) Cloud-assisted dynamic resilience for cyber-physical control systems. IEEE Wirel Commun 25(1):76–82
3. Alcaraz C, Lopez J (2017) Secure interoperability in cyber-physical systems. In: Ferrag MA, Ahmim A (eds) Security solutions and applied cryptography in smart grid communications. IGI Global, Hershey, pp 137–158
4. Alcaraz C, Lopez J (2018) A cyber-physical systems-based checkpoint model for structural controllability. IEEE Syst J 12:1–12
5. Alcaraz C, Roman R, Najera P, Lopez J (2013) Security of industrial sensor network-based remote substations in the context of the Internet of Things. Ad Hoc Netw 11(3):1091–1104
6. Bakhshi Z, Balador A, Mustafa J (2018) Industrial IoT security threats and concerns by considering Cisco and Microsoft IoT reference models. In: 2018 IEEE Wireless Communications and Networking Conference Workshops (WCNCW), Apr 2018. IEEE, pp 173–178
7. Barrett MP (2018) Framework for improving critical infrastructure cybersecurity version 1.1. Technical report
8. Caselli M, Zambon E, Kargl F (2015) Sequence-aware intrusion detection in industrial control systems. In: Proceedings of the 1st ACM Workshop on Cyber-Physical System Security – CPSS'15. ACM Press, New York, pp 13–24
9. Choi J, Shin Y, Cho S (2018) Study on information security sharing system among the industrial IoT service and product provider. In: 2018 International Conference on Information Networking (ICOIN), Jan 2018. IEEE, pp 551–555
10. Esfahani A, Mantas G, Matischek R, Saghezchi FB, Rodriguez J, Bicaku A, Maksuti S, Tauber M, Schmittner C, Bastos J (2017) A lightweight authentication mechanism for M2M communications in industrial IoT environment. IEEE Internet Things J 6(1):288–296
11. Gill H (2008) From vision to reality: cyber-physical systems. In: HCSS national workshop on new research directions for high confidence transportation CPS: automotive, aviation, and rail
12. Hassanzadeh A, Modi S, Mulchandani S (2015) Towards effective security control assignment in the industrial Internet of things. In: 2015 IEEE 2nd World Forum on Internet of Things (WF-IoT), Dec 2015. IEEE, pp 795–800
13. He D, Ma M, Zeadall S, Kumar N, Liang K (2017) Certificateless public key authenticated encryption with keyword search for industrial Internet of Things. IEEE Trans Ind Inf 14(8):3618–3627
14. Jeschke S, Brecher C, Meisen T, Özdemir D, Eschert T (2017) Industrial Internet of things and cyber manufacturing systems. In: Jeschke S, Brecher C, Song H, Rawat D (eds) Industrial Internet of Things. Springer, Cham, pp 3–19
15. Kalpakjian S, Schmid S (2006) Manufacturing engineering and technology, 5th edn. Prentice Hall, Upper Saddle River
16. Katsikeas S, Fysarakis K, Miaoudakis A, Van Bemten A, Askoxylakis I, Papaefstathiou I, Plemenos A (2017) Lightweight & secure industrial IoT communications via the MQ telemetry transport protocol. In: 2017 IEEE Symposium on Computers and Communications (ISCC), July 2017. IEEE, pp 1193–1200

17. Khalili A, Sami A (2015) SysDetect: a systematic approach to critical state determination for industrial intrusion detection systems using Apriori algorithm. J Process Control 32:154–160
18. Kim D, Yang H, Chung M, Cho S (2017) Squeezed convolutional variational autoEncoder for unsupervised anomaly detection in edge device industrial Internet of Things, Dec 2017. arXiv:1712.06343v1
19. Langner R (2011) Stuxnet: dissecting a cyberwarfare weapon. IEEE Secur Priv Mag 9(3):49–51
20. Lee S-Y, Chung T-M (2017) A study on the fast system recovery: selecting the number of surrogate nodes for fast recovery in industrial IoT environment. In: 2017 International Conference on Information and Communications (ICIC), June 2017. IEEE, pp 205–207
21. Lesjak C, Hein D, Winter J (2015) Hardware-security technologies for industrial IoT: TrustZone and security controller. In: IECON 2015 – 41st Annual Conference of the IEEE Industrial Electronics Society, Nov 2015. IEEE, pp 002589–002595
22. Lesjak C, Druml N, Matischek R, Ruprechter T, Holweg G (2016) Security in industrial IoT – quo vadis? e & i Elektrotechnik und Informationstechnik 133(7):324–329
23. Manzoor A (2016) Securing device connectivity in the industrial Internet of Things (IoT). Springer, Cham, pp 3–22
24. Mitrokotsa A, Rieback MR, Tanenbaum AS (2010) Classifying RFID attacks and defenses. Inf Syst Front 12(5):491–505
25. Mouratidis H, Diamantopoulou V (2018) A security analysis method for industrial Internet of Things. IEEE Trans Ind Inf 14(9):4093–4100
26. Pinto S, Gomes T, Pereira J, Cabral J, Tavares A (2017) IIoTEED: an enhanced, trusted execution environment for industrial IoT edge devices. IEEE Internet Comput 21(1):40–47
27. Pohl F, Schotten HD (2017) Secure and scalable remote access tunnels for the IIoT: an assessment of openVPN and IPsec performance. In: Lecture Notes in Computer Science (Including Subseries Lecture Notes in Artificial Intelligence and Lecture Notes in Bioinformatics), LNCS, vol 10465. Springer, Cham, pp 83–90
28. Ponomarev S, Atkison T (2016) Industrial control system network intrusion detection by telemetry analysis. IEEE Trans Dependable Secure Comput 13(2):252–260
29. Quarta D, Pogliani M, Polino M, Maggi F, Zanchettin AM, Zanero S (2017) An experimental security analysis of an industrial robot controller. In: 2017 IEEE Symposium on Security and Privacy (SP), May 2017. IEEE, pp 268–286
30. Sadeghi A-R, Wachsmann C, Waidner M (2015) Security and privacy challenges in industrial Internet of Things. In: Proceedings of the 52nd Annual Design Automation Conference on – DAC'15. ACM Press, New York, pp 1–6
31. Sharmeen S, Huda S, Abawajy JH, Ismail WN, Hassan MM (2018) Malware threats and detection for industrial mobile-IoT networks. IEEE Access 6:15941–15957
32. Skwarek V (2017) Blockchains as security-enabler for industrial IoT-applications. Asia Pac J Innov Entrepreneurship 11(3):301–311
33. Szymanski TH (2017) Strengthening security and privacy in an ultra-dense green 5G radio access network for the industrial and tactile Internet of Things. In: 2017 13th International Wireless Communications and Mobile Computing Conference (IWCMC), June 2017. IEEE, pp 415–422
34. Thames L, Schaefer D (2017) Cybersecurity for industry 4.0. Springer, Cham
35. Ulltveit-Moe N, Nergaard H, Erdödi L, Gjøsæter T, Kolstad E, Berg P (2016) Secure information sharing in an industrial Internet of Things, Jan 2016
36. Ulz T, Pieber T, Steger C, Haas S, Matischek R, Bock H (2017) Hardware-secured configuration and two-layer attestation architecture for smart sensors. In: 2017 Euromicro Conference on Digital System Design (DSD), Aug 2017. IEEE, pp 229–236
37. Urquhart L, McAuley D (2018) Avoiding the Internet of insecure industrial things. Comput Law Secur Rev 34(3):450–466
38. US Office of Technology Assessment (1984) Computerized manufacturing automation: employment, education, and the workplace. Washington, DC: U.S. Congress, Office of Technology Assessment, OTACIT-235

39. van Lier B (2017) The industrial Internet of Things and cyber security: an ecological and systemic perspective on security in digital industrial ecosystems. In: 2017 21st International Conference on System Theory, Control and Computing (ICSTCC), Oct 2017. IEEE, pp 641–647
40. Wurm J, Hoang K, Arias O, Sadeghi A-R, Jin Y (2016) Security analysis on consumer and industrial IoT devices. In: 2016 21st Asia and South Pacific Design Automation Conference (ASP-DAC), Jan 2016. IEEE, pp 519–524
41. Xie C, Deng S-T (2017) Research and application of security and privacy in industrial Internet of Things based on fingerprint encryption. Springer, Cham, pp 102–110
42. Yan Q, Huang W, Luo X, Gong Q, Richard Yu F (2018) A multi-level DDoS mitigation framework for the industrial Internet of Things. IEEE Commun Mag 56(2):30–36

Advanced Persistent Threats and Zero-Day Exploits in Industrial Internet of Things

Ioannis Stellios, Panayiotis Kotzanikolaou, and Mihalis Psarakis

Abstract Manufacturing industry, electricity networks, supply chain, food production and water treatment plants have been heavily depended on Industrial Automation and Control (IAC) Systems. Integration of Information and Communication Technology (ICT) played a significant role in the evolution of these systems. New emerging trends and technologies, such as Internet-of-Things (IoT) interact with traditional, isolated IAC systems. Sectors such as manufacturing, electric grids, pharmaceuticals, and water treatment facilities incorporate part of these "smart" technologies in order to increase efficiency, performance and reduce production costs. But despite of its benefits, interconnectivity between smart and legacy IAC systems also creates complex interdependencies, which in turn, make imperative the need for more safety and security countermeasures. This rapid evolution has also affected greatly the threat landscape. In order to comprehend this radical change we present and analyze recent, well documented attacks that target mission critical IAC systems, which incorporate Industrial IoT technologies. In particular, we focus on highly profiled, sophisticated attacks against interconnected automation and monitoring field devices, related software platforms and systems (e.g., Programmable Logical Controllers – PLCs, industrial robots) installed on industrial facilities and smart grid generation, transmission and distribution networks and systems.

Keywords Security · Privacy · IIoT · Advanced persistence threats · Cyberattacks

---

I. Stellios (✉) · P. Kotzanikolaou
SecLab, Department of Informatics, University of Piraeus, Piraeus, Greece
e-mail: jstellios@unipi.gr; pkotzani@unipi.gr

M. Psarakis
ESLab, Department of Informatics, University of Piraeus, Piraeus, Greece
e-mail: mpsarak@unipi.gr

C. Alcaraz (ed.), *Security and Privacy Trends in the Industrial Internet of Things*,
Advanced Sciences and Technologies for Security Applications,
https://doi.org/10.1007/978-3-030-12330-7_3

1 Introduction

Industrial Control Systems (ICSs) have been constantly evolving in terms of efficiency, productivity, quality, manageability and operational security. Rapid evolution in computer science also affected ICS: Supervisory Control and Data Acquisition (SCADA) systems enabled manufacturers to remotely control complicated production lines via Human-Machine Interfaces (HMIs) placed in central management stations. Initially developed in the early 1950s, first generation of ICSs consisted mainly of Wide Area Networks (WANs) used to communicate with Remote Terminal Units (RTUs). The second generation ICSs utilized smaller and cheaper devices that were connected via Local Area Networks (LANs) whereas in the third generation interconnectivity with third-party peripherals was introduced.

SCADA systems were mainly built from customized hardware, controlled with the use of specialized software, utilized domain specific or proprietary network protocols and, until recently, these systems were, mostly, isolated from the outside world. But the constant need for improving efficiency, interoperability, manageability and production cost reduction introduced the fourth generation of SCADA systems which included, among others, new evolutionary technologies such as Industrial Internet of Things (IIoT) technologies, thus widening significantly their attack surface [3]. Modern SCADA systems that utilize relative IIoT technologies have been widely adopted in almost every critical aspect of our modern lifestyle, ranging from manufacturing industry, power generation, transmission and distribution, water treatment and reservoir, intelligent transportation and smart city/building systems. Real cyberattacks that utilize IIoT technologies have been on the rise throughout the world during the recent decade [7, 27, 34, 35].

High impact attack scenarios usually involve refined exploitation methods named after the term *Advanced Persistence Threats* (APTs). The term *advanced*, mainly corresponds to the fact that the adversaries utilize sophisticated attack techniques, that take advantage the full spectrum of publicly available exploits against well-known vulnerabilities, as well as custom payloads and delivery methods (*zero-day*) depending on the target's response. The term *persistence*, corresponds to the continuous interaction between the adversary and the compromised systems until the goal of the attack is achieved. In short, APTs are well planned, stealthy attacks, that use advanced exploitation techniques against a particular target, designed to be effective for a large period of time.

In a recent survey paper [44], we presented how relative Internet of Things (IoT) technologies, applied in different sectors (industry, smart grid, intelligent transportation systems, medical and smart home), can be utilized by adversaries to create new, hard to identify attack paths in order to launch high impact attacks against critical infrastructures and services. In order to further understand the threat landscape of IIoT ecosystem, we herein describe the different phases that take place in an APT attack scenario and analyze the applicable exploitation techniques used on each phase for several real/PoC attacks. In particular, we analyze APT attack scenarios against Internet facing field devices (e.g., PLCs, industrial robots) [17, 30, 42] as well as indirect attacks that mostly utilize spear-phishing techniques and existing connectivity paths between corporate and industrial networks [16, 25, 34]. The latter include recent, high profile APT attack scenarios against smart grid's transmission [18] and distribution [7] IIoT devices and systems.

The outline of the chapter is as follows: Sect. 2 presents an overview of ICS software, protocols and architectures. Then, Sect. 3 categorizes and presents *zero-day* exploits found on HMI software [5], some of which can/have be used to APT attack scenarios. Furthermore, Sect. 4 defines the basic phases of an APT attack, which in turn, are used to analyze several attack scenarios against IIoT field devices and smart grid SCADA networks in Sects. 4 and 5, respectively. Finally, Sect. 6 presents an overview of the attack scenarios' characteristics and proposes security countermeasures and best practices.

2 SCADA Related Protocols and Architectures

Due to the diversity of SCADA systems there is plethora of open industrial network communication protocols, such as Modbus/TCP, Distributed Network Protocol 3 (DNP3), Profibus, IEC-104, DeviceNET, ControlNET, Ethernet/IP, wireless IEEE 802.15.4x [12, 20, 28, 45] as well as proprietary ones. Since most of these protocols have been designed with no security features in mind they are susceptible to cyberattacks such as passive/active information sniffing, message spoofing and command injection [15].

Furthermore, in order to reduce production costs, manufacturers utilize off-the-shelf hardware to build industrial field devices, in which, they incorporate vulnerable communications protocols (e.g., 802.15.4x). Most of these devices are capable of communicating directly to the Internet via embedded network protocols (e.g., 6LoWPAN [33]) and are equipped with software that facilitates the integration of IoT technologies such as commercial cloud-computing services. The latter are mainly used to improve data accessibility, reduce the operational costs and increase flexibility, optimization and scalability. On the other hand, many of these cloud platforms [46] come with a plethora of vulnerabilities [38] on both system software and Application Programming Interfaces (APIs) that adversaries may use as an enabler to attack mission critical SCADA systems.

Modern SCADA systems consist (see Fig. 1) of a large number of Intelligent Electronic Devices (IEDs), such as sensors, actuators (e.g., circuit breakers), smart meters, robotics and motors that are controlled through Programmable Logical Controllers (PLCs) and Remote Terminal Units (RTUs). Both RTUs/PLCs utilize wireless/wired network interfaces and protocols in order to communicate to each other and to Master Terminal Units (MTUs). PLCs/RTUs are used to acquire a device's status, (e.g., valve open/closed), read and re-transmit operational variables (e.g., pressure, voltage) as well as control industrial equipment (e.g., robotics) by sending commands [23].

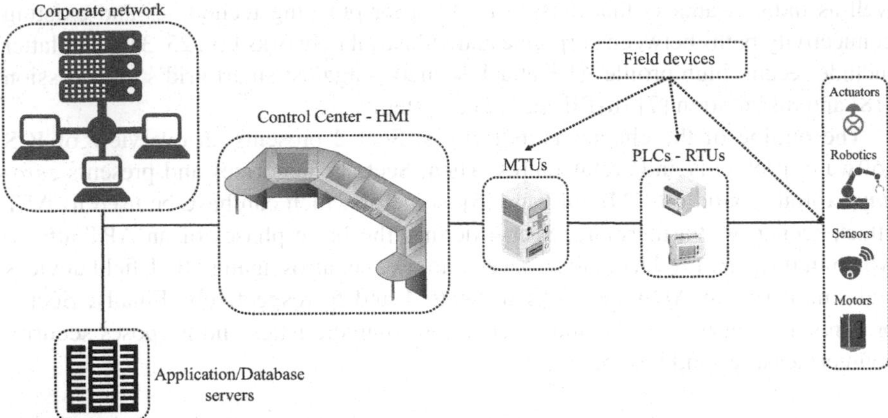

Fig. 1 A typical SCADA architecture

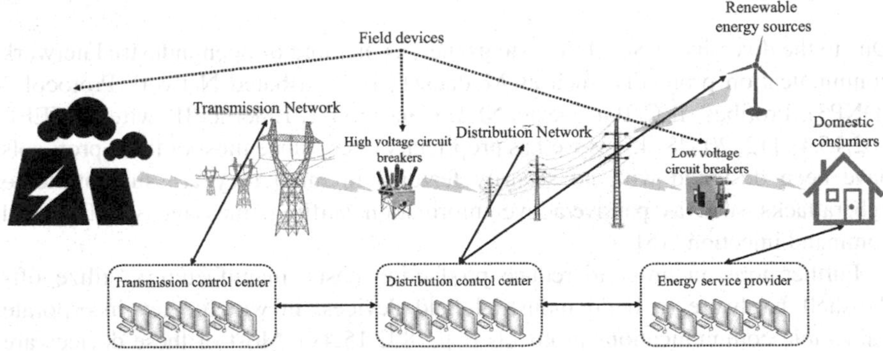

Fig. 2 A typical smart grid architecture

Typically industrial SCADA systems rely on a predefined structure. RTUs are usually interconnected using a hierarchical model to MTUs which, in turn, are connected to Command-and-Control (C&C) centers. Then, using software applications, human operators can administer, regulate and remotely control entire production lines via graphical HMIs from PC type workstations. Integration of IIoT technologies in SCADA systems, requested that C&C rooms must also be connected to the company's corporate network and/or to the Internet. Furthermore, international corporations also interconnect several C&C centers to master regional stations.

Smart grid adopts a hierarchical model similar to the industrial one, but far more complex. Its cornerstone is mainly the generation systems that produce electricity (see Fig. 2). Then, the electric current is transmitted using the backbone of smart grid, consisting of the transmission network and its substations. From there, the electricity is delivered to both home and industrial consumers through the distribution network. The latter mainly consists of the Advanced Metering Infrastructure (AMI) [31] and domestic renewable energy sources (e.g., solar panels).

Energy optimization and control are achieved with the use of Energy Management Systems (EMS) located in strategical places throughout the distribution network, whereas independent systems operators are responsible to manage the electricity flow between service providers and customers [32].

3 Zero-Day Exploits on Human-Machine Interface Applications

HMI software is considered to be the most critical application in IIoT ecosystem since it is installed on control rooms' workstations and its main purpose is to administer mission critical SCADA systems. Compromising an HMI system may lead to a series of attacks ranging from information gathering, deactivation of notification systems (e.g., alarms), notifications to operators up to physically damage industrial equipment. To make things worse, HMI vendors do not always enforce security best practices on the controlling software, thus focusing only on the managed devices. In this section we present the findings of an extensive research conducted by the *Zero Day Initiative* (ZDI) team of *Trend Micro* security company that took place throughout a two-year period (2015–2016) [5] and successfully identified 250 *zero-day* vulnerabilities on HMI applications. During the disclosure process, researchers observed that the average time period for the vendors to release a corresponding patch of a *zero-day* exploit averaged to 150 days. This actually meant that mission-critical SCADA systems were vulnerable for almost five months before a patch was available from software vendors. The exploitation techniques were classified into 4 main categories: (i) Memory corruption, (ii) Credential harvesting, (iii) Insecure installation, authentication and authorization procedures and (iv) Code injection. These exploitation techniques which can be used in various APT attack scenarios are described in detail in the following sections.

3.1 Memory Corruption

Memory corruption issues accounted for the 20% of the total number of vulnerabilities found. The majority were stack/heap-based buffer overflows [13] and out-of-bounds read/write vulnerabilities. In a particular vendor, the software Advantech WebAccess HMI Solution was proved to have a vulnerable `sprintf` function and no protection mechanisms such as stack cookies, Address Space Layout Randomisation (ASLR) [40] and SafeSEH [19]. Due to the absence of ASLR protection an adversary needs only to overwrite the return address to a controlled Return Oriented Programming (ROP) chain, in order to execute malicious code with elevated privileges. Even though the vendor issued a large number of patches these corrected only specific issues and did not address the problem globally or replaced other problematic functions.

3.2 *Credential Harvesting*

Vulnerabilities found in credential management represented the 19% of the overall vulnerabilities found. These included the use of hard-coded passwords as well as insecure storage and/or protection of passwords (e.g., stored clear text/with reversible encryption algorithms). Furthermore, in a particular case study of General Electric (GE) MDS PulseNET, a software that is used to monitor industrial equipment and communication networks deployed in energy, water, and waste water sectors globally, they managed to identify an embedded account with full privileges apart from the administrator and user account (CVE-2015–6456 [2]). By utilizing *HeidiSQL* tool they managed to extract the `ge_support` account as well as the password's MD5 hash value (`PulseNET`). Notably, even after a successful logging process of the discovered account its username did not appear in the user management screen.

3.3 *Insecure Installation, Authentication and Authorization Procedures*

This category represents the 23% of the total vulnerabilities found, including unencrypted communications, such as the transmission in plaintext of sensitive information (e.g., usernames or passwords), as well as vulnerable ActiveX controls which where marked as 'safe'. In another case study concerning Siemens SINEMA Server, a network management software for monitoring and diagnostics, a misconfiguration allowed standard authenticated users to have full access to Windows sensitive system folders (CVE-2016–6486). In addition, the binary code used to start the SINEMA service run at local system level thus allowing an adversary with local access to the workstation, to replace the legitimate binary code with a malicious one. Then, triggering a reboot allowed the adversary to execute the malicious code with system privileges.

In another case study considering Advantech WebAccess, a cross-platform user interface management based in HTML5, an authenticated user was able to retrieve the passwords of other platform's users including the administrator.

3.4 *Code Injection*

Although Structured Query Language (SQL) type and Operating System (OS) command injections occupy a small fraction (9%) of the overall vulnerabilities discovered, the impact of such threats on HMI systems is considered to be very high, especially those injections that apply to domain-specific languages for SCADA software solutions. In a particular case study, 'Cogent DataHub', a real time visualization software for complex SCADA systems, was evaluated. The application incorporates Gamma script language, a domain-specific language that has built-in features and functions for SCADA systems. Cogent DataHub also includes a database, that resides in server's memory providing interchange of data for Object Linking and Embedding (OLE) for Process Control (OPC) and other Windows applications. Researchers discovered that it is possible for an attacker to take advantage a flaw in the `EvalExpression` method of Gamma script language and enable the insecure processing mode in the Asynchronous JavaScript and XML (AJAX) web server, resulting in a remote code execution on the server.

4 APT Attack Scenarios on Industrial IoT Field Devices

Industrial systems usually attract well-funded, high-skilled and strongly motivated adversaries that seek to gain substantial economic profit (e.g., cybercriminals) or to disrupt a nation's Critical Infrastructures (CIs) (e.g., nation state adversaries). These attacks are considered of high impact due to the effect that SCADA systems have on our every day life. APTs' attack vectors, that exploit relative IIoT technologies, may include but not limited to, the following basic phases:

- **Reconnaissance/Data gathering and host discovery phase** Gathering valuable information regarding corporates' employees and executives, enumerating the targeted company's web presence and compromising corporate email accounts to launch a series of spear phishing campaigns [7, 18, 34] are considered to be the most prevailing methods in the early stages of an APT attack scenario. In addition, web search engines (e.g., Shodan) are also used to locate web exposed industrial equipment that then can be enumerated for vulnerabilities before the exploitation/ initial infection phase begins [5, 17, 30, 42].
- **Initial infection phase** Since corporate users must communicate with the outside world and, at the same time, are usually connected (directly or indirectly) to mission critical industrial control systems are considered to be the prime target for adversaries. This is usually accomplished by launching spear-phishing campaigns, which include the process of sending malware infected, office documents and malicious web links from hijacked corporate/legitimate accounts (e.g., [7, 18]). Another more direct approach is to exploit the web interfaces of modern industrial equipment, that utilize IoT enabling technologies, in order to be able to be operated, managed and updated remotely (e.g., [5, 17, 30, 42]). In addition, it is common practice for manufacturers as well as companies that provide technical support to industrial equipment, to distribute essential software components and/or updates (e.g., IIoT devices' firmware and relative management software) via vulnerable websites and unsecured methods (e.g., HTTP), with devastating consequences on IIoT ecosystem [34]. Finally, off-line exploiting techniques can be also used, as presented in [16, 25].

- **Establish and maintain remote access** Asynchronous communication, data masquerade and encryption, Intrusion Prevention/Detection System (IDS/IPS) evasion and privilege escalation are some of the techniques used in order to achieve stealthiness. To ensure access persistence, payloads are made so as to withstand power loss/reboot processes and equipped with auxiliary communication modules for redundancy.
- **Lateral movement and propagation phase** In APT attack scenarios, adversaries utilize several enumeration and pivoting techniques (e.g., probing nearby systems for open ports, connect to default drive shares, spread to different network segments) in order to locate and exploit other mission critical vulnerable ICT equipment such as control rooms' workstations and IIoT devices.
- **Remote control and device manipulation** Attackers must incorporate a series of well established and new industrial network protocols in order to remotely communicate and ultimately take control the IIoT device(s). The payloads installed on IIoT devices must be able to run with minimum resources and hide their code so as to avoid detection from machine operators.
  Functionality plays an essential role when designing payloads that target industrial equipment, since, adversaries must be able to issue arbitrary commands and even control all functions and features of the IIoT device/system. The latter enables adversaries to lock out legitimate operators thus preventing them from responding to the threat accordingly [7].

In many cases of APT attack scenarios the adversaries include payloads that are used to renter the devices and systems affected unusable and/or hide their footprints (e.g., [7, 18, 34]).

4.1 *Stuxnet*

The most well-known APT attack against SCADA systems, that managed to infect the software of at least 14 industrial sites in Iran, including a uranium enrichment plant, is considered to be *Stuxnet* [16, 25]. This 500 KB computer worm utilized four 0-day vulnerabilities to compromise two digital certificates, inject code into industrial control systems and hide the code from the operator. Its main goal was to sabotage industrial facilities by reprogramming network connected field devices to operate out of their specified boundaries. Although it required a victim to unintentionally install it in the network (e.g., via an infected external usb drive) the code was extremely stealth and sophisticated. Its main target was Siemens Steps7 software used for controlling industrial centrifuges. The worm operated autonomously by using self-replicating techniques to spread out to the internal network. It was equipped with advanced exploitation payloads that targeted Windows operating machines used to control specialized industrial equipment, thus enabling the adversaries to spy on the infected devices and even cause the destruction of the fast-spinning centrifuges. Although the authors of Stuxnet have not been officially identified, the sophistication of the discovered code indicates the involvement of nation state adversaries [6]. In particular the vector of the attack can be described as follows:

1. **Reconnaissance phase** Nation state adversaries create malware infected usb drives which, then, place in strategically chosen sites (e.g., at the Iran's industrial sites' entrances) so as to allure industrial workers to plug them to their computers.
2. **Initial infection phase** The worm is designed to infect Windows operating machines by taking advantage of auto-execution features in removable drives (Microsoft Windows Shortcut LNK/PIF Files Automatic File Execution Vulnerability – Bugtraq ID 41732). Then, it takes advantage of two *zero-day* Windows vulnerabilities to perform privilege escalation. In order to avoid detection, it utilizes a rootkit to hide its binaries so as to evade antivirus products.
3. **Lateral movement and propagation phase** Module *Export 22* was the main payload responsible for network communications and propagation. In particular:
   - Infects any newly inserted removable drives.
   - Utilizes peer-to-peer networks in order to connect to C&C servers.
   - Uses hardcoded credentials to infect WinCC devices [4].
   - Connects to all available default network shares.
   - Exploits a *zero-day* vulnerability (MS10-061) in Microsoft Windows print spooler service.
   - Exploits MS08-067 Windows Server Service Vulnerability.
4. **Establish and maintain remote access** Adversaries utilize peer-to-peer networks for communicating to C&C centers and updating purposes, whereas during the final infection process the malware was designed to hide its code on PLCs using a specially crafted rootkit.
5. **Remote control and device manipulation** The adversaries were able to remotely adjust the spinning rate of the network enabled centrifuges and, at the same time, falsify the information sent back to the operators. The latter enabled them to increase the spinning rate at a level where centrifuges started to fail without anyone noticing.

Although its main target were Iran's enrichment uranium plants the worm managed to spread throughout the world. In September 2010, approximately 100,000 hosts were infected (40,000 unique external IP addresses from over 155 countries), 60% of which, were located in Iran.

4.2 *Dragonfly*

A group of well-funded, highly-skilled adversaries launched a cyber-espionage campaign, the first advanced attack after Stuxnet that targeted ICS equipment [34]. The group behind the attack was named 'Dragonfly' by Symantec or 'Energetic

Bear' by other security firms. Initially, the targeted systems were aviation and defense industries located in the US/Canada but afterwards the attacker's group showed interest for industries of the energy sector. Using the *watering hole* attack technique [9] the adversaries managed to infect with malware several company networks. Furthermore, they managed to inject malicious payloads on available ICS vendor software found on official websites. The attack was staged in three phases: Firstly, spear-phishing campaigns were launched and remote access was established via a Remote Access Trojan (RAT) horse. Then, Havex software was used in *watering hole* attacks against official vendor websites thus redirecting users to servers with malware infected ICS software.

1. **Reconnaissance phase** Retrieval of corporate information from aviation, defense and energy industries' web presence.
2. **Initial infection phase** Via spear-phishing techniques 'Dragonfly' group infected employees' workstations with HAVEX malware. Initially, the malware harvested data, such as emails, contact lists and documents.
3. **Establish and maintain remote access** HAVEX malware served as a means for installation of other malware sent from Dragonfly servers (e.g., Karagany RAT, password stealer module, etc.). It consisted of a remote access Trojan and a server module written in PHP. After installation, the malware communicated with C&C server in order to download and execute other malicious payloads, such as an OPC scanning module, that utilized specific TCP ports used by Siemens and Rockwell automation systems, to retrieve information for ICS equipment.
4. **Propagation phase** By exploiting vulnerabilities in the vendors' websites the Dragonfly group was able to place its payloads in three major ICS vendor websites. In the case of the first vendor's website (eWon) the adversaries managed to change a download link so as to point to a modified package of a VPN application (Talk2Me) that provided access to PLCs. The second compromised website belonged to a European manufacturer of PLC devices whereas the third website was owned by a company that manufactured ICS for energy sector, including wind turbines. None of the websites affected, enforced any authorization mechanisms for accessing the ICS software.
5. **Remote control and device manipulation** Havex's main target was ICS communication interfaces and especially OLE for Process Control information. In all three cases described the attackers successfully managed to inject malicious code into the vendor's driver package. Investigators were able to identify 88 different versions of Havex, 146 C&C centers (mainly vulnerable blog websites) and 1,500 IP addresses of potential victims, most of which, in Europe.

Although Dragonfly attack did not disturb any industrial control process or lead to a severe energy outage, the adversaries manage to collect a large amount of valuable information that could potentially assist them in launching future attacks [34]. Furthermore, the OPC scanning module could be used to compromise ICS maintenance suppliers' services such as eWon, which utilizes approximately a million remote connections in order to provide remote support on ICS equipment. Based on later investigations, it was discovered that the Dragonfly group had also targeted the pharmaceutical industry aiming at stealing valuable information such as medicine recipes, batch production sequence steps as well as manufacturing plant volumes and capabilities.

4.3 Attacking Internet Facing PLCs: PLC-Blaster

In BlackHat 2015, security researchers presented a malware that targets network enabled PLCs [24], while a similar attack was presented the next year at the same conference [42]. The latter was mainly consisted of a self-replicating worm that could infect specific manufacturer PLCs such as Siemens SIMATIC. The malware was able to probe port 102/TCP in order to identify PLC devices (in this case S7-1200). Then, after establishing connection with the target the exploitation phase begins. The worm mimicked TIA-portal, a platform supported by Siemens for remote management, to implement the manufacturer's proprietary binary protocol named `S7CommPlus`. The latter utilized both TPKT and ISO8073 [14], to remotely infect and control the PLC. The functionality features of the protocol included configuration of the device, start/stop its operation, modify its processes' variables, uploading/downloading of programs as well as debugging & alerting. In order to issue commands from one PLC to another an analysis of the protocol's message structure was conducted and vulnerabilities such as insufficient integrity protection mechanisms, password originated from hash encryption keys and disabled default access protection settings were found. In summary the attack vector included the following steps:

1. **Reconnaissance phase** The adversary scans the Internet using Shodan (or similar) search engine and locates vulnerable PLCs.
2. **Initial infection phase** Since the PLC lacks of integrity protection mechanisms, the malware mimics TIA portal to issue commands and transfer its malicious payload in order to take over the PLC(s).
3. **Propagation phase** The warm sends messages to 102/TCP communication port. Then, by using the proprietary Siemens protocol (`S7CommPlus`), tests the target and tries to download a copy of itself. If no connection is established after 200 prob cycles the IP address is incremented.
4. **Establish and maintain remote access** Using an embedded Socks4proxy the worm communicates to an external C&C center. The worm is stored on the device so it can survive a restart or even a power loss.
5. **Remote control and device manipulation** The worm can alter any outputs of the compromised PLCs as well as force them to enter an endless loop thus triggering an error condition.

4.4 Attacking Industrial Robots

According to the International Federation of Robotics forecast [30], 1.3 million industrial robot units will be installed in factories located all over the globe until the end of 2018. Robots are used in almost every critical industrial sector such as automotive, aerospace, defense, plastics, electronics and electrical, metal fabrication, pharmaceutical, railway and many more. Several security firms and researchers have pointed out vulnerabilities in both domestic and industrial robots [8, 30]. The latter are usually of large volume used in complex manufacturing processes and play an essential role in production lines. Industrial robots are exceptionally complex cyber-physical systems that include actuators, sensors, human-robot interfaces and are constantly connected to computer networks primarily for operation, programming and maintenance purposes. In [30] researchers mainly focused on industrial robots by analyzing protocols and relative software. The impact of a single software vulnerability could have serious consequences, since, it could enable an adversary to inflict a massive financial damage and/or even threat human lives. After the Industry 4.0 [26] was introduced, almost all new models of industrial robots tend to incorporate IoT technologies such as connectivity and operational features that expose them to a much broader attack surface. The researchers in [30] utilized an actual robot (ABB six-axis IRB140) in order to demonstrate a series of attacks such as alter or introduce minor defects in the manufactured products, physically damage the robot, steal industrial secrets and/or cause human injuries.

Using well known search engines (Shodan, ZoomEye and Censys) they managed to discover multiple industrial robots' network interfaces connected directly to the Internet. As of late March of 2017, researchers discovered approximately 84,000 of industrial robots that were exposed to the Internet, 5105 of which did not require any authentication, 59 had known vulnerabilities whereas the researchers were able to identify 6 totally new (*zero-day*) ones. These included the usage of a self-signed certificate for multiple devices, network service banners that disclosed sensitive information (vendor's name, MAC address, firmware version, CPU model, CPU frequency, etc.), outdated software components (application & cryptography libraries, compilers, kernel), default credentials or no/poor authentication mechanisms, static VPN private keys on publicly available firmware images, adoption of symmetric cryptography schemes in VPNs, the use of plain HTTP web interfaces with no/poor input sanitization, default 'as is' use of open software (e.g., REST layer in PHP) and publicly available unstripped firmware images. A realistic APT attack vector against industrial robots includes the following steps:

1. **Reconnaissance phase** Adversaries use search engines to discover and enumerate Internet-exposed robot interfaces by searching for specific strings in the HTTP header (e.g., 'eWON', 'Westermo', etc.). Then, they manage to locate several software vulnerabilities by reading freely available technical documentation, reverse engineering publicly available software (e.g., firmware files, controller software) and even run exploitation tests using available simulation software (e.g., ABB's software suite).

2. **Initial infection phase** Using the vulnerabilities found in previous phases adversaries establish a connection with the device (e.g., authentication bypass in ABB's eWON industrial cellular router, FTP static credentials to access the command driver, memory errors found in the RobAPI). Since no security mechanisms are present and the Internet interface is used, the attack will remain undetected from any IDS/IPS equipment installed in the internal network.
3. **Establish and maintain remote access** Through FTP access, attackers upload custom, malicious software and trigger a reboot using the command shell reboot FTP function. The malicious files are executed and all robot features are now remotely controlled via a C&C center.
4. **Propagation phase** Utilizing connectivity features installed in robot's main computer (e.g., FlexPendant, RobotStudio) attackers manage to discover and attack other robot network interfaces that are connected on the company's internal network.
5. **Remote control and device manipulation** Adversaries are able to launch a series of attacks, which the researchers categorized into five classes, evaluating the potential impact of each one individually. The categorization was made under the assumption that a robot must be able to at least read accurately from its sensors and execute its control logic, perform precise movements, and not harm humans in any circumstance. In particular:
    (a) **Altering the Control-Loop Parameters** This attack includes the modification of the configuration control closed/open loop parameters used to control robot movements. Implications of such attack can lead to safety boundary violation and even breakage of robot parts.
    (b) **Tampering with Calibration Parameters** Repeatedly manipulation of the controller's calibration parameters at runtime could lead to Denial-of-Service (DoS) attacks.
    (c) **Tampering with the Production Logic** In the case where the controller does not enforce *end-to-end* integrity checks a program task could be altered thus leading to the manufacturing of defective products or fully compromising a factory's manufacturing process.
    (d) **Altering the User-Perceived Robot State** In this case the robot's user interface is manipulated in order to hide/misinform the operator of the true robot status so as to fool him/her into making wrong risk evaluations. This kind of attack can put operators at risk and even lead to human injuries.
    (e) **Altering the Robot State** Changing the robot's true state may have major impact especially when combined with other attacks (e.g., manufacture a large amount of defective products).

Realistic threat attack scenarios may include sabotage of an entire production line via product's characteristics alteration followed by a ransomware campaign in order to reveal which product batch was affected, physical damage to industrial equipment, human injuries and/or the use of the device as a means to exfiltrate sensitive industrial data (e.g., industrial secrets such as calibration parameters).

4.5 PLC Ransomware: LogicLocker

In 2017 researchers of Georgia Institute of Technology [17] presented a hypothetical ransomware attack scenario in which, adversaries target network connected PLCs located in a water treatment plant. The targeted PLCs were used to control the valves which, in turn, control the amount of chlorine that is added into the water. In particular, they developed a framework named 'Logiclocker' that then used to attack some of the most popular PLCs in the market such as Schneider Modicon M221, Allen Bradley MicroLogix 1400, and Schneider Modicon M241. The phases described in order to launch a successful ransomware campaign included initial infection, lateral movement within internal SCADA networks, reconnaissance and target discovery, locking and encrypting process and finally the negotiation for the ransom. In their PoC attack the researchers managed to retrieve the device's credential (in this case Modicon M241) either by stealing or using brute force attack techniques. A typical ransomware attack scenario consists of the following phases:

1. **Reconnaissance phase** Adversaries locate Internet facing PLCs via search engines (e.g., Shodan).
2. **Initial infection phase** Using stealing, brute force and dictionary attack techniques they manage to recover authentication information (e.g., user/system credentials) from the discovered Internet facing PLCs.
3. **Propagation phase** Embedded payloads enable the malware to scan the internal SCADA networks of the water treatment plant in order to infect other vulnerable PLCs.
4. **Establish and maintain remote access** Adversaries remotely reprogram the infected PLCs with new passwords thus locking the legitimate operators out.
5. **Remote control and device manipulation** The attackers remotely encrypt the PLCs' software using well known encryption algorithms (e.g., AES) with a newly generated key.
6. **Ransomware phase** Via the LogicLocker framework an email is sent to the water treatment plant that threatens to release chlorine in the water and cause massive human fatalities.

5 APT Attacks on Smart Grid SCADA Networks and Field Devices

Smart grids are not always engineered having in mind the security-by-design principle, thus making them vulnerable to various novel cyber threats. In this section we analyze recent, high profile attacks that utilize APT techniques.

5.1 Attacks on Generation Systems: The Aurora Attack

In 2007, an attack scenario that targeted electric power generators, was demonstrated at the Idaho US National Labs [43, 47]. Network enabled PLCs (circuit breakers) were forced to open and close in a very fast rate (4 times per second) in order to force the affected power generator to desynchronize thus resulting in its physical destruction. In a potential attack scenario described in [47] an attacker compromises the company's corporate network to propagate to the facility's main control center and take advantage of an existing communication link that is used to remotely administer the PLCs.

In order to launch an Aurora-like attack, the attacker would have to overcome intentional delays in switching on and off and synchronization checks that exist to ensure the smooth operation of the system. Assuming that the attacker has compromised a sufficient number of devices, it is possible to inject falsified commands to trip and reclose a circuit breaker in a rapid repetitive way. In particular, an hypothetical Aurora attack scenario can be described as follows:

1. **Reconnaissance phase** Adversaries manage to collect corporate information (e.g., email accounts) that then use to launch a spear-phishing campaigns.
2. **Initial infection phase** Using known and *zero-day* exploits they manage to elevate privileges and install a RAT tool in order to control the infected workstations remotely.
3. **Establish and maintain remote access** Using network pivoting techniques they manage to navigate the facility's internal network and infect a workstation located in the control center. Moreover, using similar exploitation techniques they establish remote access to the workstation and through it to the target PLCs.
4. **Lateral movement and propagation phase** Using NMAP or similar tools they fingerprint the relays' brand name and model (Ethernet and/or Modbus). Then, via passive eavesdropping and vulnerability exploitation techniques (e.g., false data injection attacks [29]) they manage to remotely control the circuit breakers and bypass protection relays.
5. **Remote control and device manipulation**
   - **Step 1** The circuit breaker(s) are opened isolating the generator from the grid.
   - **Step 2** The generator starts to speed up and the frequency of the generator increases.
   - **Step 3** The frequency difference between the grid and the generator increases.
   - **Step 4** After a particular amount of time the circuit breakers are closed, thus connecting back the generator to the grid.
   - **Step 5** The generator is forced into synchronization with *out-of-sync* conditions thus causing substantial electrical and mechanical transients.
   - **Step 6** Steps 1–5 are repeated in a timely manner until the generator is permanently damaged.

Adversaries (e.g. terrorists, nation state) could launch concurrent attacks against multiple generators, in order to destabilize large areas of a country's smart grid thus maximizing the potential impact of the attack.

5.2 Attack on the Ukraine's Smart Grid Distribution Network (2015)

One of highest impact, highly coordinated, stealthy APT attack against the smart grid is considered to be the one that took place on December 23, 2015 against an Ukraine regional electricity company named "Kyivoblenergo". The attack resulted in massive outages that affected approximately 225,000 customers for several hours [7], whereas substation control (e.g., circuit brakers) was switched to manual for weeks.

The adversaries utilized a variety of attack techniques including the use of spear-phishing campaigns (they impersonated an email message from the Ukrainian parliament), variants of *BlackEnergy* 3 and *KillDisk* malware as well as the manipulation of Microsoft Office documents in order to gain an initial foothold to the company's internal network. The attackers possessed specialized knowledge of ICS network connected devices such as Uninterruptible Power Supplies (UPSs), HMI interfaces, credential harvesting techniques, and SCADA client software. The attack vector can be described as follows:

1. **Reconnaissance phase** Nation-state adversaries launched a spear-phishing campaign with malware-infected Microsoft Office documents against corporate users.
2. **Initial infection phase** By exploiting Windows well known and *zero-day* vulnerabilities they managed to install key-loggers and retrieve user credentials.
3. **Lateral movement and propagation phase** Initially, the adversaries performed a reconnaissance of internal SCADA network and devices. Then, pivoting throughout different network segments enabled them to locate and infect SCADA dispatch workstations and servers. In particular, they managed to gain access to operators' workstations, located in control rooms, that run HMI software.
4. **Establish and maintain remote access** Using existing, legitimate remote administration tools, installed on operators' workstations, they managed to remotely connect to the aforementioned workstations and lock the legitimate operators out. In addition they uploaded malicious firmware in field communication devices to prevent any recovery attempts.
5. **Remote control and device manipulation** In order to magnify the impact of the attack the adversaries proceeded with the following actions:
    (a) Remotely opened multiple circuit breakers to cause massive outages. (attack's main target)
    (b) Reconfigured UPS systems to cause outages in company's buildings.

(c) Launched a remote telephonic denial of service on the energy company's call center to frustrate the impacted customers.
(d) Utilized a modified version of *KillDisk* malware to destroy forensic evidence and render workstations inoperable.

5.3 Attack on the Ukraine's Kiev Transmission Station (2016)

In December 2016, the Ukrainian's smart grid SCADA systems were targeted for a second year in a row [18]. The target of the attack was a 200 Megawatt transmission station located near the city of Kiev. Similar to the previous attack, the adversaries launched spear phishing campaigns in which they wrapped in a word document attachment the malware *CrashOverride/Win32/Industroyer* [27] in order to infect the employees' workstations. This time the attack techniques used were far more sophisticated and stealthier than the first attack. The malicious code was capable of being preprogrammed to launch an attack against multiple targets, at a future time, without any intervention from the attackers. The malware was modular and included, among others, the main program that ensured communications with C&C centers and IIoT equipment, four different malicious payloads that correspond to industrial control protocols IEC 101, IEC 104, IEC 61850, OPC Data Access (OPC DA) and a DoS tool that targeted a particular family of protection relays (Siemens SIPROTEC). Figure 3 depicts the basic functionality of the malware. A more detailed description of the software components as well as a walkthrough of the attack [10] is presented here:

1. **Reconnaissance phase** Using publicly available information found on the Internet (e.g., YouTube) the adversaries were able to enumerate substation's ICS. Having selected their target, then they launched a spear-phishing campaign (July 2016) against corporate users.
2. **Initial infection phase** Using advanced exploitation techniques they managed to gain a foothold to the substation's internal network. In particular, after they managed to infect corporate workstations and/or servers, the malware installed the main backdoor program responsible also to control all other SCADA modules. The latter could be programmed to communicate with the attackers at a specific time every day via C&C servers (active TOR nodes). Initially, it authenticated with a local proxy (TCP port 3128) and then utilized an HTTPS channel to connect to external C&C servers. After a successful privilege escalation process, the backdoor was masqueraded as a legitimate windows service program to avoid any detection.
3. **Lateral movement and propagation phase** The adversaries incorporated highly customized, sophisticated SCADA communication modules in order to interact with IIoT equipment. The purpose of the SCADA communication modules was twofold: Initially, they were used in the enumeration/propagation phase, in which specific commands were issued to fingerprint IIoT devices, and as a means of launching the main attack by issuing the necessary control commands.

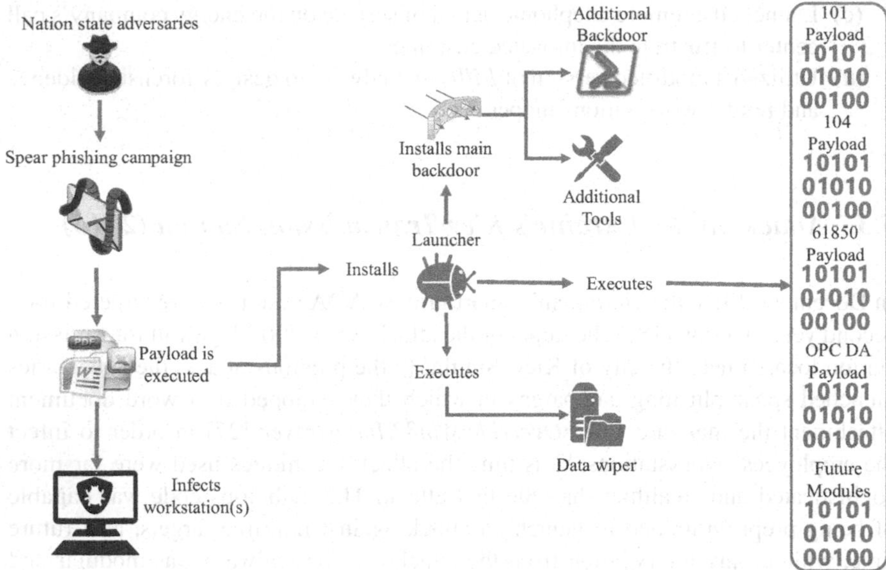

Fig. 3 Attack vector of the malware (CrashOverride/Win32/Industroyer) on Ukraine's smart grid (December 2016)

- **IEC 60870-5-101 module** It utilized the file 101.dll to implement the IEC 101 protocol so as to communicate with compatible RTUs. Upon execution, the payload located and terminated the legitimate process used to communicate with IEC 101 devices. Then, a new process was started in order to take over the control of the RTUs.
- **IEC 60870-5-104 module** Since IEC 104 extends the IEC 101, the module utilized TCP/IP network as its main communication channel. It also supported a configuration file for customization and operated in a similar way as the IEC 101 payload.
- **IEC 61850 module** Unlike the previous modules this one consisted of both an executable file (61850.exe) as well as a DLL file. When executed, the malicious program enumerated all IP addresses and tried to connect to TCP port 102. Then, Manufacturing Message Specification (MMS) commands were used to enumerate and control all discovered devices, such as circuit breakers.
- **OPC DA module** OLE, Component Object Model (COM) and Distributed Component Object Mode (DCOM) are Microsoft technologies that are used for real-time data exchange, based on a client/server model. Similar to IEC 61850 payload, the malicious program consisted of a .EXE and a .DLL file that, incorporated both 61850 and OPC DA functionalities. Upon execution, enumeration of all OPC servers and devices was performed (ABB solutions). Then, the OPC's state was altered using the `IOPCSyncIO` interface.

  - **Port scanner and DoS tools** Additionally, a custom-made port scanning program and a DoS tool were included in the malware. The latter could be used against SIPROTEC Siemens devices by utilizing a known vulnerability (CVE-2015-5374).

4. **Establish and maintain remote access** Aside the main backdoor, the attackers utilized a trojanized version of the Windows notepad application, to serve as a back-up persistence mechanism, in order to regain access in the case of the main backdoor was found and disabled. To avoid detection, the embedded malicious code was heavily obfuscated and utilized different C&C servers than the one used from the main backdoor program.
5. **Remote control and device manipulation** To launch the attack, the adversaries utilized the 'Launch' module in which they had embedded specific time and dates (17 and 20 December). Once one of the dates was reached the module was programmed to execute two processes in high priority. In particular:

  - **Payload.DLL** The actual name of the DLL file that contained the main payload was not hardcoded into the module but had to be supplied from the adversaries along with a configuration file. Upon execution, the payload used the functionality embedded in aforementioned modules to issue commands to located RTUs and PLCs, such as turn the device off or change their status (e.g., open/closed).
  - **Data wiper module** This payload was scheduled to launch with a delay of 1–2 h from the first payload. It included the file `haslo.exe/dat` that when executed, it modified the registry value `ImagePath` with an empty string thus rendering the system unusable. In addition, it deleted specific files by overwriting them twice and terminated all running process in order to make the system crash. The list with the file extensions for deletion included, among others, Windows binaries as well as MS SQL server and ICS configuration (ABB PCM600) files.

6 Conclusions

In this chapter, we analyzed recent, high impact, real APT attacks on IIoT ecosystem, as well as PoC attack scenarios that utilize APT techniques based on related work of security researchers. In the case of real APT attack scenarios, malicious actors mainly focused in exploitation techniques that took advantage of indirect attack paths, that exist between corporate and industrial networks. From the analysis of the attack techniques used in real cyberattacks against Ukraine's smart grid [7, 18], one can ascertain that adversaries are constantly evolving their attack techniques in terms of customization, stealthiness and user interaction. Adversaries mainly target systems that are being used by corporate users, such as mail servers, to infiltrate to the company's internal network. Then, via privilege escalation techniques, they manage to fully control the infected workstations and propagate throughout the network so as to locate and exploit IIoT equipment such as SCADA HMI software. Finally, using vulnerabilities found on legacy/new network industrial protocols they infect and remotely control field devices (e.g., circuit breakers).

On the other side the majority of PoC attack scenarios utilize search engines, such as Shodan, to locate and enumerate exposed web interfaces of IIoT equipment. Such attack scenarios are quite realistic. Indications of real attacks on Internet facing IIoT equipment can also be found in [22]. Another interesting finding is that a vast amount of publicly available industrial software, such as firmware files, is available to adversaries. Such information may allow them to extract valuable information and to refine their exploitation methods. In the attack scenarios examined [8, 17, 30, 42], security researchers managed to successfully exploit *zero-day* and well known vulnerabilities of the IIoT ecosystem (e.g., hardcoded credentials on firmware files, remote code execution on web interfaces) to remotely control and manipulate mission critical industrial equipment such as sensors, actuators and robotics.

In order to mitigate the risks that involve relative IIoT technologies, organizations/companies should always ensure that their mission critical ICSs can survive a large scale APT attack. Disaster recovery plans should always include well defined incident response procedures that correspond to several attack scenarios which, in turn, are thoroughly planned and tested. Furthermore, specialized security equipment, such as IDS/IPS equipped with advanced detection techniques (e.g., YARA rules [21]), should be applied throughout industrial networks. Developers of HMIs and other relative SCADA applications should adopt the secure software life cycle practices used by operating system and other application developers for over a decade. Security-by-design [36] should also be adopted by IIoT equipment manufacturers and industrial software developers to ensure strong authentication, integrity protection and authentication mechanisms are in place. Various critical procedures, such as the over-the-air updating process of IIoT equipment, should be properly implemented. In addition, network administrators should always protect industrial web interfaces via dedicated firewall devices, properly segment and sometimes isolate mission critical SCADA systems, especially those that utilize outdated, vulnerable industrial network protocols. Finally, other security best practices that promote defense-in-depth include anomaly detection systems [1], as well as message authentication, integrity and encryption [11, 37, 39, 41].

References

1. Alves T, Das R, Morris T (2018) Embedding encryption and machine learning intrusion prevention systems on programmable logic controllers. IEEE Embed Syst Lett 10:99–102
2. Andrea M (2015) GE MDS PulseNET hidden support account remote code execution vulnerability. https://www.zerodayinitiative.com/advisories/ZDI-15-440/
3. Antón SD, Fraunholz D, Lipps C, Pohl F, Zimmermann M, Schotten HD (2017) Two decades of scada exploitation: a brief history. In: 2017 IEEE Conference on Application, Information and Network Security (AINS). IEEE, pp 98–104

4. Berger H (2014) Automating with SIMATIC S7-400 inside TIA portal: configuring, programming and testing with STEP 7 Professional. Wiley
5. Brian G, Fritz Sands TTMZDI Hacker machine interface: the state of scada HMI vulnerabilities. White paper, Trend Micro
6. Broad WJ, Markoff J, Sanger DE (2011) Israeli test on worm called crucial in Iran nuclear delay. NY Times 15:2011
7. Case DU (2016) Analysis of the cyber attack on the Ukrainian power grid. Electricity Information Sharing and Analysis Center (E-ISAC)
8. Cerrudo C, Apa L (2017) Hacking robots before Skynet1. IOActive Website
9. Chen P, Desmet L, Huygens C (2014) A study on advanced persistent threats. In: IFIP International Conference on Communications and Multimedia Security. Springer, pp 63–72
10. Cherepanov A (2017) Win32/industroyer: a new threat for industrial control systems. White paper, ESET, June 2017
11. Cherifi T, Hamami L (2017) A practical implementation of unconditional security for the IEC 60780-5-101 scada protocol. Int J Crit Infrastruct Prot 20:68–84
12. Clarke GR, Reynders D, Wright E (2004) Practical modern SCADA protocols: DNP3, 60870.5 and related systems. Newnes
13. Cowan C, Wagle F, Pu C, Beattie S, Walpole J (2000) Buffer overflows: attacks and defenses for the vulnerability of the decade. In: DARPA Information Survivability Conference and Exposition, 2000, DISCEX'00. Proceedings, vol 2. IEEE, pp 119–129
14. Devarajan G (2007) Unraveling scada protocols: using sulley fuzzer. In: Defon 15 Hacking Conference
15. Drias Z, Serhrouchni A, Vogel O (2015) Taxonomy of attacks on industrial control protocols. In: 2015 International Conference on Protocol Engineering (ICPE) and International Conference on New Technologies of Distributed Systems (NTDS). IEEE, pp 1–6
16. Falliere N, Murchu LO, Chien E (2011) W32. stuxnet dossier. White paper, symantec corporation. Secur Response 5(6):29
17. Formby D, Durbha S, Beyah R (2017) Out of control: ransomware for industrial control systems. http://www.cap.gatech.edu/plcransomware.pdf
18. Goodin D (2017) Hackers trigger yet another power outage in Ukraine. https://arstechnica.com/security/2017/01/the-new-normal-yet-another-hacker-caused-power-outage-hits-ukraine/
19. Gruber E (2014) Verifying ASLR, DEP, and safeSEH with powershell. Blog, NetSPI 23
20. Gutierrez JA, Naeve M, Callaway E, Bourgeois M, Mitter V, Heile B (2001) IEEE 802.15. 4: a developing standard for low-power low-cost wireless personal area networks. IEEE Netw 15(5):12–19
21. Hurd CM, McCarty MV (2017) A survey of security tools for the industrial control system environment. Technical report, Idaho National Laboratory, Idaho Falls, ID
22. Israel B, Ross R (2018) ICS threat broadens: nation-state hackers are no longer the only game in town. https://www.cybereason.com/blog/industrial-control-system-specialized-hackers
23. John KH, Tiegelkamp M (2010) IEC 61131-3: programming industrial automation systems: concepts and programming languages, requirements for programming systems, decision-making aids. Springer, Heidelberg
24. Klick J, Lau S, Marzin D, Malchow JO, Roth V (2015) Internet-facing PLCs-a new back orifice. Black Hat USA, pp 22–26
25. Kushner D (2013) The real story of stuxnet. IEEE Spectr 50(3):48–53
26. Lasi H, Fettke P, Kemper HG, Feld T, Hoffmann M (2014) Industry 4.0. Bus Inf Syst Eng 6(4):239–242
27. Lee RM, Assante, MJ, Conway T (2017) CRASHOVERRIDE: analysis of the threat to electric grid operations. Dragos Inc. https://dragos.com/wp-content/uploads/CrashOverride-01.pdf
28. Lian FL, Moyne JR, Tilbury DM (2001) Performance evaluation of control networks: ethernet, controlnet, and devicenet. IEEE Control Syst 21(1):66–83
29. Liang G, Weller SR, Zhao J, Luo F, Dong ZY (2017) The 2015 Ukraine blackout: implications for false data injection attacks. IEEE Trans Power Syst 32(4):3317–3318

30. Maggi F, Quarta D, Pogliani M, Polino M, Zanchettin AM, Zanero S (2017) Rogue robots: testing the limits of an industrial robot's security. Technical report, Trend Micro, Politecnico di Milano
31. Mohassel RR, Fung A, Mohammadi F, Raahemifar K (2014) A survey on advanced metering infrastructure. Int J Electr Power Energy Syst 63:473–484
32. Momoh J (2012) Smart grid: fundamentals of design and analysis, vol 63. Wiley, Hoboken
33. Mulligan G (2007) The 6LoWPAN architecture. In: Proceedings of the 4th Workshop on Embedded Networked Sensors. ACM, pp 78–82
34. Nelson N (2016) The impact of dragonfly malware on industrial control systems. SANS Institute, Bethesda
35. Pagliery J (2015) The inside story of the biggest hack in history. CNN, 5 Aug 2015
36. Radvanovsky R, Brodsky J (2016) Handbook of SCADA/control systems security. CRC Press, Boca Raton
37. Rrushi JL (2017) Defending electrical substations against 0-day malware through decoy I/O in protective relays. In: Dependable, Autonomic and Secure Computing, 15th International Conference on Pervasive Intelligence & Computing, 3rd International Conference on Big Data Intelligence and Computing and Cyber Science and Technology Congress (DASC/PiCom/DataCom/CyberSciTech), 2017 IEEE 15th International. IEEE, pp 486–493
38. Sadeghi AR, Wachsmann C, Waidner M (2015) Security and privacy challenges in industrial Internet of things. In: Proceedings of the 52nd Annual Design Automation Conference. ACM, p 54
39. Saxena N, Grijalva S (2017) Efficient signature scheme for delivering authentic control commands in the smart grid. IEEE Trans Smart Grid 9:4323–4334
40. Shacham H, Page M, Pfaff B, Goh EJ, Modadugu N, Boneh D (2004) On the effectiveness of address-space randomization. In: Proceedings of the 11th ACM Conference on Computer and Communications Security. ACM, pp 298–307
41. Shahzad A, Lee M, Lee C, Xiong N, Kim S, Lee YK, Kim K, Woo SM, Jeong G (2016) The protocol design and new approach for scada security enhancement during sensors broadcasting system. Multimed Tools Appl 75(22):14641–14668
42. Spenneberg R, Brüggemann M, Schwartke H (2016) PLC-blaster: a worm living solely in the PLC. Black Hat USA, Singapore
43. Srivastava A, Morris T, Ernster T, Vellaithurai C, Pan S, Adhikari U (2013) Modeling cyber-physical vulnerability of the smart grid with incomplete information. IEEE Trans Smart Grid 4(1):235–244
44. Stellios I, Kotzanikolaou P, Psarakis M, Alcaraz C, Lopez J (2018) A survey of IoT-enabled cyberattacks: assessing attack paths to critical infrastructures and services. IEEE Commun Surv Tutorials 20:3453–3495
45. Tovar E, Vasques F (1999) Real-time fieldbus communications using profibus networks. IEEE Trans Ind Electron 46(6):1241–1251
46. Wan J, Tang S, Shu Z, Li D, Wang S, Imran M, Vasilakos AV (2016) Software-defined industrial Internet of things in the context of industry 4.0. IEEE Sens J 16(20):7373–7380
47. Zeller M (2011) Myth or reality does the aurora vulnerability pose a risk to my generator? In: 2011 64th Annual Conference for Protective Relay Engineers. IEEE, pp 130–136

Part II
Secure Interconnection Mechanisms

A Survey on Lightweight Authenticated Encryption and Challenges for Securing Industrial IoT

Megha Agrawal, Jianying Zhou, and Donghoon Chang

Abstract Embedded systems are widely deployed nowadays in various domains like smart cards, automobiles, telecommunications, home automation systems, computer networking, digital consumer electronics, defense and aerospace. IoT is the technology enabling the inter-connection of these embedded devices (composed of sensors, actuators etc.) through the internet to exchange data, optimize processes, monitor devices in order to generate benefits for the industry, the economy, and the end user. These operations typically consists of sensitive or critical information that needs to be protected against outside world. Therefore their security comes as a primary concern. However the main challenges while providing security for these devices are resource constrained environment in terms of computing power, memory capacity, chip area and the power usage. The limited capabilities of these devices necessitate the adoption of Lightweight Cryptography (LWC). Lightweight cryptography is a field dealing with cryptographic algorithms or cryptographic protocols specially designed for the usage in constrained environments which includes RFID tags, contactless smart cards, sensors, embedded systems, health-care devices and so on. This work provides a survey of existing lightweight authenticated encryption algorithms. We surveyed 17 lightweight AE schemes (LWAE), out of which 9 schemes are from the ongoing CAESAR competition.

Keywords Symmetric cryptography · Lightweight cryptography · Authenticated encryption · Embedded system security · Internet of Things (IoT) · Industrial IoT (IIoT)

M. Agrawal (✉) · D. Chang
Indraprastha Institute of Information Technology, Delhi, India
e-mail: meghaa@iiitd.ac.in

J. Zhou (✉)
Singapore University of Technology and Design, Singapore, Singapore
e-mail: jianying_zhou@sutd.edu.sg

C. Alcaraz (ed.), *Security and Privacy Trends in the Industrial Internet of Things*, Advanced Sciences and Technologies for Security Applications,
https://doi.org/10.1007/978-3-030-12330-7_4

1 Introduction

Communication security is needed for many devices especially when we are moving to the age of Internet of Things (IoT). The Internet of Things is a network of connected devices including everything like cellphones, wearable devices, household devices and everything else, which aims to make our life more comfortable by managing most of things their own. IoT architectures are supposed to deal with an estimated population of billions of objects, which will interact with each other and with other entities, such as human beings or virtual entities. And all these interactions must be secured somehow, protecting the information and service provisioning of all relevant actors and limiting the number of incidents that will affect the entire IoT [65]. According to the Ericsson Mobility report [21] "there are now about 4.6 billion connected devices excluding phones, tablets and laptops. A number that is expected to increase to 15.3 billion in the next five years". All these advanced developments are supposed to lead huge benefits to consumers of those technologies. Application of IoT in health care devices will enable patients suffering from serious diseases to consult with their doctors to manage their condition while sitting at home. Home automation systems will allow consumers to turn on the lights, television and air-condition and heat up their dinner before they are at home. Connected cars will inform the rescue team in case of an accident. Overall, the Internet of Things may bring comfort beyond our expectations.

There is a growing interest in using IoT technologies in various industries. The part of IoT which deals with the industrial application is called the Industrial IoT (abbreviated as IIoT) which is used across several industries such as manufacturing, healthcare, logistics, oil and gas, transportation, energy/utilities, mining and metals, aviation and other industrial sectors. The industrial IoT market is estimated to reach $123.89 Billion by 2021 [72]. The IIoT can greatly improve connectivity, efficiency, scalability, time savings, and cost savings for industrial organizations. However, the major concerns surrounding the IIoT is interoperability between devices and machines that use different protocols and have different architectures and the security of these protocols and data generated from these devices.

These connected devices or objects will gather, store and transmit or share enormous amount of this highly sensitive consumer data. For an example, power usage information gathered from electricity meter at home or workplace by your utility provider opens up the window for a number of threats. An instance include an attacker tracking the power usage to find your presence at home or workplace in an order to plan a theft attack or an attacker could tamper the data being transmitted to the utility provider and modify the information. All of these attacks threaten the trust in the information gathered and transmitted, which results in less confidence in the overall system. So, it becomes extremely essential to implement the basic security features like confidentiality and integrity. Other challenges to IIoT applications include lack of standardization and lack of skilled workers.

However the main challenges while providing security for IoT and IIoT applications are the constrained on devices. These devices are usually constrained in computing power, storage capacity and the power usage. Therefore it is a challenge to use existing cryptographic algorithms which generally require more resources than these resource constrained devices. Hence developing new cryptographic algorithms for providing confidentiality and integrity altogether as well as satisfying the device constraints is the need of the time. A cryptographic algorithm providing confidentiality and integrity together is known as Authenticated Encryption (AE). The Authenticated Encryption for the resource constrained devices is known as Lightweight Authenticated Encryption (LAE). Encryption processes the plaintext and produces the ciphertext and tag pair. Decryption is provided in a single step along with integrity verification, the plaintext is returned if a tag passes the verification, otherwise, an error is produced. AE is required in communication protocols like Secure Socket Layer (SSL)/Transport Layer Security (TLS) and online applications to prevent attackers from tampering, intercepting, or giving ciphertexts to the receiver. If these attacks are launched, messages can be decrypted and the communicating data is completely revealed to the attacker. For mainstream applications, AES-GCM is widely used authenticated encryption scheme. The pervasive and ubiquitous computing develops relevant lightweight schemes for constrained environment. Three approaches implement the AE over associated data

- **Encrypt-and-MAC (E&M)** The ciphertext and Message Authentication Code (MAC) is generated individually by applying encryption and authentication on plaintext respectively.
- **Encrypt-then-MAC (EtM)** The MAC is computed over the encrypted plaintext.
- **MAC-then-encrypt (MtE)** The MAC is produced over plaintext and then it is encrypted along with the plaintext to generate ciphertext.

E&M produces the most efficient result, as the ciphertext and the MAC can be computed simultaneously. However, it is also considered the least secure, as the statistical correlations of the plaintext can be reflected in its MAC. Combinational attacks are performed that exploit the MAC's vulnerabilities to disclose the encryption key and recover the whole communication. EtM and MtE are slower, as the two primitives are executed sequentially. Both approaches destroy the statistical features of the plaintext, with EtM producing the most secure results.

For LWC, many AE schemes have been proposed. In the following sections we will present a survey of Lightweight authenticated encryption schemes for embedded systems as well as their implementations (in hardware and software).

Roadmap: Sect. 2 provides some of the key applications of IIoT followed by Sect. 3, which gives the idea about target devices. Section 4 presents the current Lightweight Crytography standards. Section 5 provides the evaluation metrics for the target devices followed by Sects. 6 and 7 which present the design classification and security requirements for authenticated encryption schemes. Section 8 gives the overview of lightweight AE schemes and Sect. 9 provides comparison among them. Section 10 raises some open challenges to IoT and IIoT followed by the conclusion in Sect. 11.

2 Key IoT Applications in Industries

The use of IoT is rapidly evolving and growing. However, only a few IoT applications are being developed and/or deployed in various industries including environmental monitoring, healthcare service, inventory and production management, food supplychain, transportation, workplace and home support, security, and surveillance. Here, we will briefly discuss few IoT applications in industries.

- Using IoT in healthcare industry [2, 62]: IoT provides new opportunities to improve healthcare industry. Enabled by its global connectivity, all the healthcare related information like patient's diagnosis, therapy, medication can be tracked and managed efficiently [2]. For example, a patient's heart rate can be collected by sensors from time to time and then sent to the doctor's office, which can be diagnosed remotely. However, security and privacy of patient's data are major concerns here.
- Using IoT in transportation and logistics: IoT will play an increasingly important role in transportation and logistics industries [3]. As more and more physical objects are equipped with bar codes, RFID tags or sensors, transportation and logistics companies can conduct real-time monitoring of the move of physical objects from an origin to a destination across the entire supply chain including manufacturing, shipping, distribution, and so on [50]. Furthermore, IoT is expected to offer promising solutions to transform transportation systems and automobile services [82]. As vehicles have increasingly powerful sensing,networking, communication, and data processing capabilities, IoT technologies can be used to enhance these capabilities and share under-utilized resources among vehicles in the parking space or on the road. An intelligent informatics system (iDrive system) developed by BMW used various sensors and tags to monitor the environment such as tracking the vehicle location and the road condition to provide driving directions [63].
- Using IoT in energy and utilities [74]: Spikes in energy consumption around major TV broadcasts and weather events have long troubled utility firms. But with effective energy demand management through the IIoT, the need for investment in both energy networks and power plants is reduced. Smart meters are one example of the industry's move towards IoT technologies, although at the moment they only record usage amounts and timings. Utility firms could potentially provide price information to these meters, which could in turn interact with other IoT devices to use energy at the most efficient time.

3 Target Devices

Lightweight cryptography aims to design a cryptographic schemes that can be implemented on a low end devices. By low end devices we mean the devices having limited amount of memory and low processing power such as embedded systems,

Table 1 Target devices

Conventional cryptography	Lightweight cryptography
Desktops & servers	Embedded systems & CPS systems
Tablets & smartphones	RFID & sensor networks
	Contactless smart cards

CPS systems, RFID devices, smart cards and sensor networks. On the other hand, high end devices are with unlimited amount of storage and high processing power such as servers, desktop computers, tablets and smartphones. These traditional cryptographic algorithms perform good on the high end devices; hence, there is no need of lightweight algorithms. However, conventional cryptography can not work on low end devices due to limited memory and processing power capabilities. Therefore, lightweight cryptography primarily targets the low end devices that works under highly constrained scenario (Table 1).

Now a days, wide variety of microcontrollers are available with various performance characteristics. Though 8-bit, 16-bit and 32-bit microcontrollers are the most famous these days, 4-bit microcontrollers also finds significant usage for some ultra-low cost applications. A large number of a instruction set containing only a few easy instructions can be found for these microcontrollers. These instructions requires a large number of CPU cycles to implement those common cryptographic algorithms, which in turn results them extremely slow and consume lots of energy for some application. This is a typical issue for the applications where it requires to satisfy real-time constraints with a significant amount of energy.

For devices like RFID tags that does not have their own power source, only a restricted supply is accessible from the surroundings. These devices demands the cryptographic algorithms with strict timing and power requirements along with the one that requires small number of Gate Equivalents (GEs).

Lightweight algorithms may work in various other constraints. Therefore, above mentioned examples are not supposed to be an exhaustive list. These example are just to demonstrate the scenarios where conventional cryptographic algorithms can not be applied, in order to clearly determine the need of lightweight cryptography.

4 Lightweight Cryptography Standards

Lightweight Cryptography standardization has been discussed in ISO/IEC 29192. It consists of six part that discusses the lightweight cryptographic algorithms for various security properties. Part 1 [38] specifies the "security terms and definitions and sets the security requirements, classification requirements and implementation requirements". Part 2 [39] discusses the suitable block ciphers for lightweight cryptography. It mentions PRESENT [10] and CLEFIA [69] as a standard, later SIMON and SPECK [6] also added to this list during an amendment in 2014. Part

3 [40] specifies "Enocoro and Trivium [15] as two dedicated keystream generators for lightweight stream ciphers". Part 4 [41] mentions "three asymmetric techniques, namely (i) a unilateral authentication mechanism based on discrete logarithms on elliptic curves [79] (ii) an authenticated lightweight key exchange (ALIKE) [75] and (iii) an identity based signature mechanism [55]". Elliptic Curve-based authentication scheme called ELLI is added to this list during an amendment [43] to this part. Part 5 [42] mentions "3 hash-functions PHOTON [25], SPONGENT [11] and Lesamnta-LW [27]" suitable for applications requiring lightweight cryptographic implementations. Part 6 [44] dedicated to Message Authentication codes (MACs) are currently in development phase.

ISO/IEC 29167 standard defines the security services for RFID air interface communications. Part 1 [28] defines "the architecture for security services for the ISO/IEC 18000 air interfaces standards for radio frequency identification (RFID) devices. Its purpose is to provide a common technical specification for optional security services for RFID devices that may be used by ISO committees developing RFID application standards". Additional parts define the crypto suites. Currently, there are eight suites that specify the use of AES-128 [30], PRESENT-80 [29], ECC-DH [31], Grain-128A [32], AES OFB [33], ECDSA-ECDH [34], cryptoGPS [35], and RAMON [36] security services for air interface communication.

Recently NIST published the NISTIR 8114 Report on Lightweight Cryptography [59] that summarized the results of the lightweight cryptography project and explained NIST's plans for standardization of lightweight algorithms. NIST lightweight cryptography project aims to include all cryptographic algorithms required in constrained environments. However, initially the project focuses on only symmetric key cryptography and hashing. For secret-key cryptography, the main goals are to provide entity authentication, confidentiality and data authentication. All three goals can be satisfied by providing an algorithm for Authenticated Encryption with Associated Data (AEAD). In this context, they published a call for submission draft [58] for two AEAD profiles.

- First profile includes call for Authenticated Encryption with Associated Data (AEAD) and hashing for constrained software and hardware environments.
- Second profile includes AEAD for hardware constrained environment only.

5 Evaluation Metrics

While designing cryptographic algorithm, trade-off between the performance and the resources requirements for achieving a given security level is always an issue. Performance of an cryptographic algorithm can be measured in terms of latency, throughput, power, energy consumption and efficiency. For hardware implementations, resources can be given in terms of gate area or gate equivalents, slices and efficiency. And the resources for software implementation can be summarized in no. of registers, RAM (Random Access Memory) and ROM (Read Only Memory) usage. Resource requirements are usually measured as cost, as adding a more gates or memory results in increasing the manufacturing cost of a device.

5.1 *Performance Metrics*

- Throughput: It is defined as the rate at which new outputs are produced. It is calculated as number of output bits divided by the time i.e., by the number of needed cycles and multiplied by frequency. Most of the lightweight cryptographic schemes use frequencies of 100 KHz and 4 MHz for the hardware and the software implementation respectively.
- Latency: Latency is an important metric for real-time applications where very quick response is needed. For an example in an automobile applications very quick response is require for brakes, steering and airbags. Latency can be understand as the amount of time or no. of clock cycles required to produce the final output from the initial request.
- Power consumption: For memory constrained devices, power and energy is the important metrics to consider. Power may be of particular importance for the devices which don't have their own power sources, therefore, they depends on their surroundings to get power for their functioning. For instance, RFID tags that draws their power from the electromagnetic field generated by a reader to power up their circuit. It is estimated on the gate level.
- Energy consumption: It denotes power consumption over a certain time period. It can be determined by multiplying power consumption with required time of operation.

5.2 *Hardware-Specific Metrics*

For hardware implementation of the schemes, resource requirements are measured in terms of gate area or GEs. The area depends on the technology and the standard libraries used for the implementation. It is measured in μm^2. For FPGA and ASIC implementations, area is given in terms of slices and GE respectively.

For ASICs implementation, one GE is considered to be same as the area required by a two-input NAND gate. Hence, the area in GE is given by dividing the area in μm^2 by the area required by the NAND gate. However it is very difficult to compare the schemes just based on the number of GEs as it is very specific to the technology. The same implementation for a given scheme may require different number of GEs over a different technology.

For FPGAs, a slice is the basic configurable logical cell. A slice consists of number of flip-flops and look-up tables (LUTs), which varies based on the different FPGAs. For an instance a slice on Virtex-II consist of 2 flip-flop and 2 LUTs whereas slice on Virtex-5 has 4 flip-flops and 4 LUTs.

5.3 Software-Specific Metrics

Resource requirements for software implementation of the schemes can be given in terms of no. of registers required and the memory (RAM and ROM) usage. Implementation using a few registers have a lower overhead as few replacement needs to be done on the stack. Intermediate values generated during the computations are stored in the RAM while the actual code is stored in the ROM. Accessing intermediate values from RAM may result in the additional trade-off.

6 Design Classification

Authenticated Encryption Scheme (with Associated Data) Π is defined as a set of 3 algorithms $\Pi = (\mathcal{K}, \mathcal{E}, \mathcal{D})$, where $\mathcal{K}$ represents a key-generation algorithm that returns a randomly chosen secret key $K \in \{0,1\}^k$, $\mathcal{E}_K$ is a deterministic encryption algorithm that takes three inputs nonce $N \in \{0,1\}^n$, associated data $A \in \{0,1\}^*$ and message $M \in \{0,1\}^*$ and always outputs a ciphertext-tag pair (C, T) where $C \in \{0,1\}^*, T \in \{0,1\}^\tau$. $\mathcal{D}_K$ is its inverse decryption algorithm that takes nonce N, associated data A and ciphertext tag pair (C, T) as an input and outputs the plaintext M that corresponds to C if the tag is valid, otherwise the bot symbol $\perp$.

$$\mathcal{E}_K : N \times A \times M \rightarrow C \times T$$

$$\mathcal{D}_K : N \times A \times C \times T \rightarrow M \; or \perp$$

6.1 Underlying Constructions

This section describes the basic constructions in brief that can be used as a basic primitive of an AE schemes.

- **Block Cipher** It is defined as a keyed permutation that performs a encryption on fixed-length message blocks using a secret key K which is shared among both the parties prior to the communication.
- **Stream Cipher** A stream cipher works on bit level where ciphertext is produced by XORing a plaintext with a keystream generated by pseudo-random bit generator (PRBG) using a fixed-length secret key. Authenticated encryption can be designed using secure stream ciphers to attain confidentiality and integrity.
- **Key-Less Permutation** It can be defined as bijective mapping on the fixed-length inputs. Now a days, sponge construction is the most famous used keyless permutation. It is an iterated function which takes a variable-length input and produces variable length outputs. The permutation itself works on a fixed-

length intermediate state. Many of the submissions competing in a CAESAR competition are based on key-less permutation.

- **Hash Function/Compression Function** It can be defined as a mapping of a arbitrary length strings to fixed length outputs. A cryptographic hash functions should be preimage resistant, collision resistant and second preimage resistant. A compression function is defined in a same way as a hash function, except that in addition to fixed size message, it also takes a fixed size chaining value as input and generates a new fixed size value.
- **Hybrid** It is a combination of block and stream cipher. The hybrid designs can achieve the given security level even with the small input size. Hence they are able to satisfy the strict time and power consumption requirements for various lightweight applications.

6.2 Underlying Modes

We explicitly mention the underlying encryption mode for the block cipher based AE candidate discussed in Table 3. Here, we only list the underlying modes for the schemes that we consider only.

- CFB: Ciphertext Feedback Mode
- EME: Encrypt Mix Encrypt
- LEX: Leakage Extraction Mode
- OFB: Output Feedback Mode
- TAE: Tweakable Authenticated Encryption

6.3 Functional Characteristics

In this section, we enlists the certain functional characteristics of the AE schemes that enhances the performance.

- Online: An AE scheme is said to be online if the encryption of its ith block of plain text M_i depends only on the previous $(i - 1)$ blocks $(M_1, \ldots, M_{i-1})$ or in other words encryption of ith block is independent of the upcoming blocks M_{i+1} and so on. An AE schemes that does not fulfill this criteria are called off-line or two pass.
- Parallelizable: A scheme is said to be parallelizable for its encryption/decryption operation if the encryption/decryption of the ith input block is independent of the processing of the rest of the other jth blocks where $i \neq j$.
- Inverse-free: A scheme that uses either encryption or decryption function only requires less resource like memory and area. An AE scheme is said to be inverse-free if the underlying primitive is used in only one direction. For example, a block

cipher based AE using its encryption operation only, doesn't require decryption function.

- Intermediate tags: An intermediate tags allows the receiver to discard the message early if the initial blocks of a decrypted message doesn't match the tag. This results in saving the computations and processing time while working with long messages as message is discarded early without further processing.
- Incremental AE/AD: An AE scheme is said to satisfy incremental AE (or AD) property, if, given a ciphertext tag pair (C, T) corresponding to message M, computing (C', T') for a message M' which looks alike M except a few additional blocks, can be done significantly faster in comparison to computing (C', T') for M' individually.

7 Security

As defined by Bellare et al. in [7], an authenticated encryption scheme is considered to be secure if it satisfies following two security notions: one is privacy or confidentiality (IND-CPA) which is defined in terms of indistinguishability from a random oracle and other is authenticity (INT-CTXT) which is defined as an adversary's ability to produce a valid ciphertext tag pair which has not been earlier generated by an encryption oracle. Formally, an AE scheme is secure for any nonce-respecting adversary if both IND-CPA and INT-CTXT advantages are negligible, where IND-CPA and INT-CTXT advantages for a given AE $\Pi = (\mathcal{K}, \mathcal{E}, \mathcal{D})$ are defined as follows:

Definition 1 For a computationally bounded adversary A who has access to an oracle $\mathcal{O}$ which responds to the query either with real encryption using $\mathcal{E}$ or a random permutation \$, the IND-CPA advantage over Π is defined as:

$$Adv_{\Pi}^{IND-CPA}(A) \leq |Pr[K \xleftarrow{\$} \mathcal{K} : A^{\mathcal{E}_K} \Rightarrow 1] - Pr[A^{\$} \Rightarrow 1]|$$

Definition 2 For a computationally bounded adversary A having access to an encryption and decryption oracle $\mathcal{E}$ and $\mathcal{D}$ respectively, the IND-CTXT over Π is defined as:

$$Adv_{\Pi}^{IND-CTXT}(A) \leq Pr[K \xleftarrow{\$} \mathcal{K} : A^{\mathcal{E}_K(),\mathcal{D}_K()} \Rightarrow forges]$$

Nonce Misuse Resistant Rogaway and Shrimpton in [64] defined a new model called nonce-misuse resistance. Informally, nonce-misuse resistance schemes ensure that a repeated random nonce doesn't result in plaintext compromise. The scheme achieving nonce misuse property does not affect the authenticity whereas privacy is only affected to the extent that an adversary can see the repetitions of (N, A, M) triplets, where N represents the nonce, A represents Associated Data (AD) and M is plaintext.

8 Overview of LWAE Candidates

Following the announcement of the CAESAR competition [13], many lightweight authenticated encryption schemes came in existence as a participant for the same. Among them, several are based on dedicated block ciphers, some based on sponge function and others based on stream cipher and compression function. Basic summary of stream-cipher based, sponge based and hybrid AE candidates are given in Table 2. Basic summary for block cipher based AE candidates are given in Table 3.

8.1 Stream-Cipher Based AE Candidates

- ACORN [76]: ACORN design is based on a bit-based stream cipher where the difference is injected into the state for generating authentication tag. ACORN uses 6 LFSR's having a total length of 293 bits and 2 Boolean functions when computing the keystream bit and the feedback bit. ACORN-128 is intended to provide 128 bits of security for both encryption and authentication. It's design is inspired by that of hardware oriented stream ciphers like Grain [26] and Trivium [15]. According to the designers, "the hardware cost of ACORN-128 is slightly higher than that of Trivium, which makes it efficient in hardware". Currently ACORN is competing in a third round of CAESAR competition.

Table 2 AE schemes based on stream ciphers, sponge and hybrid

Construction	Candidate	Year	Parameters (in bits)			Primitive
			Key size	Tag size	Nonce/ IV size	
Stream-cipher based	ACORN	2014	128	128	128	ACORN
	Sablier	2014	80	32	80	Sablier
	Grain-128a	2011	128	32	96	Grain-128
Sponge based	ASCON	2014	96	96	96	ASCON
			128	128	128	
	FIDES	2013	80	80	80	AES
			96	96	96	
	Ketje	2014	128	128	128	Keccak-f
			96	96	80	
	NORX8 and NORX16	2015	80	80	32	NA
			96	96	32	
Hybrid	ALE	2014	128	128	128	AES & LEX
	ASC-1	2012	128	128	0	AES & LEX
	Humminbird-2	2011	128	64	64	AES & LEX
Hash function based	C-QUARK	2012	256	64	64	QUARK

Table 3 Block-cipher based AE candidates

Candidate	Year	Parameters(in bits)			Mode	Primitive	Analysis/Attacks
		Key size	Tag size	Nonce/ IV size			
AES-JAMBU	2014	128	64	64	OFB	AES	None
iSCREAM	2014	128	128	96	TAE	iSCREAM	Forgery attack [70], Nonlinear invariant attack [73]
Joltik $\neq$	2014	64	64	32	TAE	Joltik-BC	None
		80	64	56			
		96	64	48			
		128	64	32			
Joltik $=$	2014	64	64	64	EME	Joltik-BC	None
		80	64	64			
		96	64	64			
		128	64	64			
LAC	2014	80	64	64	LEX	L-Block	Forgery attack [52]
SCREAM	2014	128	128	96	TAE	SCREAM	Forgery attack [70], Nonlinear invariant attack [73]
SILC	2014	128	64	96/64	CFB	AES	Fault based forgery [22]
		80	32	48			

- Sablier [80]: Sablier is a hardware-efficient stream cipher with bulit-in authentication. Unlike the traditional LFSR-based stream ciphers and the usual nonlinear/linear shift registers combined structure in Grain and Trivium, Sablier adopts a new internal structure to generate the keystream from a 80-bit key and a 80-bit IV. It can be efficiently implemented in constrained hardware environments and "the encryption speed is expected to be 16 times faster than Trivium" [15] in hardware. "A practical key recovery attack and a forgery attack against Sablier" has been shown in [23].
- Grain-128a [1]: The Grain-128a AE is based on the eSTREAM finalist Grain-128 cipher [26]. It improves the security of Grain-128 and provides additional support for the authentication. The scheme uses 128-bit keys with 96-bit IVs and tag size is variable up to 32 bits. The cipher outputs one bit per cycle and the AE outputs one bit per two cycles. Except from the general attacks on the Grain family ciphers, differential fault attacks [67] are also feasible on Grain-128a. "A key recovery attack on Grain-128a, in a chosen IV related Key setting" is also presented in [5].

8.2 *Hybrid Candidates*

- Hummingbird-2 [20]: It is one of the first AE schemes for LWC and a benchmark for newer proposals. It was developed for lightweight hardware and software applications. Hummingbird-2 produces a hybrid structure of block and stream cipher. It uses 128-bit keys with 64-bit IVs and operates on 16-bit words. Its main design disadvantage is the lengthy initialization phase due to its stream nature. The size of the hardware implementation with both encryption and decryption is about 70% larger than the encryption-only version. Although the encryption of Hummingbird-2 is fast after the initialization process, the production of the 64-bit MAC is almost seven times slower. However, related-key attacks are performed on the full cipher [66, 68].
- ASC-1 [48]: ASC stands for authenticated stream cipher. In ASC-1, ciphertext is generated by XORing the plaintext with the leak extracted from different AES rounds and tag is computed in CFB manner over a ciphertext. The construction is based on "the concept of Leak-safe Almost XOR Universal (LAXU) hash function" [48] which can allow the construction of provably secure authenticated ciphers.
- ALE [12]: ALE is an online, single-pass, nonce based authenticated encryption scheme that have an optional support for associated data. It supports the plaintext length upto 245 bytes. The overall structure of ALE is similar to Authenticated Stream-Cipher (ASC) [48]. AES is utilized due to the high security and the performance of the AES-NI assembly instruction set. "A forgery attack on ALE which further turned into a state recovery and a universal forgery attack" has been shown in [51]. Later "by exploiting the state information leaked from the

encryption", Leaked-State-Forgery Attack (LSFA) has been shown in [78], which reduces the security level to 97 bits.

8.3 Sponge Based Candidates

- ASCON [18]: ASCON is a online, single pass, nonce-based AE scheme using a sponge construction, which uses a stronger keyed initialization and keyed finalization phase than usual. It is designed to be lightweight in both hardware and software performance. One of the main feature of ASCON is to support the efficient implementation of side-channel resistance. Several result for cryptanalysis of ASCON has been published in [17, 19, 53, 54, 71].
- FIDES [9]: It functions as an on-line single-pass nonce-based scheme. It follows the design principles of AES with a duplex sponge construction structure. Two variants are proposed, with 80- and 96-bit keys requiring 793 GE and 1001 GE, respectively. Defense against side-channel attacks was one of the design criteria and a custom masking scheme is provided which does not increase the area footprint too much. "Internal state recovery attacks on both versions of FIDES" has been shown in [16].
- Ketje [8]: Ketje is a lightweight version of the Keccak, SHA-3 competition winner. Like Keccak, it is also based on the sponge structure, more precisely the MonkeyWrap mode. To support the memory constrained environment, Ketje uses the internal state of 200 and 400 bits, giving raise to Ketje Jr and Ketje Sr respectively. Both of these versions uses the variant of Keccak permutation.
- NORX8 and NORX16 [61]: These are the lightweight 8-bit and 16-bit versions of the existing authenticated cipher NORX [60]. These lightweight versions are designed to satisfy the limited resource scenarios and inherit the same security properties as NORX.
- C-QUARK [4]: It is a new Quark instance with state size b = 384 bits, and a dedicated AEAD mode based on the SpongeWrap construction. It uses the heavier version of the sponge used in the QUARK family of lightweight hash functions and provide higher security.

8.4 Block Cipher Based Candidates

- JAMBU [77]: JAMBU is an authenticated encryption mode designed for the lightweight cryptography. It uses a AES-128 and SIMON block cipher as an underlying primitive. Apart from the use of existing block cipher, it just uses an simple XOR operation to satisfy the lightweight constraints. It provides the CFB encryption security when nonce is misused.
- Joltik [49]: Joltik is based on a 64 bit tweakable block cipher called Joltik-BC. It has two mode of operations one for nonce-respecting adversary (Joltik$\neq$) and

another for nonce misusing adversary (Joltik =). Joltik-BC structure is similar to AES except that it uses the internal state of 64 bits only. It uses the same S-box as Piccolo (4 × 4), however, the MDS matrix is different, it is also involutory and non-circulant.
- LAC [81]: LAC is a lightweight authenticated encryption cipher based on a similar structure of ALE [12] and a simplified version of the lightweight block cipher LBlock as underlying primitive. "Differential Forgery Attack against full LAC" has been shown in [52].
- SCREAM & iSCREAM [24]: These are based on "Tweakable Authenticated Encryption (TAE) mode with the new tweakable block ciphers Scream and iScream" [24]. Scream and iScream are both 128-bit ciphers and uses 8 bit Sbox and 16 bit Lbox, however both uses the different Sbox and Lbox. Unlike Scream, iScream is an involutive cipher. Practical Forgery Attacks on SCREAM and iSCREAM has been shown in [70]. iSCREAM has been completely broken in [73] using Invariant subspace attacks.
- SILC [47]: SILC which stands for simple lightweight CFB, is built upon CLOC [45, 46]. The primary focus of SILC is to target the resource constrained devices which has small hardware settings like RFID. Thus, it has been designed by optimizing the hardware implementation cost of CLOC. The Security proof for the SILC is given using the pseudorandomness of the underlying blockcipher. Later, fault based forgery attack on it has been shown in CAESAR mailing list [22].

9 Comparisons

In this section we present the comparison of various lightweight AE candidates based on different functional characteristics and security discussed above in Sect. 6. Results are summarized in Table 4. Table 5 shows the result for hardware implementation of the different lightweight AE candidates. Comparison is done based on the performance characteristics defined in the Sect. 5.1 and the hardware specific metrics defined in Sect. 5.2. We have summarized the hardware implementation results from the existing implementations of the various ciphers. Table 6 presents the software implementation results for the different lightweight AE candidates.

10 Open Challenges

Interoperability and security are probably the two biggest challenges surrounding the implementation of IIoT. An interoperability is a major concern between the devices and machines that use different protocols and have different architectures. Another important concern is the security of the data generated from these devices and machines. Lightweight authenticated encryption(LAE) is among the major

Table 4 Comparison of functional characteristics of the various lightweight AE schemes: Entries for nonce misuse resistance are: None: when no security is claimed by the designers if nonces are repeated, LCP: an adversary can observe the longest common prefix of messages for repeated nonces

	Functional characteristics					Security	
Candidate	Online	Parallelizable	Inverse-free	Intermediate tag	Incremental AE/AD	Security proof	Nonce misuse resistant
AES-JAMBU	✓	–/–	✓	–	–	✓	LCP
iSCREAM	✓	✓/✓	✓	–	–	–	None
≠/=	✓	✓/✓	–	–	–	✓	LCP
LAC	✓	✓/✓	–	–	–	–	None
SCREAM	✓	✓/✓	✓	–	–	–	None
SILC	✓	–/✓	✓	–	–	✓	None
ACORN	✓	✓/✓	✓	–	–	–	None
Sablier	✓	✓/✓	✓	–	–/✓	–	None
Grain-128a	✓	–/–	✓	–	–	–	None
ASCON	✓	–/–	✓	–	–	✓	
FIDES	✓	–/–	✓	–	–	–	None
Ketje	✓	–/–	✓	✓	–	✓	None
NORX8 and NORX16	✓	✓/✓	✓	–	–	✓	None
ALE	✓	–/–	✓			✓	None
ASC	✓	–/–	✓	–	–	✓	
Hummingbird-2	✓	–/–	–	–	–	–	LCP
C-QUARK	✓	–/–	✓	–	–	–	None

Some of the data is obtained from https://aezoo.compute.dtu.dk/doku.php

Table 5 Hardware implementation results of the LWAE candidates

Candidate	Area (GE)	Frequency (kHz)	Throughput (Mbps)	Power (μW)	Technology
ACORN (8 steps)	499 LUTs	–	3400	–	Virtex 7
ACORN (32 steps)	979 LUTs	–	11300	–	Virtex 7
JAMBU	1254 LUTs	434×10^3	385	–	Virtex 7
ALE	2579	20×10^3	243.8	94.87	STM 65 nm
ALE e/d	2700	20×10^3	243.8	102.32	STM 65 nm
ASC-1	4793	20×10^3	69.18	169.11	STM 65 nm
ASC-1 e/d	4964	20×10^3	69.18	193.71	STM 65 nm
ASCON (1 round)	7080	1000	5524	43	UMC 90 nm
ASCON (6 round)	24,930	1000	13218	184	UMC 90 nm
C-QUARK (serial)	3125	100	8.33	–	TSMC 90 nm
C-QUARK (paralel)	7100	100	266.67	–	TSMC 90 nm
FIDES-80	793	100	10.64	–	NXP 90 nm
FIDES-96	1001	100	12.77	–	NXP 90 nm
Grain-128a	2867	–	–	–	–
Hummingbird-2	3220	100	400	5.10	TSMC 0.13 μm
Joltik $\neq 64 - 64$	2100	–	–	–	–
Joltik $= 64 - 64$	1600	–	–	–	–
Ketje					
LAC	2030	–	–	–	–
LAC (compact)	1300	–	–	–	–
Sablier	1925	–	–	–	–
SCREAM 1R	6230	751	457–7	–	STM 65 nm
SCREAM 2R	8310	446	5190	–	STM 65 nm
SILC (aggressive)	3109	–	–	–	STM 90 nm
SILC (conservative)	4114	–	–	–	STM 90 nm
NORX8	1368	–	–	–	–
NORX16	2880	–	–	–	–

For ACORN and JAMBU area is given in LUTs

building blocks providing security in these systems. However, there doesn't exist any standardization for it, which is another major issue. There has been some standards setup for other lightweight cryptographic primitives like block cipher, hash functions etc. as mentioned in Sect. 4. However there is not any standardization yet for lightweight authenticated encryption schemes which is the current requirement of the time. Ongoing CAESAR competition goal is to find a portfolio of authenticated cipher that can be used as a replacement for currently standardized AES-GCM and suitable for widespread uses. Though CAESAR competition requirements do not focuses on lightweight perspective, many of the candidates competing in third round have lightweight properties.

Table 6 Software performance of the LWAE candidates (in cycles per byte cpb)

	Message length (in bytes)							
Schemes	64	128	256	512	1024	2048	4096	Platform
ACORN	38.2	23.2	15.4	11.8	11.2	8.8	8.2	Intel Skylake processor
AES-JAMBU	24.41	17.08	13.41	11.57	10.65	–	9.98	Intel Core i7-4770 Haswell
ALE	17.15	11.85	9.21	7.88	6.20	6.89	6.72	Intel Sandy Bridge (AES-NI)
ASC-1	37.25	30.19	26.66	24.89	24.01	23.57	23.35	Intel Sandy Bridge (AES-NI)
ASCON	10.5	–	–	7.3	7.1	–	6.9	Intel haswell
LAC	256	206	174	152	144	140	138	Intel Core i7-3612QM @2.10 GHz
SCREAM	–	–	–	–	–	23.5	23.5	ARM Cortex A15
SILC	7.81	–	–	–	–	2.82	2.82	Intel (R) Core (TM) i5- 6600 (Skylake)
Joltik	–	510.99	510.99	510	510	510	–	Skylake processor
Ketje	–	56.84	47	42	40	35	–	Skylake processor

With the rapid movement in IoT technology in last few years, security of these IoT devices are the current time requirement of the lightweight cryptography field. IoT uses the wireless communication which is susceptible to various attacks like eavesdropping, DoS, man in the middle, replay that tears down the security of the system. A conventional security algorithms are not suitable here due to the resource constrained nature of these devices like limited memory, bandwidth, battery. There exists a number of lightweight authenticated encryption schemes in literature as discussed in this work. Some of them works well in hardware and others in software. On the other hand various cryptanalytic attacks has been shown on some of them. However among all these schemes, not any works well for very small messages.

There exists numerous applications where the message packet size is very small and fixed, even there is not enough space for storing tags. For an example data frames transmitted over CAN [37] bus in vehicular network is very small and data payload is limited to 8 bytes only. In these scenarios it is challenging to apply existing AE schemes as there is not enough space for storing tags. Similar is the case with various IoT devices where verification the tag for the continuously streaming data at a very short time interval is overhead. There has been some work in [56] where this has been achieved by maintaining the sequence number and assuming a stateful receiver. However the authors did not provide any security proof.

To address all these issues, designers must need to understand the requirements of the current scenario. And while designing the new schemes they should address all these requirements. They should efficiently run on a wide variety of platforms (8-, 16-, 32-bit micro-controllers, ASICs, FPGAs, 32- and 64-bit processors) and provide security against common attacks. At last there should be some standardization for the lightweight AE like lightweight block cipher and hash function so that it can be easily adapted for the use.

11 Conclusion

In this work, we surveyed the 17 lightweight authenticated schemes, out of which 9 were submitted to ongoing CAESAR competition. Among these 9, 4 AE schemes namely ACORN, ASCON, JAMBU and Ketje are currently competing in a third round. All of the lightweight AE candidates are proposed after 2010, therefore it is a comparatively new direction in lightweight cryptography. It is gaining more popularity these days because of their applications in various IoT devices. These IoT devices have strict resource constraint, therefore conventional AE schemes can not be used there. They require the secure and efficient AE schemes satisfying the various resource constraints. In this paper, we discuss the existing lightweight AE schemes and raise the open challenges for the future research.

Acknowledgements This work was supported by SUTD start-up research grant SRG-ISTD-2017-124. The first author's work was done during her internship in SUTD.

References

1. Ågren M, Hell M, Johansson T, Meier W (2011) Grain-128a: a new version of grain-128 with optional authentication. IJWMC 5(1):48–59
2. Alemdar H, Ersoy C (2010) Wireless sensor networks for healthcare: a survey. Comput Netw 54(15):2688–2710
3. Atzori L, Iera A, Morabito G (2010) The Internet of Things: a survey. Comput Netw 54(15):2787–2805
4. Aumasson J-P, Knellwolf S, Meier W (2015) Heavy Quark for secure AEAD. https://131002.net/data/papers/AKM12.pdf
5. Banik S, Maitra S, Sarkar S, Turan MS (2013) A chosen IV related key attack on grain-128a. In: Boyd C, Simpson L (eds) Information Security and Privacy – 18th Australasian Conference, ACISP 2013, Brisbane, 1–3 July 2013. Proceedings. Volume 7959 of Lecture Notes in Computer Science. Springer, pp 13–26
6. Beaulieu R, Shors D, Smith J, Treatman-Clark S, Weeks B, Wingers L (2013) The SIMON and SPECK families of lightweight block ciphers. IACR Cryptol ePrint Arch 2013:404
7. Bellare M, Namprempre C (2008) Authenticated encryption: relations among notions and analysis of the generic composition paradigm. J Cryptol 21(4):469–491
8. Bertoni G, Daemen J, Peeters M, Van Assche G, Van Keer R (2016) Ketje v1 http://competitions.cr.yp.to/round1/ketjev11.pdf
9. Bilgin B, Bogdanov A, Knezevic M, Mendel F, Wang Q (2013) Fides: lightweight authenticated cipher with side-channel resistance for constrained hardware. In: Bertoni G, Coron J-S (eds) 15th International Workshop on Cryptographic Hardware and Embedded Systems – CHES 2013, Santa Barbara, 20–23 Aug 2013. Proceedings. Volume 8086 of lecture notes in computer science. Springer, pp 142–158
10. Bogdanov A, Knudsen LR, Leander G, Paar C, Poschmann A, Robshaw MJB, Seurin Y, Vikkelsoe C (2007) PRESENT: an ultra-lightweight block cipher. In: Paillier P, Verbauwhede I (eds) 9th International Workshop on Cryptographic Hardware and Embedded Systems – CHES 2007, Vienna, 10–13 Sept 2007, Proceedings. Volume 4727 of lecture notes in computer science. Springer, pp 450–466
11. Bogdanov A, Knezevic M, Leander G, Toz D, Varici K, Verbauwhede I (2011) Spongent: a lightweight hash function. In: Preneel B, Takagi T (eds) 13th International Workshop on Cryptographic Hardware and Embedded Systems – CHES 2011, Nara, 28 Sept – 1 Oct 2011. Proceedings. Volume 6917 of lecture notes in computer science. Springer, pp 312–325
12. Bogdanov A, Mendel F, Regazzoni F, Rijmen V, Tischhauser E (2014) ALE: AES-based lightweight authenticated encryption. In: Moriai (2014), pp 447–466
13. CAESAR (2014) Competition for authenticated encryption: security, applicability, and robustness. http://competitions.cr.yp.to/caesar.html
14. Cid C, Rechberger C (eds) (2015) 21st International Workshop on Fast Software Encryption, FSE 2014, London, 3–5 Mar 2014. Revised Selected Papers. Volume 8540 of lecture notes in computer science. Springer
15. De Cannière C (2006) Trivium: a stream cipher construction inspired by block cipher design principles. In: Katsikas SK, Lopez J, Backes M, Gritzalis S, Preneel B (eds) 9th International Conference on Information Security, ISC 2006, Samos Island, 30 Aug – 2 Sept, 2006, Proceedings. Volume 4176 of lecture notes in computer science. Springer, pp 171–186
16. Dinur I, Jean J (2014) Cryptanalysis of FIDES. In: Cid and Rechberger [14], pp 224–240
17. Dobraunig C, Eichlseder M, Mendel F, Schläffer M (2015) Cryptanalysis of ascon. In: Nyberg K (ed) Topics in Cryptology – CT-RSA 2015, the Cryptographer's Track at the RSA Conference 2015, San Francisco, 20–24 Apr 2015. Proceedings. Volume 9048 of lecture notes in computer science. Springer, pp 371–387
18. Dobraunig C, Eichlseder M, Mendel F, Schläffer M (2016) Ascon v1.2. submission to the CAESAR competition: http://competitions.cr.yp.to/round3/asconv12.pdf

19. Dwivedi AD, Kloucek M, Morawiecki P, Nikolic I, Pieprzyk J, Wójtowicz S (2017) Sat-based cryptanalysis of authenticated ciphers from the CAESAR competition. In: Samarati P, Obaidat MS, Cabello E (eds) Proceedings of the 14th International Joint Conference on e-Business and Telecommunications (ICETE 2017) – vol 4: SECRYPT, Madrid, 24–26 July 2017. SciTePress, pp 237–246
20. Engels DW, Saarinen M-JO, Schweitzer P, Smith EM (2011) The hummingbird-2 lightweight authenticated encryption algorithm. In: Juels A, Paar C (eds) 7th International Workshop on RFID. Security and Privacy, RFIDSec 2011, Amherst, 26–28 June 2011, Revised Selected Papers. Volume 7055 of lecture notes in computer science. Springer, pp 19–31
21. Ericsson Mobility report (2016). https://www.ericsson.com/res/docs/2016/ericsson-mobility-report-2016.pdf
22. Fault Based Forgery on CLOC and SILC (2015). https://groups.google.com/forum/#!topic/crypto-competitions/qxORmqcSrY
23. Feng X, Zhang F (2014) Cryptanalysis on the authenticated cipher sablier. In: Au MH, Carminati B, Jay Kuo C-C (eds) 8th International Conference on Network and System Security, NSS 2014, Xi'an, 15–17 Oct 2014, Proceedings. Volume 8792 of lecture notes in computer science. Springer, pp 198–208
24. Grosso V, Leurent G, Standaert F-X, Varici K, Durvaux F, Gaspar L, Kerckhof S (2014) SCREAM and iSCREAM: side-channel resistant authenticated encryption with masking. http://competitions.cr.yp.to/round1/clocv1.pdf
25. Guo J, Peyrin T, Poschmann A (2011) The PHOTON family of lightweight hash functions. In: Rogaway P (ed) 31st Annual Cryptology Conference on Advances in Cryptology – CRYPTO 2011, Santa Barbara, 14–18 Aug 2011. Proceedings. Volume 6841 of lecture notes in computer science. Springer, pp 222–239
26. Hell M, Johansson T, Meier W (2007) Grain: a stream cipher for constrained environments. IJWMC 2(1):86–93
27. Hirose S, Ideguchi K, Kuwakado H, Owada T, Preneel B, Yoshida H (2012) An AES based 256-bit hash function for lightweight applications: Lesamnta-lw. IEICE Trans 95-A(1):89–99
28. Information technology – automatic identification and data capture techniques – Part 1: security services for RFID air interfaces (2014). https://www.iso.org/standard/61128.html
29. Information technology – automatic identification and data capture techniques – Part 11: crypto suite PRESENT-80 security services for air interface communications (2014). https://www.iso.org/standard/60441.html
30. Information technology – automatic identification and data capture techniques – Part 10: crypto suite AES-128 security services for air interface communications (2015) https://www.iso.org/standard/60440.html
31. Information technology – automatic identification and data capture techniques – Part 12: crypto suite ECC-DH security services for air interface communications (2015). https://www.iso.org/standard/60442.html
32. Information technology – automatic identification and data capture techniques – Part 13: crypto suite grain-128A security services for air interface communications (2015). https://www.iso.org/standard/60682.html
33. Information technology – automatic identification and data capture techniques – Part 14: crypto suite AES OFB security services for air interface communications (2015). https://www.iso.org/standard/61130.html
34. Information technology – automatic identification and data capture techniques – Part 16: crypto suite ECDSA-ECDH security services for air interface communications (2015). https://www.iso.org/standard/61321.html

35. Information technology – automatic identification and data capture techniques – Part 17: crypto suite cryptoGPS security services for air interface communications (2015). https://www.iso.org/standard/61942.html
36. Information technology – automatic identification and data capture techniques – Part 19: crypto suite RAMON security services for air interface communications (2016). https://www.iso.org/standard/63176.html
37. Introduction to the Controller Area Network (CAN) (2016). http://www.ti.com/lit/an/sloa101b/sloa101b.pdf
38. ISO (2012) ISO/IEC 29192-1:2012, information technology – security techniques-lightweight cryptography – Part 1: general. https://www.iso.org/standard/56425.html
39. ISO (2012) ISO/IEC 29192-1:2012, information technology – security techniques-lightweight cryptography – Part 2: bloch ciphers. https://www.iso.org/standard/56552.html
40. ISO (2012) ISO/IEC 29192-1:2012, information technology – security techniques-lightweight cryptography – Part 3: stream ciphers. https://www.iso.org/standard/56426.html
41. ISO (2013) ISO/IEC 29192-1:2012, information technology – security techniques-lightweight cryptography – Part 3: mechanisms using asymmetric techniques. https://www.iso.org/standard/56427.html
42. ISO (2016) ISO/IEC 29192-1:2012, information technology – security techniques-lightweight cryptography – Part 3: Hash-functions. https://www.iso.org/standard/67173.html
43. ISO (2016) ISO/IEC 29192-1:2012, information technology – security techniques-lightweight cryptography – Part 3: mechanisms using asymmetric techniques Amendment I. https://www.iso.org/standard/64591.html
44. ISO, ISO/IEC 29192-1:2012, information technology – security techniques-lightweight cryptography – Part 3: message authentication codes. https://www.iso.org/standard/71116.html
45. Iwata T, Minematsu K, Guo J, Morioka S (2014) CLOC: authenticated encryption for short input. In: Cid and Rechberger [14], pp 149–167
46. Iwata T, Minematsu K, Guo J, Morioka S (2014) CLOC: compact low-overhead CFB. http://competitions.cr.yp.to/round1/clocv1.pdf
47. Iwata T, Minematsu K, Guo J, Morioka S, Kobayashi E (2015) SILC: SImple lightweight CFB. https://competitions.cr.yp.to/round2/silcv2.pdf
48. Jakimoski G, Khajuria S (2011) ASC-1: an authenticated encryption stream cipher. In: Miri A, Vaudenay S (eds) 18th International Workshop on Selected Areas in Cryptography, SAC 2011, Toronto, 11–12 Aug 2011, Revised Selected Papers. Volume 7118 of lecture notes in computer science. Springer, pp 356–372
49. Jean J, Nikolic I, Peyrin T (2015) Joltik v1.3. https://competitions.cr.yp.to/round2/joltikv13.pdf
50. Karakostas B (2013) A DNS architecture for the Internet of Things: a case study in transport logistics. Proc Comput Sci 19:594–601. The 4th International Conference on Ambient Systems, Networks and Technologies (ANT 2013), the 3rd International Conference on Sustainable Energy Information Technology (SEIT-2013)
51. Khovratovich D, Rechberger C (2013) The LOCAL attack: cryptanalysis of the authenticated encryption scheme ALE. In: Lange T, Lauter KE, Lisonek P (eds) 20th International Conference on Selected Areas in Cryptography – SAC 2013, Burnaby, 14–16 Aug 2013, Revised Selected Papers. Volume 8282 of lecture notes in computer science. Springer, pp 174–184
52. Leurent G (2015) Differential forgery attack against LAC. In: Dunkelman O, Keliher L (eds) 22nd International Conference on Selected Areas in Cryptography – SAC 2015, Sackville, 12–14 Aug 2015, Revised Selected Papers. Volume 9566 of lecture notes in computer science. Springer, pp 217–224
53. Li Z, Dong X, Wang X (2017) Conditional cube attack on round-reduced ASCON. IACR Trans Symmetric Cryptol 2017(1):175–202
54. Li Y, Zhang G, Wang W, Wang M (2017) Cryptanalysis of round-reduced ASCON. SCIENCE CHINA Inf Sci 60(3):38102
55. Liu JK, Baek J, Zhou J, Yang Y, Wong JW (2010) Efficient online/offline identity-based signature for wireless sensor network. Int J Inf Sec 9(4):287–296

56. McGrew D (2015) Low power wireless scenarios and techniques for saving bandwidth without sacrificing security. In: NIST Lightweight Cryptography Workshop, vol 2015
57. Moriai S (ed) (2014) 20th International Workshop on Fast Software Encryption, FSE 2013, Singapore, 11–13 Mar 2013. Revised Selected Papers. Volume 8424 of Lecture Notes in Computer Science. Springer
58. National Institute of Standards and Technology (2017) Profiles for the lightweight cryptography standardization process. https://csrc.nist.gov/CSRC/media/Publications/white-paper/2017/04/26/profiles-for-lightweight-cryptography-standardization-process/draft/documents/profiles-lwc-std-proc-draft.pdf
59. National Institute of Standards and Technology (2017) Report on lightweight cryptography (U.S. Department of Commerce, Washington, D.C.), National Institute of Standards and Technology Internal Report 8114. http://nvlpubs.nist.gov/nistpubs/ir/2017/NIST.IR.8114.pdf
60. Neves S, Aumasson J-P, Jovanovic P (2014) NORX: parallel and scalable AEAD. https://norx.io/
61. Neves S, Aumasson J-P, Jovanovic P (2015) NORX8 and NORX16: authenticated encryption for low-end systems. https://eprint.iacr.org/2015/1154.pdf
62. Pang Z, Chen Q, Tian J, Zheng L, Dubrova E (2013) Ecosystem analysis in the design of open platform-based in-home healthcare terminals towards the Internet-of-things. In: 2013 15th International Conference on Advanced Communications Technology (ICACT), Jan 2013, pp 529–534
63. Qin E, Long Y, Zhang C, Huang L (2013) Cloud computing and the Internet of Things: technology innovation in automobile service. In: Yamamoto S (ed) Human interface and the management of information. Information and interaction for health, safety, mobility and complex environments. Springer, Berlin/Heidelberg, pp 173–180
64. Rogaway P, Shrimpton T (2006) Deterministic authenticated-encryption: a provable-security treatment of the key-wrap problem. IACR Cryptol ePrint Arch 2006:221
65. Roman R, Zhou J, Lopez J (2013) On the features and challenges of security and privacy in distributed Internet of Things. Comput Netw 57(10):2266–2279
66. Saarinen M-JO (2014) Related-key attacks against full hummingbird-2. In: Moriai [57], pp 467–482
67. Sarkar S, Banik S, Maitra S (2015) Differential fault attack against grain family with very few faults and minimal assumptions. IEEE Trans Comput 64(6):1647–1657
68. Shi Z, Zhang B, Feng D (2015) Practical-time related-key attack on hummingbird-2. IET Inf Secur 9(6):321–327
69. Shirai T, Shibutani K, Akishita T, Moriai S, Iwata T (2007) The 128-bit blockcipher CLEFIA (extended abstract). In: Biryukov A (ed) 14th International Workshop on Fast Software Encryption, FSE 2007, Luxembourg, 26–28 Mar 2007, Revised Selected Papers. Volume 4593 of Lecture Notes in Computer Science. Springer, pp 181–195
70. Sim SM, Wang L (2017) Practical Forgery Attacks on SCREAM and iSCREAM. http://www1.spms.ntu.edu.sg/~syllab/m/images/b/b3/ForgeryAttackonSCREAM.pdf
71. Tezcan C (2016) Truncated, impossible, and improbable differential analysis of ASCON. In: Camp O, Furnell S, Mori P (eds) Proceedings of the 2nd International Conference on Information Systems Security and Privacy, ICISSP 2016, Rome, 19–21 Feb 2016. SciTePress, pp 325–332
72. The Industrial Internet of Things (IIoT): the business guide to Industrial IoT. https://www.i-scoop.eu/internet-of-things-guide/industrial-internet-things-iiot-saving-costs-innovation/
73. Todo Y, Leander G, Sasaki Y (2016) Nonlinear invariant attack – practical attack on full scream, iscream, and midori64. In: Advances in Cryptology – ASIACRYPT 2016 – 22nd International Conference on the Theory and Application of Cryptology and Information Security, Hanoi, 4–8 Dec 2016, Proceedings, Part II, pp 3–33
74. Top 5 applications for the industrial Internet of Things. https://www.raconteur.net/technology/top-5-applications-for-the-industrial-internet-of-things
75. Wu TD (1998) The secure remote password protocol. In: Proceedings of the Network and Distributed System Security Symposium, NDSS 1998, San Diego. The Internet Society

76. Wu H (2016) ACORN: a lightweight authenticated cipher (v3). https://competitions.cr.yp.to/round3/acornv3.pdf
77. Wu H, Huang T (2014) JAMBU lightweight authenticated encryption mode and AES-JAMBU (v1). http://competitions.cr.yp.to/round1/aesjambuv1.pdf
78. Wu S, Wu H, Huang T, Wang M, Wu W (2015) Leaked-state-forgery attack against the authenticated encryption algorithm ALE. IACR Cryptol ePrint Arch 2015:159
79. Yeh H-L, Chen T-H, Liu P-C, Kim T-H, Wei H-W (2011) A secured authentication protocol for wireless sensor networks using elliptic curves cryptography. Sensors 11(5):4767–4779
80. Zhang B, Shi Z, Xu C, Yao Y, Li Z (2014) Sablier v1. https://competitions.cr.yp.to/round1/sablierv1.pdf
81. Zhang L, Wu W, Wang Y, Wu S, Zhang J (2014) LAC: a lightweight authenticated encryption cipher. http://competitions.cr.yp.to/round1/lacv1.pdf
82. Zhou H, Liu B, Wang D (2012) Design and research of urban intelligent transportation system based on the Internet of Things. In: Wang Y, Zhang X (eds) Internet of Things. Springer, Berlin/Heidelberg, pp 572–580

Access Control in the Industrial Internet of Things

Stavros Salonikias, Antonios Gouglidis, Ioannis Mavridis, and Dimitris Gritzalis

Abstract The Industrial Internet of Things (IIoT) is an ecosystem that consists of – among others – various networked sensors and actuators, achieving mainly advancements related with lowering production costs and providing workflow flexibility. Introducing access control in such environments is considered to be challenging, mainly due to the variety of technologies and protocols in IIoT devices and networks. Thus, various access control models and mechanisms should be examined, as well as the additional access control requirements posed by these industrial environments. To achieve these aims, we elaborate on existing state-of-the-art access control models and architectures and investigate access control requirements in IIoT, respectively. These steps provide valuable indications on what type of an access control model and architecture may be beneficial for application in the IIoT. We describe an access control architecture capable of achieving access control in IIoT using a layered approach and based on existing virtualization concepts (e.g., the cloud). Furthermore, we provide information on the functionality of the individual access control related components, as well as where these should be placed in the overall architecture. Considering this research area to be challenging, we finally discuss open issues and anticipate these directions to provide interesting multi-disciplinary insights in both industry and academia.

S. Salonikias · I. Mavridis
Department of Applied Informatics, School of Information Sciences, University of Macedonia, Thessaloniki, Greece
e-mail: salonikias@uom.gr; mavridis@uom.gr

A. Gouglidis
School of Computing and Communications, InfoLab21, Lancaster University, Lancaster, UK
e-mail: a.gouglidis@lancaster.ac.uk

D. Gritzalis (✉)
Department of Informatics, Athens University of Economics and Business, Athens, Greece
e-mail: dgrit@aueb.gr

C. Alcaraz (ed.), *Security and Privacy Trends in the Industrial Internet of Things*, Advanced Sciences and Technologies for Security Applications,
https://doi.org/10.1007/978-3-030-12330-7_5

Keywords Industrial internet of things · IIoT · Access control · Models · Mechanisms · Architectures · Requirements · Fog · Mist · Information security

1 Introduction

The Internet of Things (IoT) is a term widely used to describe the existence of an ecosystem where pervasive and ubiquitous computing technologies are used to provide connectivity to physical things and make them part of a network where people, devices and things coexist and interact. IoT was greatly benefited from the development of underlying technologies in wireless and mobile networks, which in turn enabled the evolution of both the cloud and Wireless Sensor Networks (WSNs). WSNs provide things with sensors and actuators that are used to sense and produce, as well as consume data and interact with the environment. Advances in the IoT domain are so rapid that although the estimation of 50 billion devices in 2020 seems optimistic, the trend is inambiguous and a number of 20–30 billion seems feasible [21]. IoT is currently used in a number of domains, such as smart homes, smart cities, medical applications and the industry.

Over time, there have been some significant advances in technology that were acknowledged as milestones for the industry development, even characterizing the whole era: In the nineteenth century, steam provided the means for machine development and made the first industrial era possible. Afterwards, the significant development, that started the second era, was the deployment of electricity and its impact in the industry. The third era was characterized by the adoption of Information and Communication Technologies (ICT) that allowed for the development of Programmable Logic Controllers (PLCs) and Supervisory Control and Data Acquisition (SCADA) systems. Today, we witness the transition into the fourth industrial era that is aided by the integration of a whole ecosystem of networked sensors and actuators into every aspect of the production stage. This integration between legacy industrial information systems and IoT, was initially described by the Industrie 4.0 initiative, mainly developed in Germany to provide competitive advantages by lowering production cost and providing workflow flexibility [27]. The outcome of the aforementioned integration is known as the Industrial Internet of Things (IIoT).

As in many emerging technologies, the adaption of ICT technologies in IIoT introduced issues with regards to standardization and security. Thus, a number of commercial entities have created the Industrial Internet Consortium (IIC) which has published a number of publicly accessible white papers on architecture and security [13]. In ICT, as well as in the IoT, information security (i.e., confidentiality, integrity and availability) is of major concern. However, in IIoT, additional concepts should be taken into consideration regarding the applicability in the application environment as well as the need for safety. Controlling access to resources for ecosystem stakeholders is crucial to fulfill both targets.

Introducing access control in IIoT is considered to be a challenging task stemming from the diversity that characterizes these industrial environments. The diversity is mainly introduced by the great variety of technologies and protocols supported by the IIoT devices and networks. Access control in Cyber-Physical Systems (CPSs) has been examined in [20], where access control models are compared and a set of requirements is examined. Yet, we anticipate that further investigation may be required to cope with access control challenges in IIoT. The aim of this chapter is to provide additional information about the most promising access control models for IIoT, examine access control mechanisms able to support the described models and propose an access control architecture for IIoT based on virtualization technologies.

Specifically, in the following of this chapter, in Sect. 2, we provide background information on IIoT architectural trends, which are necessary to gain visibility to the ecosystem and extract access control requirements. In Sect. 3, major families of access control models and mechanisms are extensively presented. Access control approaches proposed in the literature for application in IIoT are examined in Sect. 4. The various components that constitute an access control architecture for IIoT are investigated in Sect. 5. Finally, in Sect. 6, we briefly elaborate on a set of open issues with regards to access control and IIoT and provide concluding remarks.

2 Background

The IoT is defined by the pervasive presence of things that are uniquely identified and are able to interact among them and with the rest of the network [3]. Initially introduced by Radio-Frequency Identification (RFID) tagging to provide Electronic Product Code (EPC), today IoT includes a number of heterogeneous devices interconnected using various protocols and technologies to provide the most efficient means of connectivity and interoperation.

Specifically, IoT describes a network of objects that may collect and share data in an autonomous manner and without requiring assistance by humans. Examples of such objects are considered to be various type of sensors that monitor and measure the temperature or humidity of the environment, the acceleration or position of an object, etc. The application scenarios of IoT are considered to be numerous, ranging from smart appliances (e.g., smart lighting and heating devices) to fitness devices (e.g., Fitbit).

The International Telecommunications Union (ITU) has released ITU-T Y.2060 [16], which is a recommendation that provides an overview of IoT. According to the recommendation, IoT adds a third axis in the already existing "anytime" and "anyplace" communication that could be even provided by legacy ICT systems. The new axis is called "anything" and represents communication not only between computer devices, but also between human to human, human to thing and thing to thing. Things are objects that exist in the physical world and can be sensed and identified. The identification can be performed utilizing virtual entities which can exist without the presence of the physical ones.

Due to IoT great success and adoption rates, IoT technologies are also embraced by the industry and introduced in industrial environments as a means to improve operational efficiency [5]. Therefore, IIoT, "IoT Version 4.0" or "Manufacturing IoT" are expressions frequently used to denote the use of IoT for industrial purposes. By the end of 2020, it is estimated that more than 10 billion devices will account for the IIoT and represent the 57% of IoT spending [21].

IoT has already been a part of everyday life, including, but not limited to, smart cities, health-care, agriculture, leisure (smart homes), construction, intelligent transportation systems, etc. There are many initiatives aim to exploit IoT in industrial environments, such as smart factories, Industrial Internet, Factories of the Future, etc. [27]. Although IoT and underlying technologies are well established and evolving constantly, adoption in the industry is a challenging task considering both the different environment and the fact that there are already well-established ICT systems in place (e.g., Distributed Control Systems (DCS) and SCADA systems) that control and monitor production process.

Industrie 4.0 is an initiative to support manufacturing in optimizing production efficiency and increase product quality. The initiative's underlying concept is to integrate IoT into legacy production field industrial information systems, thus being able to create a new concept, the IIoT. IIoT is enabled by the advances on Machine to Machine (M2M) communication, network efficiency and simplicity induced by 4G and 5G development and of protocols like 6LoWPAN and LoRaWAN and faces all challenges that exist in the IoT, such as resource constrained devices, heterogeneity, limited connectivity, etc. In the industrial environment an important factor is also the requirement for safety [27]. Although safety is not directly concerned with information security, being a key objective in IIoT operation, it must be taken under consideration to prevent accidents that could potentially threat the integrity of humans and machinery, as well as the availability of services. Access control models do not take safety under consideration as an inherent design feature, so safety provision should be considered, if possible, when creating access control policies.

2.1 IIoT Architecture

In March 2014, AT&T, Cisco, General Electric, Intel, and IBM co-founded the Industrial Internet Consortium (IIC) with the aim to promote the growth of IIoT. IIC has released version 1.8 of the IIoT Reference Architecture [18] where an IIoT analysis define four different viewpoints, i.e., business, usage, functional and implementation viewpoints. In this chapter we are mainly concerned with the implementation viewpoint where technological aspects can be revealed and examined.

With regards to the implementation viewpoint, IIC defines a three-tier architecture, namely, the edge, platform and enterprise tiers. The edge tier is where data collection is performed from industrial and other end devices such as vehicles, machinery, workstations, automations, and all other sensors representing "things" in the industrial area. Data collected from the edge tier is sent to the platform tier, which is the medium between data collection and data exploitation, with the latter taking place to the enterprise tier (upper tier). Nodes residing on these tiers are inter and intra connected using different kinds of networks. These include the proximity network, which connects assets within the edge tier, the access network that connects the edge to the platform tier, and finally the service network that connects the platform with the enterprise tier.

The edge tier includes all the ICT components that are located in the production space. Example of such components are sensors, actuators and all other legacy devices and CPSs. The evolution of IoT led to the multiplication of the number of edge nodes that are characterized by physical limitations on computing and energy resources. The platform tier includes all the necessary processing that is required for edge device provisioning and data consolidation before those are delivered to the enterprise tier where services are developed. IIC does not provide topology-related constraints so, in its simplest form, platform and enterprise tier can be physically either located in premises or be powered by the cloud. Considering the volume of collected data, the cloud can be an enabling computing paradigm since it may provide the best candidate for big data processing. Connecting edge nodes directly to the cloud though, can be challenging considering the resource restrictions of many edge devices and latency induced by logical distance. The latter can be a potential threat to service provisioning as well as to system safety since the delay induced can lead to delayed actions that may cause damage. To overcome this issue, fog computing can be used as a middle layer between the edge and the cloud, thus reducing both distance and latency.

Fog was initially proposed by Cisco Systems [4] to provide a location aware and low latency virtualized layer between the edge and the cloud, thus bringing services nearer to the actual stakeholder. A fog layer is populated by private, community, public or hybrid [14] fog nodes that process information from edge devices and communicate with the cloud when necessary. In the fog concept, all information processing is performed in the fog nodes and little or none in the edge devices. Nevertheless, since nowadays network fabric can also provide the means to integrate processing into the network itself, a new layer can be developed between the edge and the fog. This is created by low-resources microcontroller-based devices with low-resources and is known as the mist layer [14]. Mist nodes are actually embedded in the same environment with the edge device, providing more accurate context information and enabling processing at the edge of the network, which further reduce the overall latency, provide contextual accuracy, and reduce power requirements from end devices.

Although IIC edge tier is clearly matched to the edge layer, it is not so clear how to map the platform tier and the enterprise layer. Such a mapping usually depends on the specific application domain and topology. In Fig. 1 an indicative mapping is depicted.

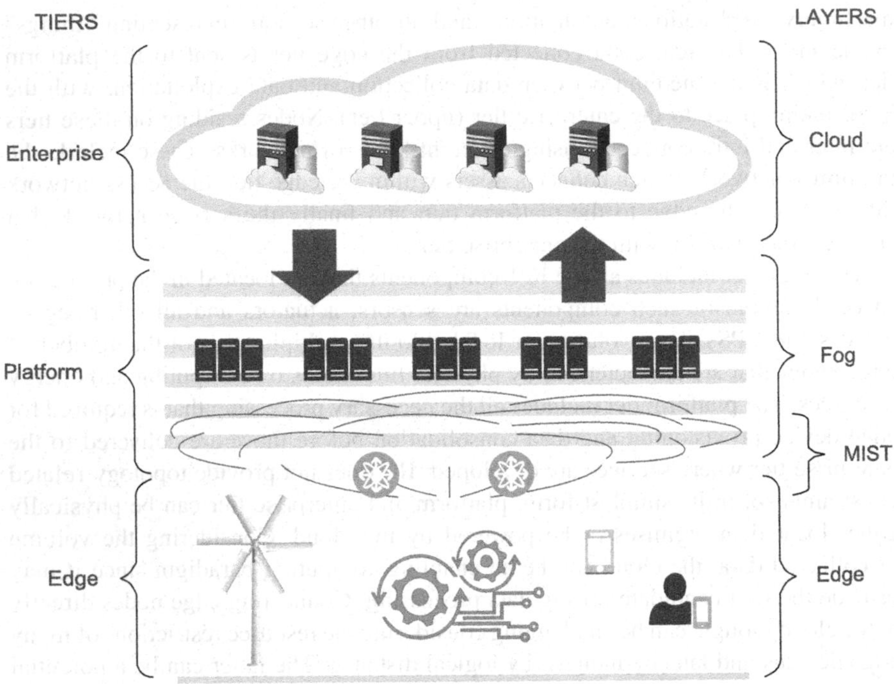

Fig. 1 IIoT Ecosystem

2.2 Access Control Requirements

Access control is essential in all systems that require to control and limit actions or operations that are performed by a user or process on a set of system resources [6]. An access control system is considered of three abstractions, namely, the access control policies, models, and mechanisms. Based on these abstractions, an access control system is made responsible for enforcing the access control policies and preventing them from subversion. Access control policies are characterized as high-level requirements that specify how and when a user, or a process, may access a resource. The access control policies are enforced through an access control mechanism, which is responsible for granting or denying access. An access control model is an abstract container of a collection of access control mechanism implementations, capable of preserving support for the reasoning of the access control policies through a conceptual framework. Thus, access control models are bridging the abstraction gap between the policies and the mechanisms in an access control system. In [26] an IoT enabled ecosystem utilizing the notion of fog computing in Intelligent Transportation Systems (ITSs) is presented. Considering the number of access control issues presented there, the following requirements can be extracted:

- **Context awareness:** Contextual information characterizes the situation of an entity and the environment [1]. Context can influence access control decision and allow for policy creation that considers factors beyond subject's and object's identity. Having visibility into the context, access control policies can also be designed with an eye on safety on top of information security.

- **Inter-domain operation:** IIoT is deployed in multiple domains supporting operation of remote sectors under the same administration authority. Any access control solution should be able to support a coherent operation among different domains.
- **Privacy assurance:** Privacy is nowadays an important factor that needs to be considered in the deployment of every ICT solution (privacy by design). Since 2018, it is also a legal obligation in the European Union, defined by the General Data Protection Regulation (GDPR). An access control mechanism should be designed in a way that no private data should be ever disclosed.
- **Resource efficiency:** Most devices on the edge are designed to perform specific tasks and consume the less power possible. This limits available resources, both in terms of processing power and storage space, so any component designed to run on those, should take these limitations under consideration.
- **Manageability:** There should be a centralized way to create, store and enforce policies that would not induce extra latency and could function over low-bandwidth networks that may even sometimes become unavailable.
- **Accountability:** Auditing should be supported to provide respective stakeholders with the ability to monitor and reveal any violations or system misuse.

The list of the above mentioned requirements is not exhaustive, but instead it operates as stepping stones in choosing a more appropriate authorization scheme. In the following, we provide more information about families of access control models and frameworks towards their investigation in the context of IIoT environments.

3 Access Control Approaches

Although there is an abundance of access control models that could be applicable in IIoT environments, we elaborate in the following on major access control family of models. This results in avoiding replication of information among models having their root on the same model family and help to describe the main characteristics offered by these models. Specifically, we provide information about the role-based, attribute-based, capability-based, and usage control family of models.

3.1 Role-Based Access Control

In Role-Based Access Control (RBAC) [17], access to resources of a system is based on user and role assignment to roles, which have predefined permissions associated with them. RBAC may support several principles, e.g., least privilege, separation of duties and separation of administrative functions, which makes it preferable for use in organizational environments.

The core RBAC model is composed of five static elements, namely, users, roles, and permissions, with the latter being composed of operations applied on objects. With regards to relationships among elements, roles are assigned to users and permissions are assigned to roles. These types of relations may be of many-to-many, i.e., one user can be assigned to many roles and many users can be assigned to a single role. The same applies for role to permission assignments. Negative permissions are not supported in RBAC.

RBAC has two different phases, i.e., the design and run-time. During the design phase, a system administrator can define a number of assignments between the elements in the computer system. At the run-time phase, the assignments in the system are enforced by the model as it is specified by the security policy, which was prescribed during the design phase. Run-time enforcements are instantiated through the concept of sessions. The latter distinguishes RBAC from other group-based mechanisms. During a session, roles for a subset of users are allowed to be activated. This means that a user could be assigned various roles during the design phase, but these roles do not need to be activated always or simultaneously. Using the latter mechanism, RBAC provides support for the principle of least privilege. A number of constraints may be also enforceable during a session.

Apart from the core model, RBAC supports also hierarchies between roles. This mechanism provides great flexibility when it comes to the management of the policies. Specifically, permissions that are assigned to a role can easily be inherited to another role, without the need to reassign the same permissions to the latter. For example, we assume two roles R1 and R2 and two permission sets PR1 = (P1,P2) and PR2 = (P3,P4), which are initially assigned to roles R1 and R2, respectively. If role R1 inherits role R2, it means that all of R2's permissions are available through R1. The available permissions to role R1 are expressed by the union of permissions on sets PR1 and PR2. When hierarchies are represented in graphs, the immediate inheritance relation is shown as $\rightarrow$. The head of the arrow or arc defines both the permissions and user membership inheritance. In the previously example, we have R1 $\rightarrow$ R2. User membership refers to the assignment of users to roles in a hierarchy. In such a case, users are authorized to access all the permissions assigned to roles either directly or through inheritance relationships. Yet, another functionality that is provided in hierarchical RBAC is the support of general and limited role hierarchies. General hierarchies comprise the most common cases in role inheritance, and they are depicted as partial order sets. However, in more restrictive environments the requirement for supporting limited hierarchies may arise. This involves usually the existence of either a single immediate ascendant or descendant role in the hierarchy tree structure.

RBAC is also capable of supporting constraints through static and dynamic separation of duty relationships. The main objective in both types of constraints is to preserve the security of the system and prevent it from being compromised. Constraints are usually used to deliver business requirements. Static separation of duty relationships copes with the enforcement of conflict of interest policies. For example, let R1 and R2 be two conflicting roles and user U1 assigned to role R1. By enforcing a static separation of duty constraint between roles R1 and R2, RBAC prohibits the assignment of user U1 with role R2 since the two roles are conflicting. This type of constraints is defined and enforced in RBAC during the design phase. In the presence of a role hierarchy, the static separation of duty constrains are enforced in the same way for all the directly assigned and inherited roles. Dynamic separation of duty relationships handles conflict of interest policies in the context of a session. In this case, the user has a set of roles activated. A dynamic separation of duty relationship is described during the design time, but it is enforced during run-time – in the context of a session – to prevent the simultaneous activation of two or more conflicting roles. In case of role hierarchies, a similar mechanism to static separation constraints is applied, but constraints are enforced only on the set of activated roles.

3.2 Capability-Based Access Control

Capability-based access control (CapBAC) is based on the concept of capabilities [29], which are known to be communicable and unforgeable tokens of authority. A capability contains entries for the resources that a subject has granted access to. Thus, in a similar way to access control lists, an access control matrix is considered that may include subjects, objects, and permissions. In CapBAC, permissions are assigned with subjects, and thus support one-to-many relationships between subjects and objects. Subjects and objects refer to the users and resources of a system (in a similar way to RBAC). Permissions are authorized operations that can be performed by a subject on an object.

CapBAC support also delegation and revocation mechanisms for capabilities. These are required to delegate access (indirectly) to other subjects and revoke access, respectively. Usually, capabilities are issued in the context of a Simple Public Key Infrastructure (SPKI) to cope with delegation of authorizations from one subject to another [25]. Such solutions may be applicable in multi-domain federated environments.

3.3 Usage Control

A representative usage control approach is UCON [24], which is based on a modern conceptual framework. The UCON conceptual framework encompasses traditional access control, trust management and digital rights management for the protection of digital resources. Nonetheless, functionalities such as administration and delegation are still absent. UCON has introduced a number of novelties compared to both RBAC and other attribute-based models, like its support for mutable attributes and continuity of access decision. Research has also been conducted regarding its usage in collaborative systems [30].

UCON is formed of eight components, namely, subjects, subject attributes, objects, object attributes, rights, authorizations, obligations and conditions. The notion of subjects and objects as well as the association with their attributes is straightforward. A subject can be an entity in a system and its definition, as well as its representation, is given by a number of properties or capabilities in the associated subject's attributes. For instance, role hierarchies similar to RBAC can be formed through the use of subject attributes. In regard to objects, they also represent a set of entities in a system. Each object can be associated with object attributes. Subjects can hold rights on objects. Through these rights, a subject can be granted access or usage of an object. This type of attributes can serve, for example, in the classification of the associated objects, by representing classes, security labels and so on and so forth. It is worth mentioning that both subject and object attributes can be mutable. This means that the values of the attributes can be modified as a result of an access. To the contrary, when an attribute is characterized as immutable, its value can be modified only by an administrative action and not by its user activity.

UCON is characterized by a number of novelties, stemming mainly from the rest of its components. The component of rights represents a number of privileges that can be held and exercised from a subject to an object. In a similar way to RBAC's roles, the UCON conceptual framework supports hierarchies among rights. Note that rights are not set a priori, but they are determined during the access. The access decision is given from a usage function by considering the following factors of subject and object attributes, authorizations, obligations and conditions. Authorizations in UCON are functional predicates, whose evaluation is used for taking decisions, namely if access to a subject is granted to an object. In a same manner to the usage function, the evaluation of authorizations is based on subject and object attributes, requested rights and a set of authorization rules. Authorizations can be characterized as pre-authorizations or ongoing-authorizations. The pre prefix refers timely before the requested right and the ongoing prefix during the time span of access.

Furthermore, obligations in UCON are used to capture the requirements that must be met from a subject requesting the usage of an object. These are expressed as functional predicates and, as already mentioned, they are used in the evaluation of access both in the usage function as well as with authorizations. Obligations are also divided into pre obligations and ongoing obligations. The former is used usually for the retrieval of history information and the latter to check whether the requested requirement is fulfilled during the time span of access. Finally, conditions in UCON are used to capture factors that are accrued from the environment of the system. The semantic difference between conditions and other variables, namely authorization and obligation, is that the former cannot be mutable since there is no direct semantic association with subjects.

3.4 Attribute-Based Access Control

Attribute-Based Access Control (ABAC) has gained a significant attention due to the development of distributed systems and networks, such as the Internet, and is considered to be a logical access control methodology [11]. In contrast to RBAC, a standardized ABAC definition is still missing, and thus several have been proposed. However, a set of guidelines are provided by NIST in [11]. ABAC can provide access decisions based on the evaluation of attribute values, policy rules and environment conditions, depending on the particular ABAC definition. One virtue of ABAC compared to other models is that its policies are expressed in terms of attributes without prior knowledge of the subjects and objects in the system. Moreover, subjects and objects in a system may be assigned with attribute values without prior knowledge of policy details. This does greatly simplify authorization management.

The ABAC model consists of the following six categories of elements: Attributes, subjects, objects, operations, policies, and environmental conditions. Attributes are characteristics of the subject, object, or environment conditions. Attributes may contain information given by a name-value pair, i.e., a tuple of the form: (Name, Value). Both subject and object attributes are able to support the use of meta-attributes. The latter provides an additional index for referring to groups of subjects and objects per se. Hierarchies in ABAC are intrinsically supported via the meta-attribute functionality. This provides ABAC with the potential to express powerful hierarchies between elements of the same type. A subject is usually interpreted as being a user or process that issues access requests to perform operations on objects. Subjects can be assigned with one or more attributes. An object can be a system resource for which access is managed by the ABAC system. These could be devices, files, records, tables, processes, programs, networks, or domains containing or receiving information. It can be the resource or requested entity, as well as any entity on which an operation may be performed by a subject including data, applications, services, devices, and networks. An operation is the execution of a function at the request of a subject upon an object. Example of operations include the read, write, edit, delete, copy, execute, and modify commands. A policy is the representation of rules or relationships that makes it possible to determine if a requested access should be allowed, given the values of the attributes of the subject, object, and possible environment conditions. An environment condition is an operational or situational context in which access requests occur. Environment conditions are detectable environment characteristics. Environment characteristics are independent of subject or object, and may include the current time, day of the week, location of a user, the current threat level, etc. The above definitions subsequently help in the provision of a reference model for ABAC and a formal specification of it.

In the following, a brief description of well-known ABAC frameworks is provided. Access control frameworks may provide useful guidelines when considering the implementation of an access control system. With regards to attribute-based approaches, the Extensible Access Control Markup Language (XACML) and the Next Generation Access Control (NGAC) appear to be the most prominent frameworks. Both provide operations to manage policies, evaluate decisions, enforce policies, etc. XACML and NGAG may facilitate the adoption of attribute-based approaches though the provision of specifications with regards to both functional operations and composition of components (e.g., policy decision point, policy enforcement point). In the following, we provide information on XACML and NGAC, so as to operate as a precursor when considering proposing access control systems applicable in IIoT environments.

3.4.1 Extensible Access Control Markup Language

XACML is an OASIS standard, currently in version 3.0, which provides a framework for deploying ABAC. To achieve this, XACML provides a data-flow model, named the XACML context, and a policy language model. The data-flow model describes the main functional components, e.g., Policy Enforcements Point (PEP), Policy Decision Point (PDP), Policy Authorization Point (PAP), etc. and interactions among them. These are used for accessing repositories – containing policies or attributes – and getting authorization decisions. The XACML context expresses access requests and responses using an XML schema, implemented by the PDP for authorization purposes. The policy language model is used for the specification of access control requirements using attributes in the context of three hierarchical components, i.e., rules, policies, and policy sets. Apart from the main components, it is also interesting to refer to the terminology differences between the XACML standard and the guidelines on ABAC provided by NIST. It is apparent that despite some terms are expressed differently, both refer to the same concepts. In the following, we briefly refer to this mapping, as identified in [12]. Subjects and actions refer to the same concept, in both XACML and ABAC. A subject refers to the entity that requests access, and an action refers to the performed operation on the requested entity. A resource in XACML is mapped to an object in ABAC – resources or objects are entities that a subject request to access. The environment in XACML is mapped to environment condition in ABAC – that is a dynamic factor, independent of subjects and objects. Lastly, while the term element is used in NIST's guidelines document to refer to subjects, objects, actions, and environment conditions, the term category is used in XACML instead to refer to subjects, resources, actions, and environments.

3.4.2 New-Generation Access Control (NGAC)

NGAC is a NIST initiative [12] for standardizing ABAC mechanism. It is able to express and enforce a wide range of policies. Defined in accordance to ABAC to meet its requirements, NGAC uses data/relations and attributes to express policies and deliver capabilities, respectively. It also provides a set of administrative operations and functions for configuring data and enforcing policies. In the following, we provide briefly information on NGAC as described in [12].

Access control data in NGAC includes elements, containers, and relations. An element may be a user, an operation or an object. These maps to ABAC's subject, action, and object, respectively. User and object attributes are supported through containers. The latter are used to administer and formulate attributes and policies. Containers are used to associate and group elements among them. Similarly, policy class containers are used to provide collection of policies. Attributes in NGAC are used in a similar way to ABAC – they represent characteristics of the user or object. For example, user attributes could express user roles, etc., while an object's attributes could express its stored data. A set of basic operations are provided by NGAC to interface with the data of objects, and administrative operations are responsible for the creation of data elements and relations.

Relations in NGAC are used to express access control policies. There is support for four different type of relations, i.e., assignments, associations, prohibitions and obligations. Assignments are used to define membership on containers. This is expressed through a tuple of the following form: (a, b) or equivalently a → b. The semantics are that element a is assigned to element b.

Associations are used to derive privileges and are expressed as 3-tuples including a user attribute ua, a set of access writes asr, and a user or object attribute at. The latter association is written as ua – ars – at with the following semantics: Users in ua can execute the ars access rights on the policy elements referenced by at.

Prohibitions are used to derive privilege exceptions. Three types of prohibitions are supported, i.e., user-deny (u_deny), user attribute-deny (ua_deny), and process-deny (p_deny). Each prohibition is expressed using a 3-tuple including a user u, a user attribute ua, and a process p, respectively, followed by an access right (ars) and a policy element (pe). A user-deny prohibition may be of the following form: u_deny(u, ars, pe). The semantics of the latter prohibition is that user u cannot execute access rights in ars on policy elements in pe. In a similar manner, attribute-deny and process-deny are expressed as ua_deny(a, ars, pe) and p_deny(p, ars, pe), respectively.

Lastly, obligations are used to dynamically alter an access state. Obligations are expressed as pairs of event patterns ep and a response r (i.e., sequence of administrative operations). The former consists of conditions, which when evaluated to true causes the response r to execute.

4 Access Control in IIoT

As stated already, access control can introduce the appropriate mechanisms in a system to restrict access of legitimate users or processes in it. IIoT can be characterized as a system of systems, and its emerging characteristics, such as automation, adaptation, high heterogeneity of devices, spatial diversity, etc. require revisiting the concept of access control. Although several works have been conducted in the context of IoT environments, access control in IIoT is still a relatively new area of research. In the rest of this section, we refer to the latest achievements in access control and IoT that appear to be prominent for application in an IIoT environment. Yet, we identify research works that has been already conducted in IIoT environments.

An extensive review of access control model and frameworks for IoT is conducted by A. Ouaddah et al., in [23]. The survey includes approaches proposed within a period of five years, starting of 2011. These approaches can be potentially applicable in IoT/IIoT environments. An interesting outcome of the survey is the compilation of a taxonomy for both access control models and frameworks. An abundance of access control models has been included, yet all of them have been grouped in representative families of models/categories, e.g., ABAC, RBAC, usage control, CapBAC, organizational-based access control models, etc. In the following, we briefly elaborate on individual models that appear to be omitted in [23] and elaborate on generic frameworks – potentially applicable in IIoT environments.

An RBAC model has been proposed in [7], which is applicable in collaborative multi-domain systems. The proposed model (domRBAC) supports all the components of the standard role-based model (ANSI INCITS 359-2004), including support for the core RBAC, hierarchical RBAC, static and dynamic separation of duties. Furthermore, domRBAC is able to enforce access control under secure interoperation, a prerequisite in multi-domain environments.

In [19] an RBAC model is proposed for application in IIoT, considering them to be multi-domain collaborative environments. Specifically, the requirements under investigation include these of resource sharing and process collaboration. The authors define RBAC policies as an authorization route optimization problem and provide a solution by proposing an algorithm for solving it. Although the proposed solution may provide optimal solutions, its performance may be restrictive in some cases, as stated by the authors (e.g., assuming excessive amounts of devices and roles). It is provided merely as an administrative tool and lacks automation, i.e., it is not applicable in a policy decision point.

An ABAC model is formally defined in [8] in adherence with NIST's recommendations in [11]. The model's main elements that can take part in the authorization process and a description of its main administrative operations and review functions are provided. ABAC approaches intrinsically support highly distributed environments due to context information conveyed through attribute values.

The UseCON model [9, 10] is a next-generation model based on the concept usage control. UseCON is able to support complex and more expressive policies com-pared to existing usage-based approaches (e.g., UCON). Although it is not explicitly defined in the context of IoT/IIoT, its main characteristics, such as continuity of decision and attribute mutability, may render it applicable in industrial environments. Although an implementation of the model is missing, formal proofs have been provided with regards to its internal functions.

Independently of the access control model and its supported policies, a set of functional components are required for an access control mechanism to be instantiated in the context of an access control system architecture. The telecommunication standardization sector of ITU provides in X.812 a recommendation of a security framework, which defines among other the main functions required in open systems to support access control services and mechanisms [15]. Thus, based on X.812 the main functions may include: An initiator (e.g., a user or process), a target (i.e., the resource access is required upon), an Access control Enforcement Function (AEF), and an Access control Decision Function (ADF). The latter is responsible for access control decision making. The decisions are made based on information applied by the access control policy rules, the context in which the access request is made, and Access control Decision information (ADI). ADI is part of the Access Control Information (ACI) function, which includes all the information used for access control purposes, including contextual information. Lastly, the responsibility of AEF is to enforce the decision taken from the ADF.

Following the core idea of X.812, existing access control frameworks as the XACML and the NGAC provide their own set of functions to support X.812 functionality. XACML main functions are a PEP, a PDP, a Policy Information Point (PIP), a PAP and a context handler (CH). Further information about the operations supported by the individual functions is provided in OASIS XACML standard documentation [22].

In a similar manner, NGAC provides its own functional architecture, too. Its main functional components are: At least one PEP; at least one PDP; zero or one Event Processing Point (EPP); one PAP; one PIP; and one or more Resource Access Points (RAPs). Further information about the operation of the individual functions in NGAC is provided in [12].

It is worth mentioning that although both XACML and NGAC frameworks share some functionality, yet they differ. For example, the PAP, PDP, and PIP appear to provide slightly different functionality in each framework. Differences apply also when it comes to their access decision process, which is logic based in XACML and enumerated in NGAC [12].

5 Components Placement

From the above it is evident that considering an access control architecture for application in an IIoT environment requires a carefully investigation of all its functional components. This will provide – depending on the applied framework (e.g., XACML, NGAC) – indications on where to place each of the functional components in respect to the layers, as depicted in Fig. 1. The placement is not just an arbitrary architectural decision since it affects both the functionality and the efficiency of the applied framework in the specific context.

The cloud is an important element in the development of IIoT. It provides a unified, ubiquitous platform for data sharing and can support various applications in the context of IIoT. Alsheri et al. [2] propose a cloud-enabled architecture for access control deployment in IoT. That architecture includes a layered environment that consists of the object layer, the application layer and the in-between middle layer(s).

Specifically, the object layer includes the things residing on the edge, whereas the middle layer includes the virtual object and cloud services layers. The virtual objects layer is an abstraction used to provide the constant presence of things including both current and historical information [28].

The cloud services layer provides resources to objects, and finally, the application layer offers an interface to communicate with the objects. In such an approach, the access control decision making is provided by a PDP placed in the cloud layer and the enforcement of access control decisions is performed by a PEP placed on the object layer. Access control administration is performed in the administration layer.

In [26] an ABAC specific deployment is proposed where cloud, fog and edge layers are used for the various components of access control system. Access control administration is provided by a PAP, which is located on the cloud along with a PIP that stores subject, object and system attributes. PDPs are in turn located in various fog nodes and interact with the PAP and PIP in the cloud. Finally, PEP is performed on the edge layer. Integrating PEP on the edge is a challenging task considering all resource limitations and the heterogeneity of objects that renders the consistent enforcement deployment to be a challenging task.

PAP is the term used by ABAC models to describe the entity that is used to create and manage policies of an access control system. Deploying PAP in the cloud makes it available enterprise-wide and it eases any consideration regarding policy exchange between remote federations, provided that all required mechanisms (e.g., authentication) are in place. The same applies to any other model implementation when it comes to policy administration.

PDP on the other hand provides time-critical services since their use is to reach into access control decisions. Access control decision making requires on the one hand resources to allow for quick processing of policies and on the other hand low latency to communicate the decision to enforcement points instantly upon making. Placing PDP on the cloud may not be the most efficient architectural decision, mainly due to the distance between stakeholders and the cloud itself. Extending cloud near the edge though, which is the case when exploiting fog computing, lowers this distance and makes the fog layer the prevalent candidate to host PDPs.

In the ABAC case, which is a suitable model in implementing context-aware access control mechanisms, thus mechanisms which use context to provide relevant information and/or services to the user, where relevancy depends on the user's task [1], the required attributes need to be retrieved from PDP to perform access decision. This information is provided to PDP upon request to PIP. To achieve this, PIP should be both aware of all available attributes but also able to both retrieve and deliver attributes in real time without stalling the whole process. Since most attributes are domain-specific in the industrial environment, considering the uniqueness of each deployment, PIP needs to have visibility to the specific domain. To achieve this fog can be utilized and host an additional "local" PIP to provide cloud functionality in close vicinity to the stakeholders.

Apart from communication between PDP and PIP, the former needs to obtain the policies to consider. Having placed PAP on the cloud and PDP in the fog may induce latency or connectivity issues between those. However, given the benefits in policy management that cloud provides, it is a matter of context handler implementation to perform propagation of policies and disconnected decisions. Access control decisions should be enforced from PEP. The enforcement usually happens in the edge, where stakeholders exist. Considering an industrial environment, main issues in this layer are resource limitation, device heterogeneity and proprietary communication methods. As a result, consistency in PEP deployment is hard to achieve. The mist layer, as introduced for IIoT deployment, can provide the area to deploy PEP.

In the IoT reference by ITU [16], edge devices can communicate either directly with the upper layers or through a gateway node. Regarding PEP deployment, it can be either integrated with the device or with the gateway. Integrating PEP in a gateway enables support for joining proprietary or other devices that cannot be natively controlled. Moreover, access control in the industrial environment can heavily rely on mist implementation directly into the edge network fabric [14], thus potentially eliminating any latency or connectivity issues. An indicative component placement is presented in Fig. 2.

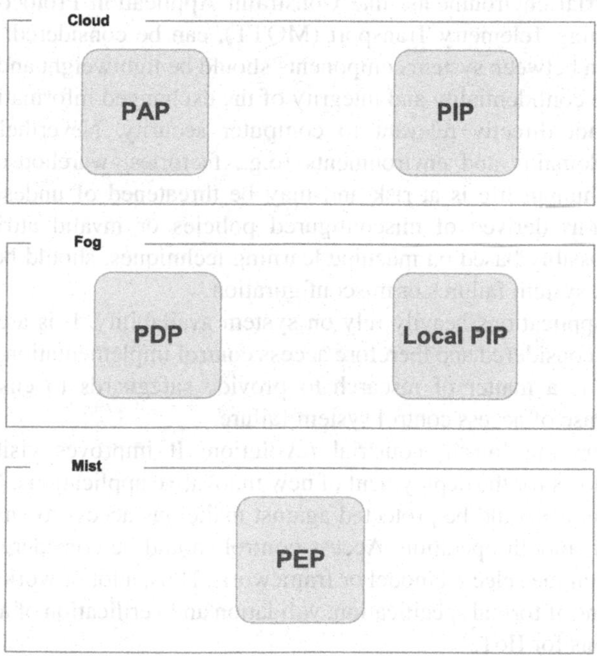

Fig. 2 Components placement

6 Open Issues and Conclusion

There are still issues to promote further research in the deployment of access control in IIoT, some of which are presented in this section.

In ABAC, stakeholder and contextual attributes are evaluated in order to allow or deny access requests. In an environment like IIoT it is challenging to limit the scope of a domain into a specific area and control interactions with other domains. While RBAC models the definition of inter-domain policies requires to exchange identities or roles [7], in ABAC based schemes there is a potential unlimited number of attributes that need to be exchanged.

Trust relationships between domains constituting federations, but also between federations, should be established. Moreover, although PIP placement in a domain's fog area is proposed as an effective approach (i.e., for retrieving attribute values), interconnectivity between PIPs and exchanging of attribute values is a matter of further research and analysis.

Communication between access control components should be optimized so that it can be secure and efficient. Working on this direction, communication protocols used in industrial environments, like Constraint Application Protocol (CoAP) or Message Queuing Telemetry Transport (MQTT), can be considered. In any case, communication between system components should be lightweight and reliable, but also ensure the confidentiality and integrity of the exchanged information.

Safety is not directly relevant to computer security. Nevertheless, IIoT is deployed in domains and environments (e.g., factories, warehouses, hospitals, roads) where human life is at risk and may be threatened of undesirable access control decisions derived of misconfigured policies or invalid attribute values. Safeguards, possibly based on machine learning techniques, should be included to protect against system failures or misconfiguration.

Industrial applications heavily rely on system availability. It is a critical factor that should be considered and therefore access control implementation should never threaten it. It is a matter of research to provide safeguards to ensure business continuity in case of access control system failure.

IIoT triggers the fourth industrial revolution. It improves visibility to the context and allows for the deployment of new innovative applications. Nevertheless, industrial systems should be protected against malicious access to ensure business continuity and smooth operation. Access control should be considered and implemented based on the selected model or framework. Thus, a lot of work still needs to be done in terms of formal specification, validation and verification of access control implementations for IIoT.

Access control appears to be a challenging research topic in the context of IIoT. In this chapter, we elaborated on the concept of IIoT and on access control models and frameworks that may be applicable in it. We anticipate these directions to provide interesting multi-disciplinary insights in both industry and academia, and to stimulate further research in this important field of study.

References

1. Abowd GD, Dey AK, Brown PJ, Davies N, Smith M, Steggles P (1999) Towards a better understanding of context and context-awareness. In: International Symposium on Handheld and Ubiquitous Computing. Springer, pp 304–307
2. Alshehri A, Sandhu R (2016) Access control models for cloud-enabled internet of things: a proposed architecture and research agenda. In: 2016 IEEE 2nd International Conference on Collaboration and Internet Computing (CIC), pp 530–538
3. Atzori L, Iera A, Morabito G (2010) The internet of things: a survey. Comput Netw 54(15):2787–2805
4. Bonomi F, Milito R, Zhu J, Addepalli S (2012) Fog computing and its role in the internet of things. In: Proceedings of the First Edition of the MCC Workshop on Mobile Cloud Computing. ACM, pp 13–16
5. Daugherty P, Banerjee P, Negm W, Alter AE (2015) Driving unconventional growth through the industrial internet of things. In: Accenture technology. https://www.accenture.com/ph-en/_acnmedia/Accenture/next-gen/reassembling-industry/pdf/Accenture-Driving-Unconventional-Growth-through-IIoT.pdf. Accessed 1 June 2018
6. Ferraiolo DF, Kuhn DR, Chandramouli R (2003) Role-based access control. Artech House Inc., Norwood
7. Gouglidis A, Mavridis I (2012) domRBAC: an access control model for modern collaborative systems. Comput Secur 31(4):540–556
8. Gouglidis A, Hu VC, Busby JS, Hutchison D (2017) Verification of resilience policies that assist attribute based access control. In: Proceedings of the 2nd ACM Workshop on Attribute-Based Access Control. ACM, pp 43–52
9. Gouglidis A, Grompanopoulos C, Mavridou A (2018) Formal verification of usage control models: a case study of UseCON using TLA+. In: International Workshop on Methods and Tools for Rigorous System Design
10. Grompanopoulos C, Gouglidis A, Mavridis I (2012) A use-based approach for enhancing UCON. In: International Workshop on Security and Trust Management. Springer, pp 81–96
11. Hu VC, Ferraiolo D, Kuhn R, Friedman AR, Lang AJ, Cogdell MM, Schnitzer A, Sandlin K, Miller R, Scarfone K et al (2014) Guide to attribute based access control (ABAC) definition and considerations. NIST Special Publication. U.S. Department of Commerce Gaithersburg, p 800
12. Hu VC, Kuhn DR, Ferraiolo DF, Voas J (2015) Attribute-based access control. Computer 48(2):85–88
13. Industrial Internet Consortium: Technical papers, publications, and white papers. https://www.iiconsortium.org/white-papers.htm/. [Online; Accessed 01 June 2018]
14. Iorga M, Feldman L, Barton R, Martin M, Goren N, Mahmoudi C (2018) Fog computing conceptual model, recommendations of the National Institute of Standards and Technology. NIST Special Publication, pp 500–325
15. ITU Information technology – open systems interconnection – security frameworks for open systems: access control framework. https://www.itu.int/rec/T-REC-X.812/en/. [Online; Accessed 1 June 2018]
16. ITU Y.2060: overview of the internet of things. https://www.itu.int/rec/T-REC-Y.2060-201206-I/. [Online; Accessed 1 June 2018]
17. Jin X, Krishnan R, Sandhu R (2012) A unified attribute-based access control model covering DAC, MAC and RBAC. In: IFIP Annual Conference on Data and Applications Security and Privacy. Springer, pp 41–55
18. Lin SW, Miller B, Durand J, Joshi R, Didier P, Chigani A, Torenbeek R, Duggal D, Martin R, Bleakley G et al (2015) Industrial internet reference architecture. Industrial Internet Consortium (IIC), Technical Report

19. Liu Q, Zhang H, Wan J, Chen X (2017) An access control model for resource sharing based on the role-based access control intended for Multi-domain manufacturing internet of things. IEEE Access 5:7001–7011
20. Lopez J, Rubio JE (2018) Access control for cyber-physical systems interconnected to the cloud. Comput Netw 134:46–54
21. Navarro-Ortiz J, Sendra S, Ameigeiras P, Lopez-Soler JM (2018) Integration of LoRaWAN and 4G/5G for the industrial internet of things. IEEE Commun Mag 56(2):60–67
22. OASIS: eXtensible Access Control Markup Language (XACML) Version 3.0. http://docs.oasis-open.org/xacml/3.0/xacml-3.0-core-spec-os-en.html/. [Online; Accessed 1 June 2018]
23. Ouaddah A, Mousannif H, Elkalam AA, Ouahman AA (2017) Access control in the internet of things: big challenges and new opportunities. Comput Netw 112:237–262
24. Park J, Sandhu R (2004) The UCON ABC usage control model. ACM Trans Inf Syst Secur (TISSEC) 7(1):128–174
25. Pesonen LI, Eyers DM, Bacon J (2006) A capability-based access control architecture for multi-domain publish/subscribe systems. In: International Symposium on Applications and the Internet, SAINT 2006, 7 pp
26. Salonikias S, Mavridis I, Gritzalis D (2015) Access control issues in utilizing fog computing for transport infrastructure. In: International Conference on Critical Information Infrastructures Security. Springer, pp 15–26
27. Serpanos D, Wolf M (2017) Internet-of-things (IoT) systems: architectures, algorithms, methodologies. Springer, Singapore
28. Welbourne E, Battle L, Cole G, Gould K, Rector K, Raymer S, Balazinska M, Borriello G (2009) Building the internet of things using RFID: the RFID ecosystem experience. IEEE Internet Comput 13(3):48–55
29. Wilkes MV, Needham RM (1979) In: Denning PJ (ed) The Cambridge CAP computer and its operating system, Operating and programming systems series. North Holland, New York City
30. Zhang X, Nakae M, Covington MJ, Sandhu R (2008) Toward a usage-based security framework for collaborative computing systems. ACM Trans Inf Syst Secur (TISSEC) 11(1):3

A Distributed Usage Control Framework for Industrial Internet of Things

Antonio La Marra, Fabio Martinelli, Paolo Mori, and Andrea Saracino

Abstract This work presents a distributed Usage Control framework designed to ensure high flexibility, performance and fault tolerance in security and safety policy enforcement. The framework has been designed for distributed Peer-to-Peer (P2P) systems, without a root of trust, being thus suitable for Industrial Internet of Things (IIoT) settings. The proposed framework benefits from the presence of a set of Usage Control Systems, logically interconnected through a DHT which enables shared and replicated memory, distributed evaluation and distributed attribute retrieval. Furthermore, being based on the Usage Control paradigm, it is able to enforce policies with mutable attributes, revoking ongoing sessions when policies are not matched anymore with the current request context. The presented framework is validated through performance experiments performed in both an emulated and real settings.

Keywords Usage control · Distributed policy enforcement · Industrial control systems · Industry 4.0 · IoT

1 Introduction

The paradigm of Internet of Things has entered pervasively in this last years in several environments of daily life, such as smart homes, smart cities, smart grids and sensor networks. More recently, the IoT has also entered industrial settings, becoming thus the enabler of two novel paradigms, namely Industrial IoT and Industry 4.0. Hence, through both these paradigms, sensors and actuators, including machines and assembly lines, become *Connected Objects* and *Smart Objects*,

A. La Marra · F. Martinelli · P. Mori · A. Saracino (✉)
Istituto di Informatica e Telematica, IIT-CNR, Pisa, Italy
e-mail: AntonioLa.Marra@iit.cnr.it; fabio.martinelli@iit.cnr.it; paolo.mori@iit.cnr.it; andrea.saracino@iit.cnr.it

C. Alcaraz (ed.), *Security and Privacy Trends in the Industrial Internet of Things*, Advanced Sciences and Technologies for Security Applications,
https://doi.org/10.1007/978-3-030-12330-7_6

communicating between them and eventually with the Internet and becoming able to take their own decisions with reduced or no-human intervention.

Connected devices, differently from traditional ones, are managed by operative systems (such as Google Android, or some specific Linux versions), which in several cases allow the installation of additional applications to add further functionalities. Moreover, the capabilities of these devices dramatically increase due to the connection among them and to the Internet. Each device has in fact, sensors, actuators and access to different kind of information which can be used collaboratively by other devices to provide additional and more complex and powerful services, which in turn might increase efficiency and improve other industrial KPIs, also related to safety of human users. In fact, classical production environments are moving to connected machines as well, and Cyber Physical Systems are replacing traditional industrial control systems (e.g., Programmable Logic Controllers) in the operating production systems of modern factories. This would allow to easily and automatically operate, monitor and optimize complex production assembly chains, composed of several machineries interacting each others. Another relevant advantage could be the ability of predicting, detecting and automatically and quickly reacting to machine failures by properly changing the assembly chain, even by outsourcing part of the productive process.

The connection capabilities of smart devices, as well as the capability of installing additional and/or third party software on them, expose devices, users and the whole environment to new potential security and safety issues. In the context of IoT, the notion of security is typically tightly related with the notion of safety, because attacking IoT devices is a mean to injury the people who directly interact with them and the people living in the environment in which these devices are placed. As an example, in 2010 a successful attack (called Stuxnet) has been executed against the control system of the Iranian nuclear facilities causing damages to the centrifuges used for the production of uranium235. Obviously, this attack could have had tragic consequences on the people living nearby. Other incidents concerning industrial control systems are reported in [8, 13].

The security and safety risks can be mitigated by introducing advanced systems for security policy enforcement in the connected devices environment. The need for the introduction of enhanced access control support, such as Usage Control mechanisms, in the IoT environment has been anticipated by relevant and recent works in literature concerning smart home (such as [3, 19] and [14]), and concerning Industry 4.0 (such as).

In this paper, we present UCIoT (Usage Control in the Internet of Things), a framework which aims at bringing the Usage Control on IoT architectures in a seamless, configurable and dynamic manner. The framework is designed for heterogeneous and distributed architectures of connected devices, evaluating and enforcing security, safety or general purpose policies by exploiting the expressive U-XACML language [5].

The contribution of this paper is summarized in the following:

- We define UCIoT, a distributed, decentralized, fault tolerant and infrastructure independent framework to implement Usage Control functionalities in IoT systems;
- We report details on the implementation of the UCIoT framework, which exploits the CHORD-based *Apache Cassandra* Distributed Hash Table for enabling the distributed communication and data storage, reporting the challenges and implementation choices to decentralize the UCON functions.
- We propose a sample industrial setting which would benefit by the presence of a distributed usage control mechanism, presenting policies, attributes and actuators which would be considered;
- We present an extension of the basic model with a hierarchical structure, to represent complex and hierarchical industrial settings;
- We present a set of experiments to measure performance of the proposed framework, to measure the performance overhead caused by the UCIoT framework on a real implementation performed on Raspberry PI-3.

Differences with previous works This work completes and extends the one presented in [12] by proposing a multi-tiered implementation of the UCIoT framework and proposes applications on an environment which is much more critical than a smart-home setting, presenting specific policies and attributes related to that specific environment. This work is thus a first attempt to bring the Usage Control in the Industrial IoT and in the Industry 4.0 setting, showing how the flexibility and generality brought by UCON might benefit an industrial setting, focusing also on issues like the presence of single point of failures, which is avoided by exploiting a distributed fault tolerant architecture for system control, such as the one presented in this work.

The remainder of the paper is organized as follows. In Sect. 2 describes the UCON components and workflow phases. Section 3 describes the UCIoT framework, detailing the components and the workflow and implementation choices to adapt UCON components to a distributed architecture, also presenting a possible use case and the hierarchical framework extension. Section 4 reports the performance evaluation of UCIoT documenting the implementation on a testbed of real devices. Section 5 reports a set of related work on access and usage control in IoT environments. Finally Sect. 6 briefly concludes proposing some future directions.

2 The Usage Control Model

In this section we are going to introduce the Usage Control model, presenting its theoretical aspects, the formal components and the workflow. Furthermore, will be introduced the Usage Control System, i.e. a framework implementing the Usage Control workflow, with six components performing the needed functionalities. The Usage Control (UCON) model [16, 18] extends traditional access control models introducing *mutable attributes* and new decision factors besides *authorizations*: *obligations* and *conditions*. Mutable attributes represent features of subjects, resources, and environment that change their values as a consequence of the normal operation of the system [17]. For instance, some mutable attributes change their values because the policy includes attribute update statements that are executed before (*pre-update*), during (*on-update*), or after (*post-update*) the execution of the access. For instance, the e-wallet balance is a subject attribute which could be decreased by the policy every time the subject performs a new access to a resource.

Since mutable attributes change their values during the usage of an object, the usage control model allows to define policies which are evaluated before *(pre-decision)* and continuously during the access to the object (*ongoing-decision*).

The continuous evaluation of the policy when the access is in progress is aimed at executing proper countermeasures (such as interrupting the access) when the execution right is no more valid, in order to reduce the risk of misuse of resources. Hence, in the Usage Control model it is crucial to be able to continuously retrieve the updated values of the mutable attributes, in order to perform the continuous evaluation of the policy and to promptly react to the attribute change by taking proper actions, e.g., by interrupting those ongoing accesses which are no longer authorized.

This paper takes into account Usage Control systems based on the XACML reference architecture [15], with particular reference to the one we presented in [2], which is shown in Fig. 1. In the XACML reference architecture, the

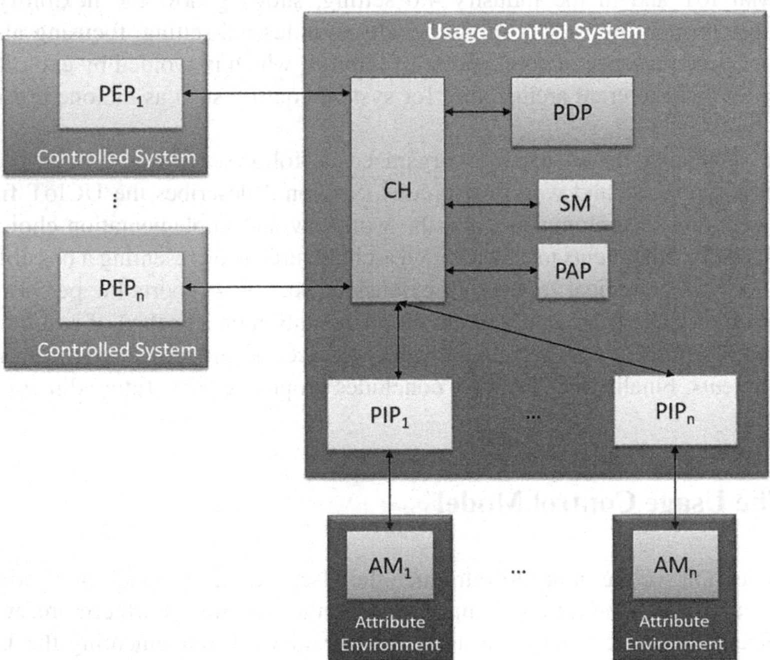

Fig. 1 Usage control system architecture

Policy Enforcement Points (PEPs) embedded in the controlled system intercept the execution of security relevant operations, and they invoke the Context Handler (CH), which is the frontend of the Usage Control system. The Policy Information Points (PIPs) are the components invoked by the CH to retrieve the attributes required by the Policy Decision Point (PDP) for the execution of the decision process, i.e., to evaluate the policy retrieved from the Policy Store (PS). Attributes are managed by Attribute Managers (AMs), sometimes called Attribute Providers or Attribute Stores, which provide the interfaces to retrieve and, in case of mutable Attributes, to update their current values. Each specific scenario where the Usage Control system is exploited requires its own set of AMs to manage the attributes required for the policy evaluation. Hence, PIPs are properly configured in order to be able to query the specific AMs adopted in the scenario of interest for retrieving and updating attributes. In particular, each PIP implements the specific protocol required to interact with the related AM and exploits the provided mechanisms for securing the communications. The Usage Control model emphasizes the role of PIPs because it introduces the continuous policy enforcement while an access is in progress to cope with mutable attributes. In particular, the PIP is also in charge to detect when the value of an attribute changes in order to trigger the policy re-evaluation for the involved ongoing accesses, which are managed by the Session Manager (SM). To detect attribute changes, the PIP could exploit the subscription mechanism provided by the AM or the PIP must emulate it if it is not supported by the AM.

The phases of the Usage Control decision process are regulated by the interactions between the PEP and the Usage Control systems as follows (derived from [21]):

- $TryAccess$: is the *pre-decision* phase, which begins when the Tryaccess message is sent by the PEP to the Usage Control system because a subject requests to execute the access. The TryAccess phase finishes when the Usage Control system sends the response to the PEP. The possible responses are: PERMIT, to allow the access, or DENY;
- $StartAccess$: is the first part of the *ongoing-decision* phase, which begins when the StartAccess message is sent by the PEP to the Usage Control system because the access has just started, and finishes when the policy has been evaluated and the response has been sent back to the PEP;
- $RevokeAccess$: this is the second part of the *ongoing-decision* phase. This phase is executed every time an attribute changes its value. This phase starts when an attribute changes its value. It finishes when the policy has been evaluate and, if a policy violation occurs, the RevokeAccess message is sent by the Usage Control system to the PEP.

When the subject s tries to execute a security relevant action a, the PEP suspends its execution and retrieves the information related to this access (subject and resource IDs, etc.). The PEP sends the *TryAccess* message with the data previously collected to the Usage Control system, which performs the *pre-decision* process and returns the result to the PEP, which enforces it. If the execution of a is permitted, the PEP sends the *StartAccess* message to the Usage Control system as soon as a

is started, to start the *on-decision* phase. Again, the Usage Control system performs the first evaluation of the Usage control policy. From this moment on, as long as the action a is in progress, the Usage Control service evaluates the Usage Control policy every time an attribute changes its value, and we call this phase *RevokeAccess*. If the policy is violated, the Usage Control system sends the *RevokeAccess* message to the PEP, in order to take proper countermeasures.

3 The UCIoT Framework

This section describes the proposed framework which implements Usage Control on the IoT devices installed in a Industrial IoT setting. Though the paradigm of Usage Control has been already successfully applied in several scenarios bringing, for example, the Usage Control System (UCS) in Cloud [2], and on Android devices [11], the present work faces the challenge of decentralizing the UCS functionalities proposing a distributed a Peer-to-Peer (P2P) architecture for constrained devices, such as the Smart Home ones. The devices we are considering are micro-controllers, embedded devices, control boards, security cameras adn smart sensors, which are optimized for very specific tasks and have generally limited computational power and storage capabilities, hence they are not suitable to handle and evaluate all the requests which might be issued in a smart home environment, nor to store needed information for policy reevaluation for all the active sessions. As will be detailed in the following, UCIoT partially replicates the functionalities of the UCS on each P2P node. Also, any policy evaluation can be performed by any of the nodes which can exploit the attributes collected from the other nodes. These two features enable respectively fault tolerance and higher complexity for enforced policies.

Another relevant advantage of such a distributed system concerns the fault tolerance aspects, because the failure of one (or of a number of) device(s) can be tolerated as long as the system can be executed on the remaining ones.

Moreover, being distributed and exploiting several devices, UCIoT allows the Usage Control framework to exploit the data collected from the sensors of all these devices, thus allowing to define more complex policies.

3.1 Architecture

The architecture of UCIoT framework, depicted in Fig. 2, is a distributed P2P framework, where each node represents a smart device logically connected to all the others through a Distributed Hash Table (DHT) [1]. In particular, the DHT exploited by UCIoT is the Cassandra protocol [9], based on a modified version of the CHORD DHT, which also includes a non-relational distributed database, used by several popular applications such as Facebook.

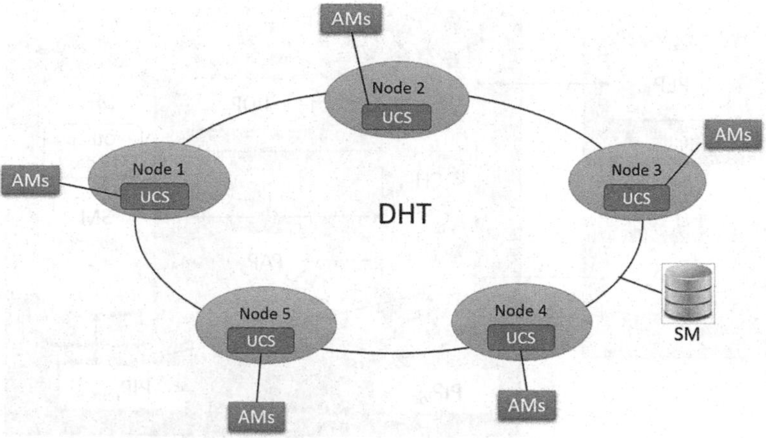

Fig. 2 Logical architecture of UCIoT

We consider as *smart device* any device such as, micro-controller, sensor, actuator, board, computer, smartphone, or other device which is running an operative system which is able to install and run externally provided applications. Every smart device will be thus considered a P2P node, whilst connected devices not matching these specifications will be considered as peripherals of a specific node, logically incorporated in the node itself. Some elements of the smart devices environment could even have also an acceptable computational power, a consistent power autonomy, still with limited storage space. Given the capabilities of smart devices, the UCS can be installed as a third party app and run on any smart device in UCIoT.

At network and datalink layer, the used communication protocol is ad-hoc wireless which, being completely distributed, does not rely on the presence of a single router node, which might represent a single point of failure in the network. Thus, thanks to the DHT protocol and to the ad-hoc wireless, communication among smart devices will still be possible if a limited subset is switched off or malfunctioning. The network protocol is completely oblivious to the applications running on the smart devices, thus smart devices will exchange messages as if they are connected to the same Wireless-LAN.

A representation of the logical architecture components of UCIoT in a single node is shown in Fig. 3. As anticipated, the functionalities of UCS are replicated on every device, which can thus decide to permit or deny the access/usage to an operation, or a resource that it controls. This request can be issued by physical users, other smart or peripheral devices and is matched against usage control policies stored locally by the UCS. More specifically, every smart device runs (i) the Policy Administration Point (PAP) to store policies, (ii) an instance of the Policy Decision Point (PDP) to match request and policies and deciding to permit or deny the resource usage, (iii) a set of Policy Information Points (PIP) to query local

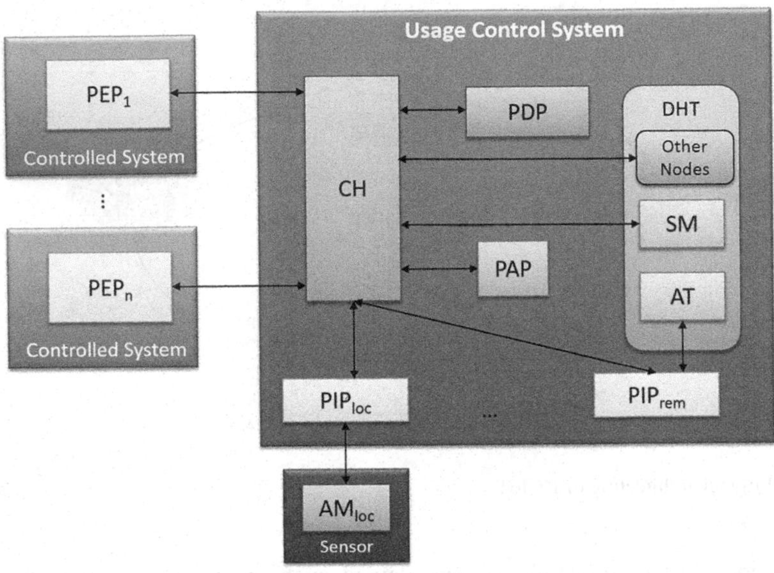

Fig. 3 Logical architecture of a UCIoT node

attributes, (iv) a Context Handler (CH) to manage the interaction among the various components. The main difference with the UCON architecture presented in our previous works lays in the Session Manager, for which UCIoT take advantage of the distributed database offered by Cassandra. In fact, since smart devices have limited memory storage and the number of managed sessions will be likely unbalanced among devices in a smart home settings, the set of active sessions are saved in the DHT, to efficiently use the global memory storage offered by all the devices. Moreover, thanks to the configurable replication factor, the SM is not prone to single point of failure issues. The other main difference with previous UCON architectures is the presence of *remote attributes*. An attribute is considered *local* to a node when the Attribute Manager (AM) is queried directly by one of the PIPs belonging to the UCS of that node (see Fig. 2). However, in UCIoT is possible that not all attributes needed to evaluate a policy are local, i.e. some attributes are local to other smart devices and are thus considered remote. To abstract this procedure to the UCS local to the evaluating device, an abstract component named *remote PIP* is added to all UCSs, and it is in charge of interacting with the remote smart device to retrieve the attribute. The remote PIP retrieves the identifier of the node physically connected to the AM by the attribute table stored in the DHT, then retrieves the attribute according to the procedure described in the following subsection. Two distinct strategies have been designed to collect the remote attributes required to perform the decision process. The first strategy, calledSMAC strategy, simply starts the evaluation of the security policy when only the local attributes have been collected, i.e., without having collected any remote attribute. The PDP tries to evaluate the policy with the local attributes only and, if one (remote) attribute is missing, the

PDP returns an error specifying the name of the missing attribute. Hence, the UCIoT framework collects the missing remote attribute communicating with the node directly connected with the interested AM. In particular, the remote PIP finds which is the node to be contacted for retrieving such attribute by accessing the remoteAttributeLocationTable table stored in the DHT.

The PD strategy, instead, starts the evaluation of the security policy only when all the attributes required for the decision process have been retrieved, including remote ones. To this aim, the CH inspects the security policy in order to determine which attributes are required for the evaluation, and it invokes the proper local and remote PIPs to perform the retrieving procedure. As in the previous case, the nodes to be contacted to retrieve the remote attributes are determined by the remote PIP by accessing the remoteAttributeLocationTable table stored on the DHT. Only when all the selected PIPs returned the attribute values, the CH invokes the PDP for the policy evaluation.

3.2 *Workflow*

The operations performed by UCIoT are equivalent to the one performed by the UCON framework described in Sect. 2. The main difference in the workflow is introduced by the presence of remote attributes and by the distributed Session Manager. UCIoT has been designed with the aim of reducing as much as possible the differences in the workflow with respect to the standard UCON workflow, despite the distributiveness of the architecture where it is applied.

As anticipated, the SM is stored on the Apache Cassandra DHT database, which brings the advantages of increased storage space, tackling thus the issue of memory constrained devices, and ensures fault tolerance thanks to the replication factor, storing data of sessions related to temporarily unavailable devices. The DHT database also masks the underlying distributed architecture of the database to the application level, leaving thus the functionalities of the SM unmodified.

On the other hand, remote attributes introduce some additional challenges, requiring thus a small alteration of the usual workflow and UCON architecture. As anticipated, the problem of remote attributes consists of collecting the value of attributes from AMs connected to a UCS node different from the one evaluating the policy. To this end, we introduce in each node a component which abstracts the procedure of collecting the remote attributes, named *PIP remote*. This PIP exploits the *Attribute Table* stored on the DHT which memorizes, for each attribute, the node whose AM is physically connected to. Hence, when a remote attribute is needed, the local CH reaches through the DHT the CH of the interested node (*CH rem* in Fig. 4), which will instruct the local PIP (*PIP loc* in Fig. 4) to retrieve the attribute value, which is then returned to the local CH.

The complete workflow of the retrieval of a remote attribute is depicted in Fig. 4. In particular, Fig. 4 represents the workflow of the TryAccess request issued by the PEP, detailing the message exchange between the various components. For the

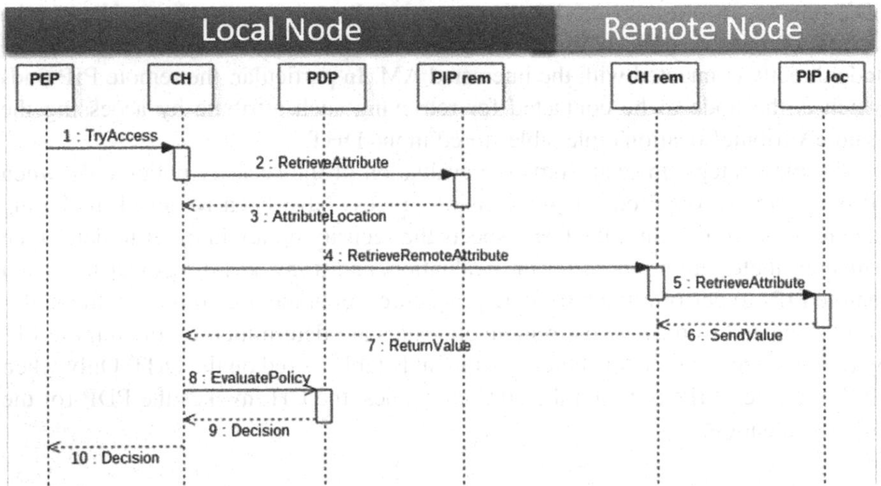

Fig. 4 TryAccess sequence diagram with remote attribute retrieval

sake of simplicity and clarity of representation, no local attributes are retrieved in the represented workflow. The remote attribute retrieval is performed also for the StartAccess, which is identical to the TryAccess in workflow, except for the addition of the started session to the distributed SM and for the subscription of the remote CH to the remote mutable attributes. Thus, if a remote attribute changes its value the remote CH is notified, and the new value is sent back to the local CH for policy reevaluation and possible revocation.

3.3 Implementation

The current implementation of the UCIoT framework consists of a Java application, shipped in the form of a `jar` file, which can be installed on any device running a JRE. When installed, the application will instantiate on the devices the DHT for the distributed database and to handle communication with other nodes belonging to the same network. The installed application includes the full UCS, where the CH and PDP are not altered with respect to the standard UCON model. The SM is installed in the distributed database handled by the DHT, which will also handle the data replication. The distributed database also hosts the attribute table, which keeps the correspondence among the attributes and the node to which the AM is physically connected, hence it can be queried and possibly updated by any node. As it happens in the standard UCON framework, PIPs and PEPs will be device specific, to be interfaced with actuators and sensors proper of the specific device. The single instances of the UCIoT application are dynamically configurable and the code related to specific PIPs and PEPs can be loaded at runtime by exploiting Java reflection.

3.4 Use Case Example

A usage control system can monitor and enforce policies separated or at the same time on several elements of the industrial architecture, spanning from the physical system to the software one. In the following we report a small set of possible attributes and conditions related to a non exhaustive list of domains of an industrial setting:

- *Physical Domain*
  - **Attributes** Machine Temperature, Power Consumption, Machine Vibrations, Rotor Speed.
  - **Actions** Machine Activation, Increase Speed, Increase Power.
- *Network Domain*
  - **Attributes** Data flows, Number of connected devices, Available bandwidth, Opened connections
  - **Actions** Opening new connection, Start data stream, Register new device
- *Controller Domain*
  - **Attributes** Device Integrity, Firmware Version, CPU load.
  - **Actions** Update firmware, Add input source, Fork new process/thread.
- *Environment Domain*
  - **Attributes** Room Temperature, Room Light, Air Quality, Number of Persons
  - **Actions** Activate air conditioning, Toggle lights, Lock/Unlock doors.
- *Software Domain*
  - **Attributes** Room Temperature, Room Light, Air Quality, Number of Persons
  - **Actions** Activate air conditioning, Toggle lights, Lock/Unlock doors.

As an example, let's consider a setting where a company has a room with one door and a window. The room hosts a controlled machine having a rotor which is rotating at variable speed. Long period of high speed rotation gradually increase the temperature of the room, monitored through an external sensor, which is part of the room air conditioning system. High temperature levels, make the environment unsuitable for human operators, hence an automated lock will forbid accesses from outside the room (it still allows eventual operators present in the room to leave it). We will consider that the machine, the air conditioning system and the door and window lock manager are three nodes of the UCIoT framework, installing thus a UCS with a single PEP attached, to enforce policies respectively on the rotation speed, the settings of the air conditioning, i.e. `cool`, `warm`, `off` and the right to open the door from outside. Considered attributes are the room temperature, the number of persons present in the room, the status of windows (i.e. opened/closed) and the rotation speed of the machine, which are collected respectively through a temperature sensor connected to the air conditioning system, a magnetic sensor, a camera connected to the door locking system and speedometer included in the machine. For each of these devices, a PIP is used as interface it with the UCS, which is responsible of collecting and monitoring the attribute values.

A set of possible policies, which are enforceable in the described setting, is the following:

- **Policy 1** It is allowed to enter and stay in the room only if the room temperature is below 32 °C.
- **Policy 2** The air conditioning will operate in any mode only if the windows are closed. It can operate in `cool` mode if temperature is higher than 15 °C and in `warm` mode if the temperature is less than 25 °C.
- **Policy 3** The machine rotor can rotate over 2000 rpm only if the temperature room is below 32° and there are no persons in the room.

Policy 1 is designed to ensure safety, by defining an Access Control condition. The policy, to be evaluated needs attribute coming from two sensors: the camera for the number of persons and the room temperature coming from the temperature sensor. Thus, the evaluation will be performed and the policy will be enforced on the UCS of the door locking system, collecting one remote attribute which comes from the air conditioning node and a local attribute. The enforcement is performed by not allowing the door to be opened from outside, moreover an audio message or alarm can be streamed to push eventual persons staying in the room to leave it. This last action is enforced through the Obligation mechanism. Policy 2 is a Usage Control policies instead, designed for energy saving. The air conditioning system, in fact, will only waste energy if operating when windows are opened or in temperature conditions where air conditioning would be ineffective. Hence, this policy checks, at the time the air conditioning system is started, that windows are closed and temperature is in range. Furthermore, if the windows are opened while the air conditioning is operating, it is stopped through a revoke operation. Also for this policy evaluation requires collection of attributes from two different nodes, namely the door control system to verify that windows are closed, and the temperature sensor of the air conditioning system. Finally, Policy 3 is a Usage Control policies which relates the allowed rotating speed for the machine and the temperature inside the room. It also exploits attributes coming from two sensors, forcing the machine to reduce the speed when the temperature in the room is too high. To this end we consider that asking to the rotor to go faster than 2000 rpm is a specific action to which should correspond a request to be performed through the PEP.

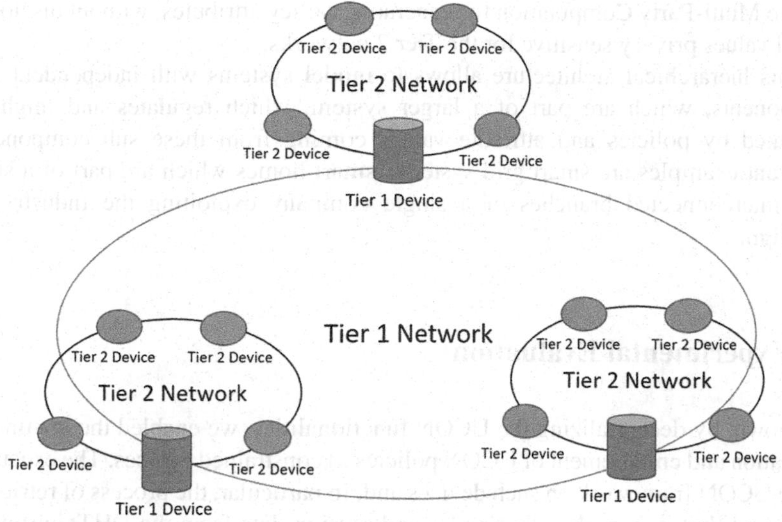

Fig. 5 Multi-tiered architecture

3.5 *Extending the Model: Multi-tier Framework*

The model of the UCIoT framework can be scaled to a multi-tier architecture to further increase flexibility and to enable the enforcement of hierarchical policies, which use attributes coming from different domains.

As shown in Fig. 5, the architecture is composed of a P2P network of Tier 1 devices, i.e., the Tier 1 network, and a set of second level networks, i.e., Tier 2 Networks where Tier 2 devices are interconnected again through a DHT (hence P2P). Every Tier 2 network can be considered as a standalone UCIoT system, except for the presence of a Tier 1 device which act as a Bridge Device. As a matter of fact, all the devices belonging to a Tier 2 Network have their own UCS and PEP(s) and are connected to a set of AMs, typically embedded in to the devices themselves, or physically connected to them, e.g., a power consumption meter, to retrieve the local attributes used for policy evaluation. To this aim, each Tier 2 Network includes a Bridge device, which is the component that connects this Tier 2 Network with the others.

One of the main advantages of the two tiers architecture is that it separate the attributes in high level attributes, i.e., attributes at the level of Tier 1 Network and low level attributes, which are specific to each Tier 2 networks. High level attribute values can be Tier 1 specific, i.e. global variables used in the Tier 1, which might also be exploited in Tier 2 policies, or can be obtained by combining the values of the attributes belonging to the Tier 2 Networks. The Bridge devices are the components embedding the AMs belonging to the Tier 1, which manage the high level attributes that are related to the system as a whole. Through Tier 1 devices it is also possible to perform privacy preserving attribute calculations, by exploiting mechanisms such as

Secure Multi-Party Computation for generating policy attributes, without disclosing actual values privacy sensitive for the Tier 2 networks.

This hierarchical architecture allows to model systems with independent sub-components, which are part of a larger system, which regulates and might be regulated by policies and attribute values coming from these sub components. Relevant examples are smart grid systems, smart homes which are part of a smart city, interconnected branches of a single company exploiting the Industry 4.0 paradigm.

4 Experimental Evaluation

As shown, by decentralizing the UCON functionalities, we enabled the continuous evaluation and enforcement of UCON policies in constrained devices. The execution of the UCON framework on such devices and, in particular, the process of retrieving remote attributes from the other nodes and session data from the DHT, introduces overhead. To quantify such overhead and to evaluate its impact on usage in real system, we conducted on our reference testbed a set of experiments aimed at measuring the performance of the proposed system and the results are discussed in the following.

4.1 The Testbed

As anticipated, the testbed consists of five Raspberries PI 3 Model B and is shown in Fig. 6. As shown, three devices are equipped with the Raspberry SenseHat module, which adds to the simple board sensing capabilities, incorporating sensors for temperature, humidity, pressure, magnetometer, accelerometer and gyroscope. The SenseHat also features a led matrix which can be used to display messages. The two remaining devices have been equipped with a Pi Camera Module v2, adding thus the capability of collecting pictures and video. The Raspberry PI 3 Model B has the following features: 1 GB RAM, 1.2 GHz ARM processor, VideoCore IV 3D graphics core and has several interfaces for wireless communication, namely 802.11n WLAN, Bluetooth 4.0 and Bluetooth Low Energy (BLE). For our set of experiments, devices are interconnected through an ad-hoc WiFi network, constituting thus a WANET with AODV routing protocol [4]. The choice of using WiFi instead of Bluetooth is due to the higher reliability and greater speed of WiFi, which also does not require pairing phases as it happens in Bluetooth. Also, being all devices connected to a power supply, in the smart home use case it is not necessary to use low consumption protocols. This hypothesis is sound since in a smart home, house appliances are generally connected to power supply and do not rely on batteries. However, it is worth noting that the UCIoT framework is completely independent from the routing and data-link protocol.

Fig. 6 Five Raspberries PI 3 used as testbed equipped with SenseHat and Pi Camera

Sensors and cameras are used in our testbed as AMs, providing the values for attributes which will be used in policies. The led matrix is instead used as actuator, hence is commanded by the PEP, showing a "P "every time a session starts, i.e., PDP returns the PERMIT decision for a StartAccess and "D "if the access is denied or revoked.

4.2 *Local Evaluation*

The first set of experiments, whose time performance is shown in Fig. 7, evaluates the time required to execute the phases of the policy evaluation process, namely: TryAccess, Startaccess, and RevokeAccess (see Sect. 2). The experiments have been performed enforcing policies including local attributes only, varying the number of the attributes required for the decision process from 2 to 50. The time of the TryAccess phase is measured from the moment when the access request is sent by the PEP to the CH to the moment when the PEP receives the response (permit/deny) to this request. This time affects the user experience because it is the delay introduced by our framework in the utilization of the smart home device. The time of the StartAccess phase is measured from the moment when the evaluation of the policy is triggered by the PEP because the access just began, to the moment when the response has been received by the PEP. The time to perform the RevokeAccess phase is measured from the moment when a given attribute changes its value to the

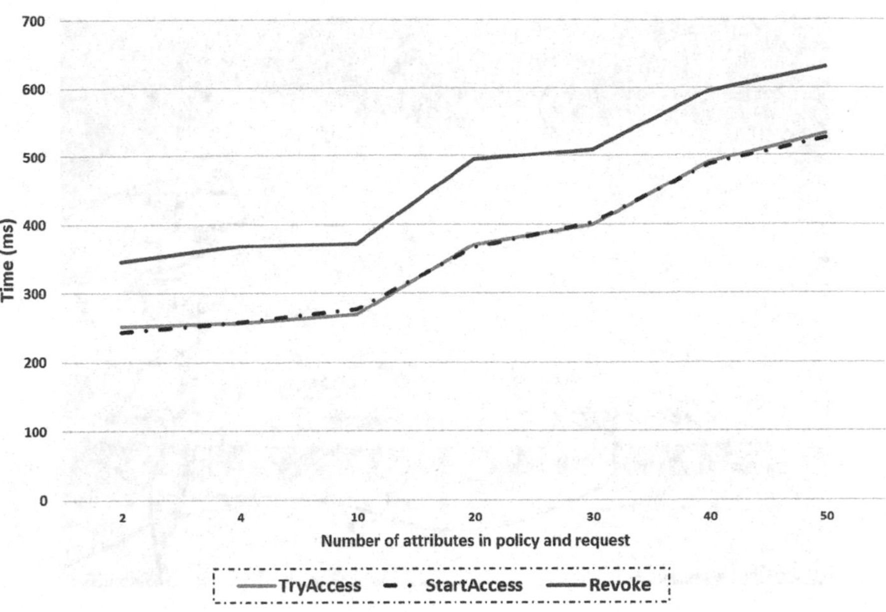

Fig. 7 Performance with local attributes

moment when the PEP receives the revocation response. Hence, it also includes the time required by the PIP to detect that an attribute changed its value. The StartAccess and RevokeAccess phases do not directly introduce any delay in the utilization of the resource because they are executed while the access to the resource is already in progress. When a policy violation is detected, the time required by the StartAccess phase or by the RevokeAccess one represent the interval while the resource has been used without holding the related right. The X axis reports the number of local attributes, nLA, while the Y axis reports the time in milliseconds. For measuring the time taken by the RevokeAccess phase, we supposed that only one of the sessions stored in the DHT requires the re-evaluation of the policy.

The results of our experiments show that, in case of 2 local attributes, the time required to perform the TryAccess phase is 252 ms, the StartAccess phase takes 247 ms while the RevokeAccess phase requires 346 ms. In case of 50 local attributes, instead, the TryAccess phase takes 533, the StartAccess phase requires 528 ms, while the time to execute the RevokeAccess phase is 630 ms. First of all, we notice that the time to execute the TryAccess phase is quite the same as the time required for the StartAccess phase. The reason is that the workflows of the two phases are very similar, i.e., they perform quite the same operations in the same order. Hence, in case of local attributes, the delay introduced by the UCIoT framework would not affect the user experience, and the usage of the resource would be revoked in a short time in case of policy violation. Moreover, the difference between the time of the RevokeAccess and of the TryAccess or StartAccess, for all the attribute values, is mainly due to the time required to detect that an attribute changes its value.

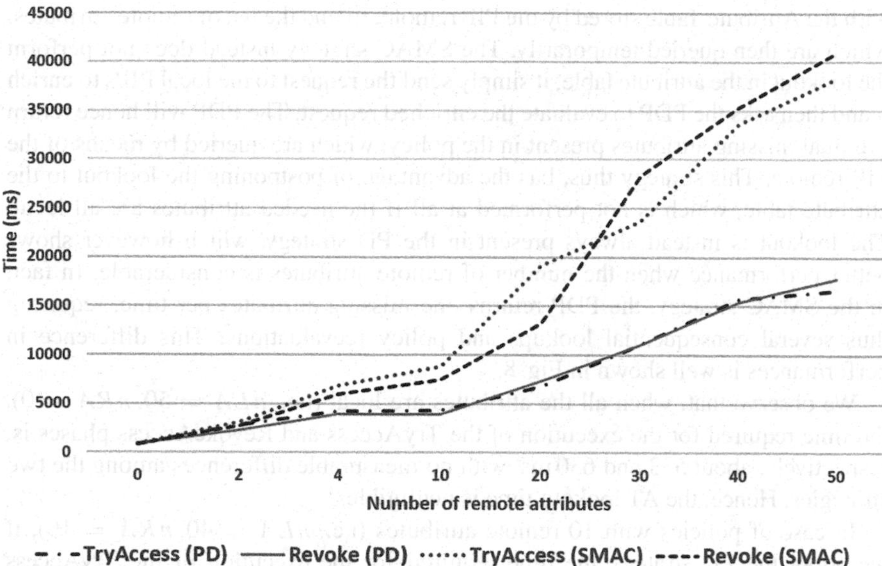

Fig. 8 Performance with local and remote attributes

4.3 Remote Attribute Retrieval

The next set of experiments, shown in Fig. 8, is aimed at evaluating the time required to execute the TryAccess and RevokeAccess phases of the policy evaluation process in case of policies including remote attributes. We omit the time required for the StartAccess phase since the results of the first set of experiments show that it is quite the same as the time of the TryAccess phase. In this set of experiments, the policy always includes 50 attributes, and we vary the ratio between local and remote attributes. The policy is written in such a way that the value of all the 50 attributes have always to be collected and evaluated to perform the decision process, i.e., the UCIoT PDP cannot perform the decision phase exploiting only a subset of the 50 attributes. We recall that real policies would embed a smaller number of attributes, hence this can be considered as a worst case test. For instance, the first policy described in Sect. 3 embeds 3 attributes only.

On the X axis of the graph in Fig. 8 we report the number of remote attributes in the enforced policy, nRA. Hence, the number of local attributes nLA is given by $(50 - X)$. In this case too, for measuring the time taken by the execution of the RevokeAccess phase, it is supposed that only one session requires the re-evaluation of the policy. The experiments have been conducted exploiting two distinct strategies for retrieving the remote attributes described, namely *Policy Driven* (PD) and *Sequential Missing Attribute Collection* (SMAC). The first strategy consists of making the CH aware of the policy to be analyzed, to have a list of the attributes to be queried by the various PIPs. Hence, the list of attributes is matched with the Attribute Table stored by the PIP remote, to find the list of remote attributes, which are then queried temporarily. The SMAC strategy instead does not perform the lookout in the attribute table, it simply send the request to the local PIPs to enrich it and then asks the PDP to evaluate the enriched request. The PDP will hence return eventual missing attributes present in the policy, which are queried by means of the PIP remote. This strategy thus, has the advantage of postponing the lookout to the attribute table, which is not performed at all if the needed attributes are all local. The lookout is instead always present in the PD strategy, which however shows better performance when the number of remote attributes is considerable. In fact, in the SMAC strategy, the PDP returns one missing attributes per time, requiring thus several consequential lookups and policy reevaluations. This difference in performances is well shown in Fig. 8.

We observe that, when all the attributes are local (i.e., $nLA = 50, nRA = 0$), the time required for the execution of the TryAccess and RevokeAccess phases is, respectively, about 533 and 630 ms, with no measurable differences among the two strategies. Hence, the AT lookout time is negligible.

In case of policies with 10 remote attributes (i.e., $nLA = 40, nRA = 10$), if we adopt the PD strategy the time required for the execution of the TryAccess phase is 4,2 s, while the time for the RevokeAccess phase is 3,9 s. Adopting the SMAC strategy the times required for the execution of the two phases are, respectively, 8,8 and 7,7 s. Instead, when the policy embeds remote attributes only (i.e., $nLA = 0, nRA = 50$), the time required for the execution of the StartAccess and RevokeAccess phases is, respectively, about 16,6 and 17,8 s if we adopt the PD strategy, while it is about, respectively, 38,3 and 40,9 s if we adopt the SMAC strategy.

4.4 Network Delay Evaluation

The aim of the third set of experiments, shown in Fig. 9, is to evaluate the overhead introduced by the communications over the network. In particular, we installed the 5 nodes on distinct Virtual Machines running on the same physical machine, and we measured the time for executing the TryaAccess and RevokeAccess phases in case of policies including remote attributes. In this experiments we adopted the PD strategy for the collection of remote attributes.

The results show that, even in case of 50 remote attributes, the time to perform the TryAccess and RevokeAccess phases is less than 2 s. Compared with the results of Fig. 8 these experiments confirm that the time required for the communications over the network is the main factor which affects the overhead introduced by the UCIoT framework.

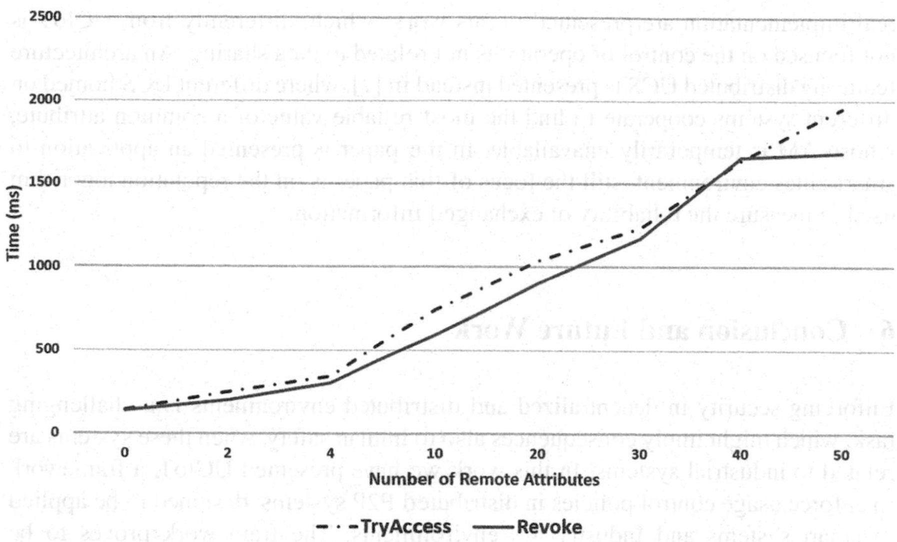

Fig. 9 Performance on emulated devices

5 Related Work

The need of introducing access control in IoT environments has recently been discussed by V. Cerf in [3], where he also proposed some directions based on the idea of introducing controlling and verification mechanisms on edge devices. UCIoT extends this vision, by introducing the more expressive UCON model and presenting an architecture which can be easily integrated in both edge and internal devices, given they have smart device requirements. The author of [20] present a model to include access control in constrained devices to be used in IoT. The work mainly focuses on the introduction in the COAP protocol of an overlay to ensure authenticated access. On the other hand, UCIoT focuses on usage control acting at application level, without modifying the standard communication protocol. An initial application of UCIoT has been presented in [12]. In this work, the UCIoT framework has been exploited to define and enforce policies for a smart home setting. In [19], a set of challenges for security and privacy in IoT is discussed. The paper is not proposing solutions to the presented issues, still it supports the claim that a distributed solution, like UCIoT would be more effective in the IoT environment. The work in [10] presents a distributed and decentralized architecture for data usage control. Differently from UCIoT, this work is focused on controlling the right to access and use data in multi-domain environment, attempting to ensure that policies are respected even after data are moved in a different environment with different control mechanisms. The work in [6] presents a methodology to enforce data usage control in IoT, to enable policy protected information sharing, exploiting semantic web technologies to derive the current context. However, not real implementation are presented in this work, which, differently from UCIoT is not focused on the control of operations not related to data sharing. An architecture featuring distributed UCS is presented instead in [7], where different UCS homed on different systems cooperate to find the most reliable value of a common attribute, whose AM is temporarily unavailable. In the paper is presented an application to smart cities environment, still the focus of this paper is on the reputation algorithm used to measure the reliability of exchanged information.

6 Conclusion and Future Work

Enforcing security in decentralized and distributed environments is a challenging task, which might imply consequences also to human safety, when these systems are related to industrial systems. In this work we have presented UCIoT, a framework to enforce usage control policies in distributed P2P systems, designed to be applied in smart systems and Industry 4.0 environments. The framework proves to be efficient and is able to ensure data replication and fault tolerance. These features are extremely important in an industrial environment, where the singular instances of UCSs can be installed directly on controlled machines. In fact, machines be temporarily unavailable due to maintenance or failure. Though we presented a plausible industrial use case, detailing attributes and policies which could be considered in a distributed UCON setting for Industry 4.0, this initial contribution has yet to be validated in a real industrial environment. In particular, implementation on a real industrial system and the consequent experimentation is left for future work.

Acknowledgements This work has been partially funded by EU Funded projects H2020 C3ISP, GA #700294, H2020 NeCS, GA #675320 and EIT Digital Trusted Cloud and IoT.

References

1. Balakrishnan H, Kaashoek MF, Karger D, Morris R, Stoica I (2003) Looking up data in P2P systems. Commun ACM 46(2):43–48. https://doi.org/10.1145/606272.606299
2. Carniani E, D'Arenzo D, Lazouski A, Martinelli F, Mori P (2016) Usage control on cloud systems. Futur Gener Comput Syst 63:37–55. https://doi.org/10.1016/j.future.2016.04.010
3. Cerf VG (2015) Access control and the Internet of Things. IEEE Internet Comput 19(5):96–c3
4. Chakeres ID, Belding-Royer EM (2004) AODV routing protocol implementation design. In: 24th International Conference on Distributed Computing Systems Workshops, Mar 2004. Proceedings, pp 698–703
5. Colombo M, Lazouski A, Martinelli F, Mori P (2009) A proposal on enhancing XACML with continuous usage control features. In: Grids, P2P and Services Computing [Proceedings of the CoreGRID ERCIM Working Group Workshop on Grids, P2P and Service Computing, 24 Aug 2009, Delft], pp 133–146

6. Das PK, Narayanan S, Sharma NK, Joshi A, Joshi K, Finin T (2016) Context-sensitive policy based security in Internet of Things. In: 2016 IEEE International Conference on Smart Computing (SMARTCOMP), pp 1–6
7. Faiella M, Martinelli F, Mori P, Saracino A, Sheikhalishahi M (2016) Collaborative attribute retrieval in environment with faulty attribute managers. In: 11th International Conference on Availability, Reliability and Security, ARES 2016, Salzburg, 31 Aug – 2 Sept 2016, pp 296–303. https://doi.org/10.1109/ARES.2016.51
8. Farwell JP, Rohozinski R (2011) Stuxnet and the future of cyber war. Survival 53(1):23–40. https://doi.org/10.1080/00396338.2011.555586
9. Featherston D (2010) Cassandra: principles and application. Department of Computer Science University of Illinois at Urbana-Champaign
10. Kelbert F, Pretschner A (2015) A fully decentralized data usage control enforcement infrastructure. In: 13th International Conference on Applied Cryptography and Network Security, ACNS 2015, New York, 2–5 June 2015, Revised Selected Papers, pp 409–430
11. Lazouski A, Martinelli F, Mori P, Saracino A (2016) Stateful data usage control for android mobile devices. Int J Inf Secur 16(4):345–369
12. Marra AL, Martinelli F, Mori P, Saracino A (2017) Implementing usage control in Internet of Things: a smart home use case. In: 2017 IEEE Trustcom/BigDataSE/ICESS, Sydney, 1–4 Aug 2017, pp 1056–1063. https://doi.org/10.1109/Trustcom/BigDataSE/ICESS.2017.352
13. Miller B, Rowe D (2012) A survey SCADA of and critical infrastructure incidents. In: Proceedings of the 1st Annual Conference on Research in Information Technology, pp 51–56. RIIT'12, ACM, New York. https://doi.org/10.1145/2380790.2380805
14. Mordeno A, Russell B (2017) Identity and access management in the Internet of things – summary guidance. https://cloudsecurityalliance.org/download/identity-and-access-management-for-the-iot/
15. OASIS: eXtensible Access Control Markup Language (XACML) version 3.0 (January 2013)
16. Park J, Sandhu R (2004) The $UCON_{ABC}$ usage control model. ACM Trans Inf Syst Secur 7(1):128–174
17. Park J, Zhang X, Sandhu R (2004) Attribute mutability in usage control. In: Research Directions in Data and Applications Security XVIII, IFIP TC11/WG 11.3 Eighteenth Annual Conference on Data and Applications Security, pp 15–29
18. Pretschner A, Hilty M, Basin D (2006) Distributed usage control. Commun ACM 49(9):39–44
19. Roman R, Zhou J, Lopez J (2013) On the features and challenges of security and privacy in distributed Internet of Things. Comput Netw 57(10):2266–2279. http://www.sciencedirect.com/science/article/pii/S1389128613000054, towards a Science of Cyber SecuritySecurity and Identity Architecture for the Future Internet
20. Sitenkov D, Seitz SL, Raza S, Selander G (2014) Access control in the Internet of Things. Master's thesis
21. Zhang X, Parisi-Presicce F, Sandhu R, Park J (2005) Formal model and policy specification of usage control. ACM Trans Inf Syst Secur 8(4):351–387

Part III
Advanced Protection Techniques

Profiling Communications in Industrial IP Networks: Model Complexity and Anomaly Detection

Mustafa Amir Faisal, Alvaro A. Cardenas, and Avishai Wool

Abstract Profiling communication patterns between devices in the Industrial Internet of Things (IIoT) ecosystems is important for deploying security measures like detecting anomalies and potential cyber-attacks. In this chapter we perform deep-packet inspection of various industrial protocols to generate models of communications between pairs of IIoT devices; in particular, we use discrete-time Markov chain models applied to four different industrial networks: (1) an electrical substation, (2) a small-scale water testbed, (3) a large-scale water treatment facility, and (4) an energy management system of a university campus. These datasets represent a variety of modern industrial protocols communicating over IP-compatible networks, including EtherNet/IP (Ethernet/Industrial Protocol), DNP3 (Distributed Network Protocol), and Modbus/TCP (Transmission Control Protocol).

Keywords IIoT · Anomaly detection · Modeling · DTMC

1 Introduction

While many critical infrastructures such as power systems have existed for over a century, it is only in the past two decades that the information exchanged between different parts of the system have migrated from serial communications to IP-compatible (and thus Internet-compatible) networks. Modern Supervisory

M. A. Faisal
The University of Texas at Dallas, Richardson, TX, USA
e-mail: mustafa.faisal@utdallas.edu

A. A. Cardenas (✉)
University of California Santa Cruz, Santa Cruz, CA, USA
e-mail: alvaro.cardenas@ucsc.edu

A. Wool
Tel Aviv University, Tel Aviv, Israel
e-mail: yash@eng.tau.ac.il

C. Alcaraz (ed.), *Security and Privacy Trends in the Industrial Internet of Things*,
Advanced Sciences and Technologies for Security Applications,
https://doi.org/10.1007/978-3-030-12330-7_7

Control and Data Acquisition (SCADA) systems are interconnected though a variety of IP-compatible industrial protocols such as DNP3, Modbus/TCP, EtherNet/IP, and Profinet. Several of these communication upgrades are motivated by the raising popularity of Industrial Internet of Things (IIot) devices. While most IIoT systems are not directly connected to the public Internet, they use IP-compatible technologies, which facilitates their interconnection and debugging functionalities.

Securing these networks is a growing concern [12]. Discerning the normal behavior of different ICS (Industrial Control System) network protocols for connecting IIoT devices is important for finding anomalies, configuration errors, and potential cyber-attacks. In particular, by looking at what type of commands are common in an ICS, and how these commands interact with each other (order, frequency, etc.) we can get a fine-grained view of the expected network behavior which can be beneficial for cyber-security purposes.

Our contributions include:

- In this chapter, we apply DTMC (Discrete-Time Markov Models) to capture the sequence of commands being exchanged between machines in industrial networks and evaluate their effectiveness for accurately representing various systems from real-world systems to testbeds.
- Various challenges of using DTMC-based models like model complexity, growth of number of states and transitions are discussed, which gives a promising exploration point for future researchers.
- We proposed an evaluation metric by combining two other metrics so that it can be used for anomaly detection.
- We also discuss practical applications of DTMC. For example, one of the challenges of anomaly-based intrusion detection systems is to generate meaningful messages for operators. From our analysis, models like DTMC can be used for this purpose as well as for visualization of communication between two individual devices.

We structure the chapter as follows. In Sect. 2, we discuss previous work. We provide a summary of the ICS protocols in this work in Sect. 3. A concise illustration of DTMC and its state generation procedure in our context is discussed in Sect. 4. After that, an overview of our datasets is presented in Sect. 5. Results from experiments measuring the model complexity are discussed in Sect. 6; Sect. 7 gives a summary of the results for anomaly detection and proposes a combined metric for model evaluation; and Sect. 8 highlights other use-cases for the use of our models.

2 Related Work

The number of companies integrating IIoT equipment to enhance sensing, processing, and computational capabilities for building a smart industrial environments [1], are increasing. IIoT networks are responsible for carrying sensitive data among various sensors, actuators, controllers, HMI (Human Machine Interface), and SCADA servers in an industrial environment.

Securing IIoT ecosystems is thus becoming a critical challenge in this sector [13]. For industrial ecosystems, Rubio et al. [12] reviewed threats existing in traditional industrial systems and recent IIoT deployments; they discuss how future IIoT devices in industrial systems can be vulnerable to different threats like availability (on routing protocols and the device itself), integrity (falsification of information), confidentiality (side channel attack), and authentication (introduction of dummy or fake nodes) threats.

Intrusion Detection Systems (IDSs) are a popular defense in these systems [15]. One of the key-observations for IoT and IIoT communications is that network traffic tends to be periodic and predictable. For example, Fu et al. [3] proposed an automata-based IDS for various heterogeneous IoT networks for three attacks like jam-attack, false-attack, and reply-attack.

The use of finite state machines to model industrial networks has been considered before. A fundamental work by Goldenberg et al. [4] introduced DFA (Discrete Finite Automata) as a modeling technique for Modbus/TCP. Follow up work by Kleinmann et al. [6] extended the same principles for another industrial control protocol, namely Siemens S7 (version 0x32). The main assumption for applying DFA to industrial communications is that machine to machine communications is very periodic. However, in many real-systems, this assumption does not hold true for a variety of reasons, including (human) operator interactions, random delays [2], multi-threading servers, etc.

To address these problems, Kleinmann et al. suggested a statechart-based approach [7] which was further automated in [8] using Fourier transform. However, these DFA-based approaches can struggle to completely capture aperiodic channels. Moreover, we cannot track which type of commands happen frequently or occasionally in the communication. To understand the full interaction among various commands, we can use a probabilistic model.

A popular probabilistic model is a Discrete Time Markov Chain (DTMC) which has been used for modeling ICS network protocols by Caselli et al. [2]. In this chapter we study further the applicability of DTMCs in various datasets from both real-systems and testbeds.

3 ICS Protocols

Among the different application-layer ICS networks, here we are considering three protocols: EtherNet/IP, Modbus/TCP, and DNP3. A brief description of each of these protocols are provided in below.

EtherNet/IP (ENIP) is an application layer protocol which adapts CIP (Common Industrial Protocol) for industrial network over standard Ethernet technologies. CIP uses object-oriented design to facilitate various services and device profiles which are required for real-time control applications. In addition, this also helps to promote uniform implementations of automation functions for a large set of diverse products. ENIP packets are encapsulated over either TCP (explicit message which can be unconnected and connected) or UDP packets (implicit or I/O which can only be

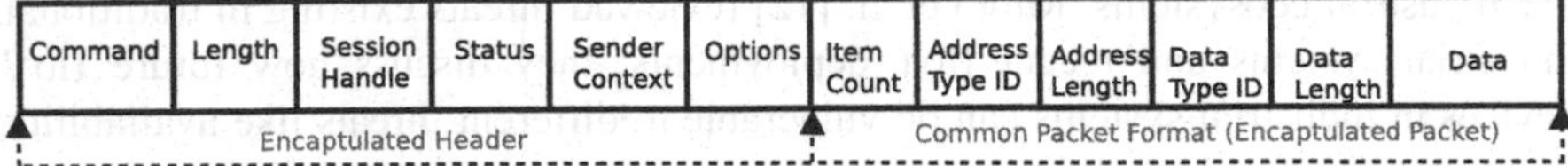

Fig. 1 A ENIP message with header and common packet format

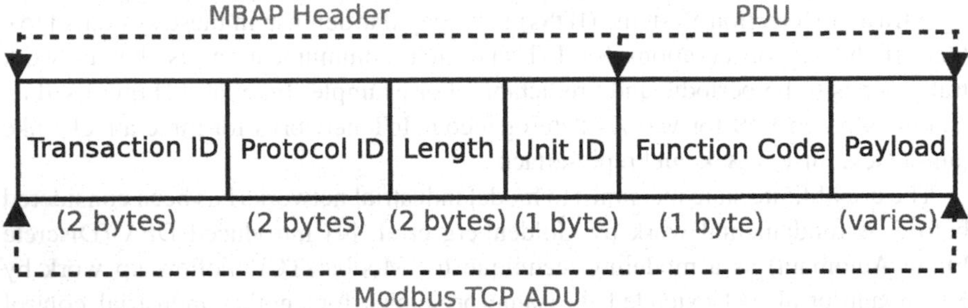

Fig. 2 Modbus TCP data packet

connected). The transport layer ports for ENIP over TCP and UDP are 44818 and 2222 respectively. Thus, ENIP allows both peer to peer or multi-casting over UDP (User Datagram Protocol). Every encapsulated message over TCP (not for UDP) is composed of a fixed-length (24 bytes) header followed by an optional data portion (up to 65,511 octets) [11]. The encapsulation header has 6 fields (Fig. 1): (i) *Command* which is allocated by the specification, (ii) *Length* of data portion, including header, and (iii) *Session Handle*, which is generated by the target and sent back to originator in response to a *RegisterSession* request command – for some commands like NOP (No Operation), this field is optional, (iv) *Status*, which indicates whether or not the receiver was able to execute the requested command (a value of 0 indicates successful status), (v) *Sender Context*, which is assigned by the sender and a receiver which returns this value without any modification (this can be used to pair request with response), and (vi) *Options*, which is set to zero by the sender and a receiver will discard packets for non-zero options values.

The Common Packet Format consist of (i) item count, (ii) address item, (iii) data item, and (iv) additional items (optional). Data and address items have three fields: (i) Type ID which indicates the type of encapsulated item (defined by the specification) (ii) Length (in bytes), and (iii) data (variable). There can be additional fields based on commands and message types (connected or unconnected). For example, for the *SendRRData* command, two additional fields are interface handle (0 for CIP) and timeout for operation.

Modbus/TCP is a variant of basic Modbus over TCP, a popular communication protocol in ICS. Modbus/TCP clients and servers listen and receive data through port 502. The protocol [10] has an ADU (Application Data Unit) which consists of two parts: MBAP (Modbus Application Protocol) header and PDU (Protocol Data Unit) (see Fig. 2). The MBAP header has four fields: (i) *TID (Transaction Identifier)*, which is used for pairing transactions because multiple messages can be transmitted through the same TCP connection by a client without waiting for a prior response; (ii) *PID (Protocol Identifier)*, which is always 0 for Modbus; (iii) *Length*, which provides the length of the remaining fields (in bytes), including the UID (Unit Identifier), function code, and data fields; and (iv) *UID*, which is used to identify a remote PLC (Programmable Logic Controller) located on a non-TCP network (for serial bridging). By default, UID is set to 00 or FF which is ignored by the server and echoed back in the response.

Three types of PDUs are defined in Modbus [9]: Modbus request, response, and exception response. PDUs have two fields, (i) *Function Code* (for Modbus request and response PDU) or exception function code (for Modbus exception response, where the most significant bit is set as 1) and (ii) *Payload* – from 1 to 252 bytes – which can be a variable RN (Reference Number), count, value, data offsets, subfunction codes, etc.

In Modbus, valid function codes range from 1 to 127 and are divided into three categories: Public (well documented), user-defined (vendor-specific functions), and reserved (used by legacy products). Public function codes [9] are {1–8, 11–12, 15–17, 20–24, 43}. There are four primary tables which are organized in series: discrete input, coils, input registers, and holding registers. Depending on the function code in the Modbus request, access is directed to a specific primary table.

DNP3 was mainly developed for standards-based interoperability between substation computers, RTUs (Remote Terminal Units), IEDs (Intelligent Electronic Devices), and master stations. The default transport layer port for DNP3 is 2000. The protocol is mainly separated in three layers [5] (Fig. 3):

The *Data Link Layer* provides an interface between transport functions and the physical media or transport layer of the OSI model (in case of IP-based network). This layer manages frame synchronization, flow control, and error handling as well as providing link status. In this layer, each frame starts with a header block (0 block), followed by optional data block(s). Each block ends with 2 bytes CRC. The header block has 5 fields excluding CRC: (i) *start* (which is $0x05$ and $0x64$ – to indicate the start of the frame), (ii) *length*, (iii) *control* (contains subfields to indicate frame direction, transaction initiator, error, flow control, and function), (iv) *destination*, and (v) *source*.

Transport Function is used to disassemble (fragments from application layer) or assemble (from data link layer) a segment to handle large sizes of data. Each segment has a header and a data part. The segment header has the following fields: FIN (final segment or not), FIR (first segment or not) and SEQUENCE (6 bits – used for order and duplication check with range 0–63).

Finally, in the *Application Layer*, data is stored as *fragments* which consist of octets of request/response information between a master and an outstation. The packets are divided into three types of fields: Application Header, Object Header, and DNP3 Object. Each packet may have multiple DNP3 objects.

The Application Header has three parts: Application Control (1 octet), Function Code (1 octet), and Internal Indications (2 octets – only in response header). Application Control has five fields: (i) FIR (1 bit – first fragment or not), (ii) FIN (1 bit – final fragment or not), (iii) CON (1 bit – application layer's confirmation will

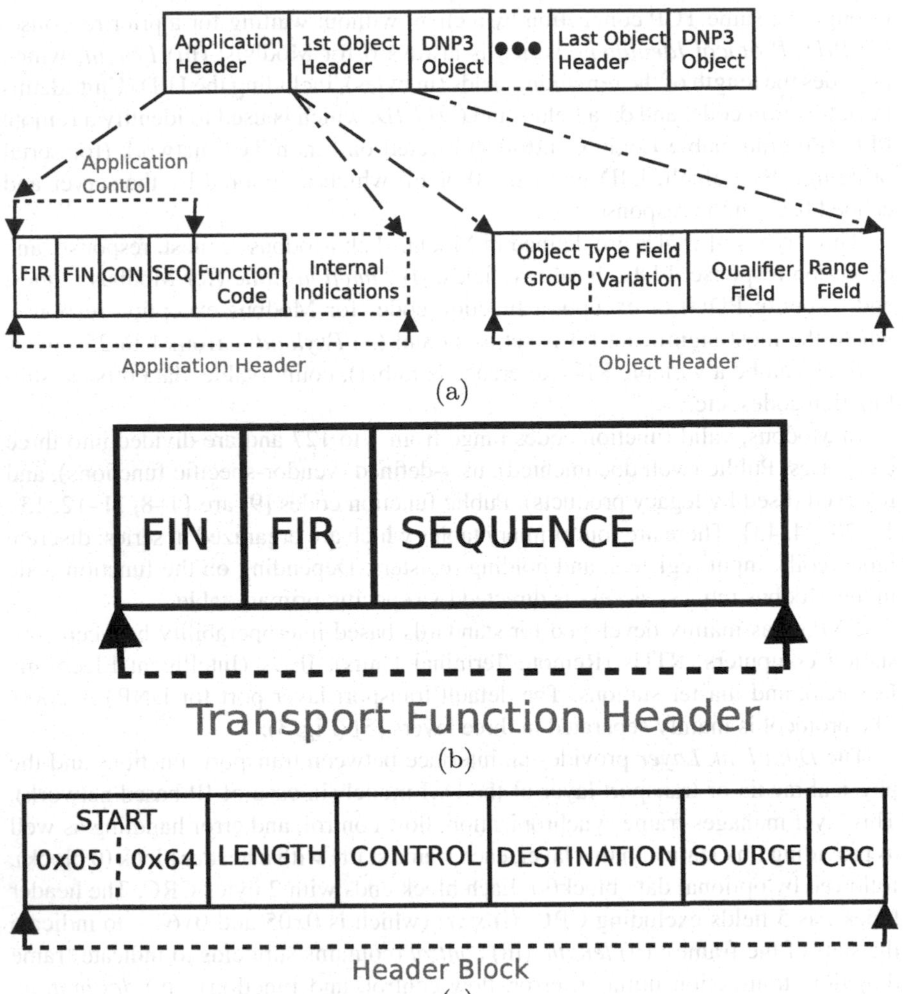

Fig. 3 Various parts of DNP3 packet. (**a**) Application layer. (**b**) Transport function. (**c**) Data link layer

be returned or not), (iv) UNS (1 bit – requested or unsolicited response message), and SEQ (4 bits – used for order and duplication check with range 0–15).

Unlike Modbus, DNP3 has a separate set of function codes for request and response. *Function codes* from 1 to 128 are used for requests among which 1–33 are currently used. On the other hand, 129–255 are used for response except 129–131 (details can be found in [5]).

Finally, the bits in Internal Indications (14 for states and 2 are kept as reserved) indicate specific states and errors in an outstation.

The Object Header and DNP3 Object provide supplementary information when the application header cannot carry all information. Requested packets do not have a DNP3 Object. The Object Header has group (data type or value – like analog, binary, etc.), variation (data format – like 16-bit or 32-bit analog etc.), qualifier (size of object prefix), and range (depends on qualifier). Note that recommended qualifiers and range are used for vendor specific master and outstations.

Data Model: Outstations can carry five types of variables (a uniquely identifiable physical or logical entity): (i) binary input, (ii) analog input, (iii) counter input, (iv) binary output, and (v) analog output. In addition, DNP3 can also carry files and other data forms. And point, a uniquely identifiable physical or logical entity, is used as a data structure to carry data.

4 Modeling Industrial Networks

Our goal is to profile each flow or channel (network communication between two devices) in ICS networks. This profile can be used for anomaly detection, visualization, meaning message generation for operators, etc. Overall, our procedure can be divided as follows: (i) identify the flows/channels from captured packets, (ii) filter packets of each flow according to the ICS protocol specification, (iii) create symbols or states from the interested fields, (iv) build the model by sequentially creating and updating states and transitions from protocol specific data.

The channel definition depends on the ICS protocol specification. In the case of Modbus/TCP, it is *(source IP address, Destination IP address, Unit ID)* [4] and for other two protocols, it is *(source IP address, Destination IP address, Destination port address)*. To create models, we use an iterative process based on [2] using DTMC with some modification for creating state data.

4.1 Discrete Time Markov Chains

A DTMC is a stochastic process where a set of random variables $X_1, X_2, X_3, \cdots, X_n$ follow the Markov property, which means the probability of moving to the future state depends on only current state not previous states. Here X_t is a random variable which takes values from a countable state space on a given time $t \in 0, 1, 2, \cdots, n$. Mathematically we can say,

$P(X_{t+1} = j|X_t = i, \cdots, X_0 = i_0) = P(X_{t+1} = j|X_t = i) = P_{ij}$.

P_{ij} is a one-step transition probability, which means that in a DTMC, the probability of moving state i to state j is P_{ij}. The square matrix $M = (P_{ij}), i, j \in S$ is called the one-step transition matrix where for each $i \in S \sum_{j \in S} P_{ij} = 1$.

In our case, we define an event according to [2] for each protocol. For example, like in [2], a Modbus/TCP sequence is defined as a time-ordered list of events, e_t where event, e is a 3-tuple $< ID, Code, Data >$ derived from a sequence of message pairs $(m_t^{Req}, m_{t+1}^{Res})$, and m is a message with type request and response at time t and $t+1$.

From each event, states and transitions are defined. A state, S, is defined by a 5-tuple: <Data, Type, number of events, First Time Seen, Last Time Seen> and every transition, T (from a source state to a destination state), is defined by a 6-tuple: <probability, number of jumps, first jump, last jump, average time elapsed, standard deviation on time elapsed>.

4.2 State Data Generation for DTMC

Construction of state data for DTMC is crucial. This mainly consists of finding relevant/interesting network fields for each ICS protocol. We find that in some protocols like EtherNet/IP (*NOP* command), and DNP3 (unsolicited response) there are no request/response pairs. Hence, instead of treating a request/response pair as a single state like in [2], we treated each packet as a single state (the type can be request, response, or not request/response). Selecting the fields of each ICS protocol for constructing state data requires domain knowledge and depends on the security requirements as well. In our state data definition, we try to focus on the fields related command functions (read or write), command type (request/response, unsolicited, confirmation, etc.), data length, data type, services, etc.

For EtherNet/IP, we selected most of the fields of the encapsulated header, common packet fields, services, class, etc. As CIP over EthenNet/IP is an object-oriented protocol, the user can define any device as an object of a class with custom function, attribute, behavior, etc.

For Modbus/TCP, state data is defined as in [4]: A 4-tuple $< Query/\ Response$ $(1\ bit),\ \ function\ code(8\ bits),\ \ reference\ number(16\ bits)\ \ bit/word$ $count(8\ bits) >$. Here *Query/Response* is a custom field derived from *transaction id* (pairing for request/response) which is set by the master and echoed by a slave. Following this field, *function code* allows us to understand the functionality of the packet, *reference number* indicates the position (start position of data to read or write) and *bit/word count* is the amount of data.

For DNP3, we focus primarily on its application layer's fields. And fields are selected from the data-link layer indicating whether the packet originated from the master or not, source, etc.

5 Datasets

We captured various datasets from real-systems and testbeds.

The PGT (Power generation testbed) dataset is gathered from a power generation SCADA (Supervisory Control and Data Acquisition) testbed. Two 1 MVA (Mega Volt Ampere) on-site generators (1 and 2) are protected by their own protective relays. The generated energy from these generators is coupled together on a generation bus which also protected by its own relay. Finally, a transformer is used to step the voltage up from 4,160 volts to 13,200 volts. This transformer is protected by a transformer relay. All the relays are connected to an automation controller, and a network sensor is deployed to collect packets from these five devices. The DNP3 protocol is used for all communications except for the channel between SCADA controller and generator 1's relay, which uses Modbus/TCP. In this testbed DNP3 uses six commands: *Confirm* ($0x00$), *Read* ($0x01$), Write ($0x02$), Freeze ($0x07$), Record current time ($0x18$), and Response ($0x81$). Among them, *Read* and *Response* are most frequently used commands. The *Confirm* command is only used for two channels after some responses for the *Read* command. The use of the rest of the commands is infrequent. For Modbus/TCP, the testbed uses Read Input Registers ($0x04$) for reading three phase current magnitudes and angles.

The WPT (Water plant testbed) dataset is collected from a water treatment testbed which has six stages: (i) raw water, (ii) pre-treatment, (iii) ultra filtration, (iv) dechlorination, (v) RO (reverse osmosis), and (iv) RO final product. Each process is managed by two PLCs: primary and secondary. Each stage of process is controlled by a PLC and all PLCs are connected through a common network line (Level 1 or L1 network). PLCs in L1 network are connected to a centralized SCADA system which contains HMI and Historian server. Moreover, each PLC is connected to local sensors and actuators through a dedicated field-bus network (Level 0 or L0 network). In this Field-bus network, sensors and actuators send and receive digital and analog signals to process-specific PLCs through RIO (Remote Input/Output) modules (which converts electrical signals into network packets). These packets are encapsulated over the L0 network protocol (Device Level Ring or DLR for EtherNet/IP in case of this testbed) which is a ring topology among the RIO, the primary and the backup PLC. These signals are sent back and forth from the PLCs. We collected traces from both L1 and some of L0 networks. From the L1 network data, we parsed the EtherNet/IP and CIP data. L1 data have both connected (over UDP) and unconnected (over TCP) messages. On the other hand, the L0 network has only connected messages (over UDP) and DLR protocol (which works in layer 2, i.e., the data link layer in OSI (Open System Interconnection) model and uses Beacon frame – a type of management frame, providing the "heartbeat" of a wireless LAN) data. For TCP/IP flows, we observed only two kinds of EtherNet/IP commands: SendRRData ($0x006f$) and SendUnitData ($0x0070$). The first one is a request and response connection. On the other hand, SendUnitData is unidirectional and is the most frequent packet.

The ESM (Energy system monitoring) dataset is an energy monitoring system of a university campus power grid using Modbus/TCP. This network has seven Modbus devices and one of them acts as a master. The only request we see from the master is the *Read Holding Registers* ($0x03$) command.

The WT (Water treatment) dataset is a network trace also over Modbus/TCP from a real-world operational large-scale water facility in Texas. The network has over one hundred controllers and some controllers have over 1000 registers. Most of the communication (about 95%) is over Modbus/TCP and the rest are general Internet protocols. The dataset has 108 Modbus devices and one of them works as a master; another one acts as a gateway and the rest of them act as slaves. Regarding commands (master to slaves), most of them (74%) are Read/Write Multiple Registers ($0x17$). And the other commands are Read Coils ($0x01$) and Read Discrete Inputs ($0x02$) (20% and 6% respectively). The network topology is shown in Fig. 4.

A summary of all used datasets is provided in Table 1. As we see, WT is largest dataset.

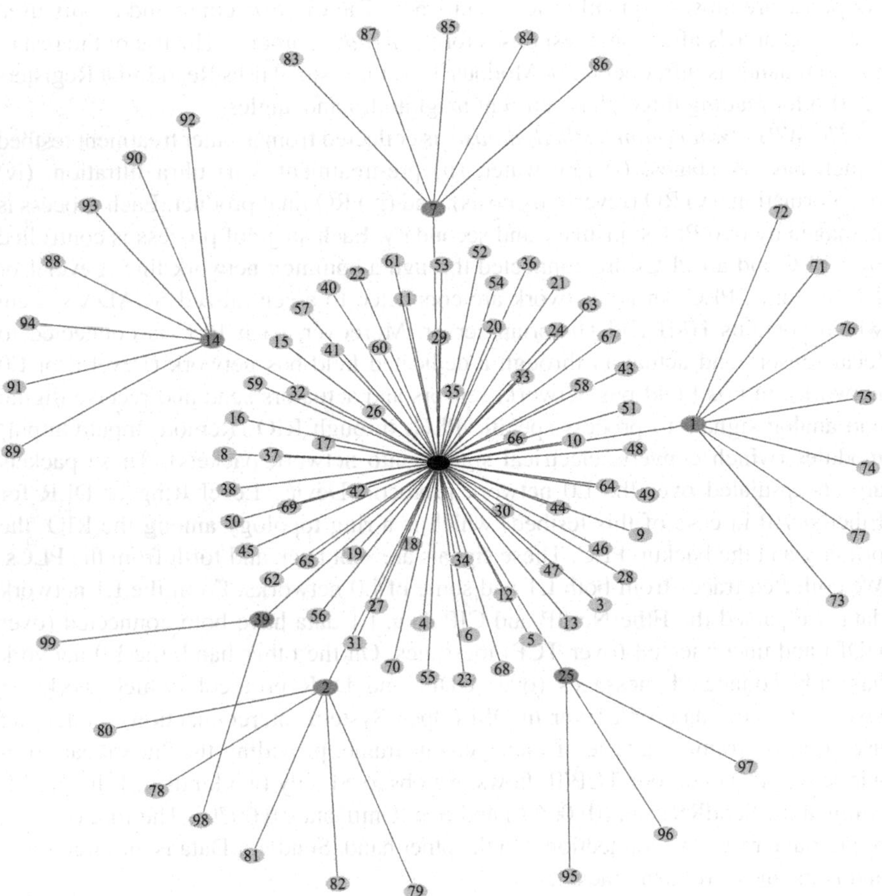

Fig. 4 Communication network for *WT* dataset

Table 1 Summary of used datasets

CPS domain	Protocol	#packets	#hours(appx.)	#channels
PGT	Modbus/TCP	58, 836	16	1
PGT	DNP3	169, 922	16	3
WPT L1 network	EtherNet/IP	10, 051, 785	0.5	57
WPT L0 (P2) network	EtherNet/IP	142, 308	0.35	4
WPT L0 (P3) network	EtherNet/IP	108, 054	0.2	5
WPT L0 (P4) network	EtherNet/IP	68, 164	0.2	5
WPT L0 (P5) network	EtherNet/IP	188, 232	0.2	9
ESM	Modbus/TCP	48, 835, 082	20	6
WT	Modbus/TCP	80, 239, 202	24	94

6 Results and Discussion

Various types of experiments are performed to understand the effectiveness of DTMC as a model. In particular, we try to understand how the DTMC model evolves as we increase the amount of data. We start from 400 network packets and increase by steps of 400 packets.

Communication over Modbus consists of request/response states of different commands on various data tables (discrete input, coil, registers, etc.). For example, Fig. 5 depicts commands' interaction between two entities for 14,000 packets in the *ESM* dataset. *Read Holding Registers* requests from address 256 for 6 words (12 bytes) and their corresponding responses happened frequently. There is a very occasional self-loops for requests which may happen because of network issues. Following these responses, a single request with its response for same function code appears with different starting address and amount of output data.

Unlike Modbus (same function code for request/response), DNP3 has a dedicated response function code ($0x81$). As it can request event data, the size of the response data may vary. For example, with 14,000 packets captured between the SCADA server and the relay in generator 2, we see that in response to *Read* requests for both static and event data, three type of responses are sent back: (i) with no data (ii) with static data, and (iii) with both event and static data. Most of the transmissions ($\approx$84%) happen due to *Read* requests for event data and their corresponding responses. After each response for event data, the SCADA server sent back a confirmation. There are some states with invalid conditions for Record Current Time ($0x18$), Write ($0x02$) and Freeze ($0x07$) commands like invalid parameters and unreported class 2 events. These invalid responses were caused by the default installation of the SCADA software, which defaults to a misconfiguration.

In EtherNet/IP, all commands do not have explicit request/response pair like *SendUnitData*, NOP, etc. For example, we observe a frequent interaction between primary PLCs of ultra filtration and dechlorination process using *SendUnitData* command. Using a customized CIP service, Read_Tag_Service, the originator requests to read a particular tag and the target confirms with the tag. There is

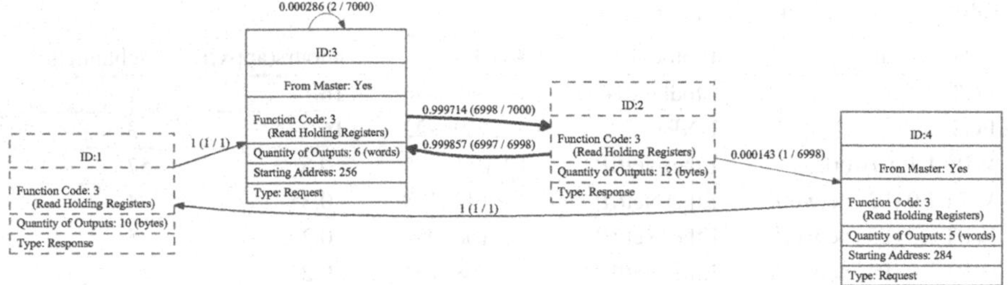

Fig. 5 DTMC with 14,000 packets for communication (over Modbus) between two components of energy monitoring dataset. Solid and dotted lines are used for request and response states respectively

an infrequent and aperiodic transition of *SendRRData* commands where the CIP service is forward open (used to establish connection with a target device) of *identity object* (used to know the identity of a particular device).

6.1 Measuring the Increase in Model Complexity Over Time

6.1.1 Extreme Cases

The simplest form of DTMC, from our observation, can be two states and two transitions between them. For example, we observed such a simple DTMC for channel between the SCADA server and the relay of generator 1 (Fig. 6) where a *Read Input Register* request is asked by the SCADA server (magnitudes and angles of 3-phase current), and a response is provided by the protection relay at generator 1.

In contrast, the largest DTMC generated from our datasets occur in the *WT* dataset (Fig. 7). This DTMC has 37 states and 277 transitions. One of the possible reason can be configuration change. That means whenever there is a change in configuration, we need to rebuild DTMC with new data. However, access to the configuration file is not always feasible (for security reason). In this case, a notification (due to configuration) can be sent to DTMC builder to collect new data and generate new DTMC. Alternatively, a set of policies can be defined for trimming stale states and sequences of security concerned transitions. Besides, a heuristic approach for pruning may be applied to make DTMC as a manageable sized.

6.1.2 State and Transition Growth

To understand, how the number of states and transitions evolve if we increase the size of the data, we perform an experiment where we increase the data size by 400 packets increments up to L_s (300,000). At each interval we build the DTMC model and get the number of states and transitions. Here, it should be noted that

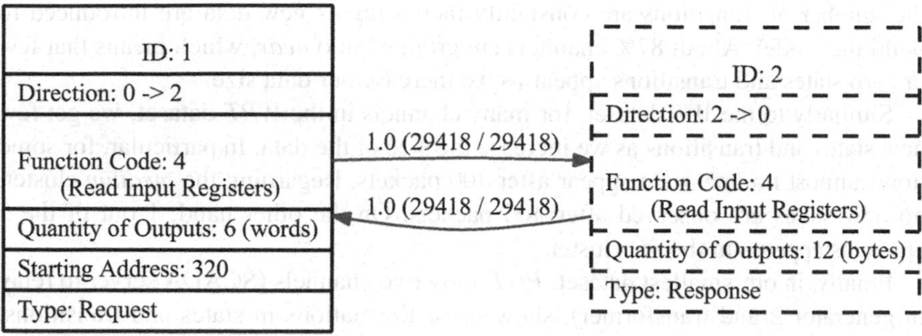

Fig. 6 One of the simple (2 states and 2 transitions) DTMCs with full data for channel controller and generator 1 (over Modbus/TCP) in power generation testbed dataset

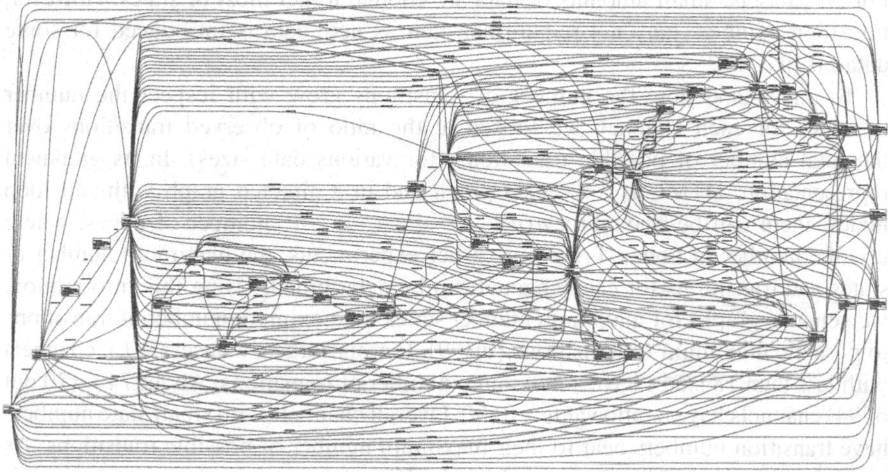

Fig. 7 One of the complex (37 states and 277 transitions) DTMCs with 40,000 packets for communication (over Modbus/TCP) between two devices in Water Treatment

some channels have less than L_s packets. We then use clustering algorithms to identify the growth trends in each dataset. We use two attributes: sum of slopes of (i) states and (ii) transitions. With these attributes, we clustered the channels into three categories according to proximity (of cluster centroid) from origin (2D plot) (i) near, (ii) medium, and (iii) far (far away from the origin).

For the *ESM* dataset, in case of two channels, we do not see new states and transitions and one has few new states and transitions. They are grouped into *near* category. In one channel (medium), only new transitions appear. And for the remaining two, appearance of new states and transitions are frequent and thus, they are clustered and labeled as *far*.

Even in the case of our largest dataset (*WT*), after a few iterations, the number of states tend to become constant with small fluctuations. However, for some channels, the number of transitions are constantly increasing as new data are introduced to build the model. About 87% channels are grouped into *near*, which means that few or zero states and transitions appear as we increase our data size.

Similarly to the *WT* dataset, for many channels in the *WPT* dataset, we get few new states and transitions as we increase the size of the data. In particular, for some flows almost no new state appear after 400 packets. Regarding the *medium* cluster, no new states are observed after 400 packets. On the other hand, 1 out of the 5 channels appears in the *far* cluster.

Finally, in our smallest dataset, *PGT*, only two channels (SCADA server to relay in generator 2 and transformer), show some fluctuations in states and transitions. The channel over Modbus/TCP has a constant number of states and transition throughout the whole dataset (Fig. 6).

Figure 8 shows the results (for state and transition growth) from all four datasets. For all datasets, small amounts of data are enough to get most of states. However, transition numbers may not remain constant as new data is introduced for some datasets like *WT*.

To understand how the number of transitions grow with respect the number of states for each channel, we looked at the ratio of observed transitions over the total number of possible transitions (for various data sizes). In its graphical representation, DTMC is a directed graph and in a directed graph with self-loop nodes, there can be maximum $n(n - 1) + n = n^2$ number of edges, where n is the number of nodes. From our earlier experiments, the maximum number of states and transitions for each channel can be determined. Using this information, we generate graphs for maximum number states vs. maximum number of transitions to possible maximum of transitions (Fig. 9). For our largest dataset (*WT*), channels with less than 18 states have a maximum number of transitions. About 74% (70 out of 94) channels have a ratio value ≥ 0.90. Overall channels with small state numbers have transition numbers near to their maximum number of possible transitions. As maximum number of states grows, the ratio decreases.

7 Results for Anomaly Detection

In this second experiment we divide the data equally for learning and testing the effectiveness of the models. We also assume that the training data does not have any malicious events. Potential anomalies generated by the DTMC model when we give them testing data can be: (i) *Unknown state*, (ii) *Unknown transition*, and (iii) *Anomalous probability* which is a dissimilarity metric between the frequency of appearances of certain transaction in the learning dataset and the testing dataset. For detecting an anomalous probability, Caselli et al. [2] define two distance measures: (i) sum of the difference (absolute) of transition probabilities of a state: $d_s = \sum_{i=1}^{T} |p_{i_{test}} - p_{i_{learning}}|$, and (ii) absolute difference between learning and detection transition probability: $d_{i_t} = |p_{i_{test}} - p_{i_{learning}}|$. Here T is the number of

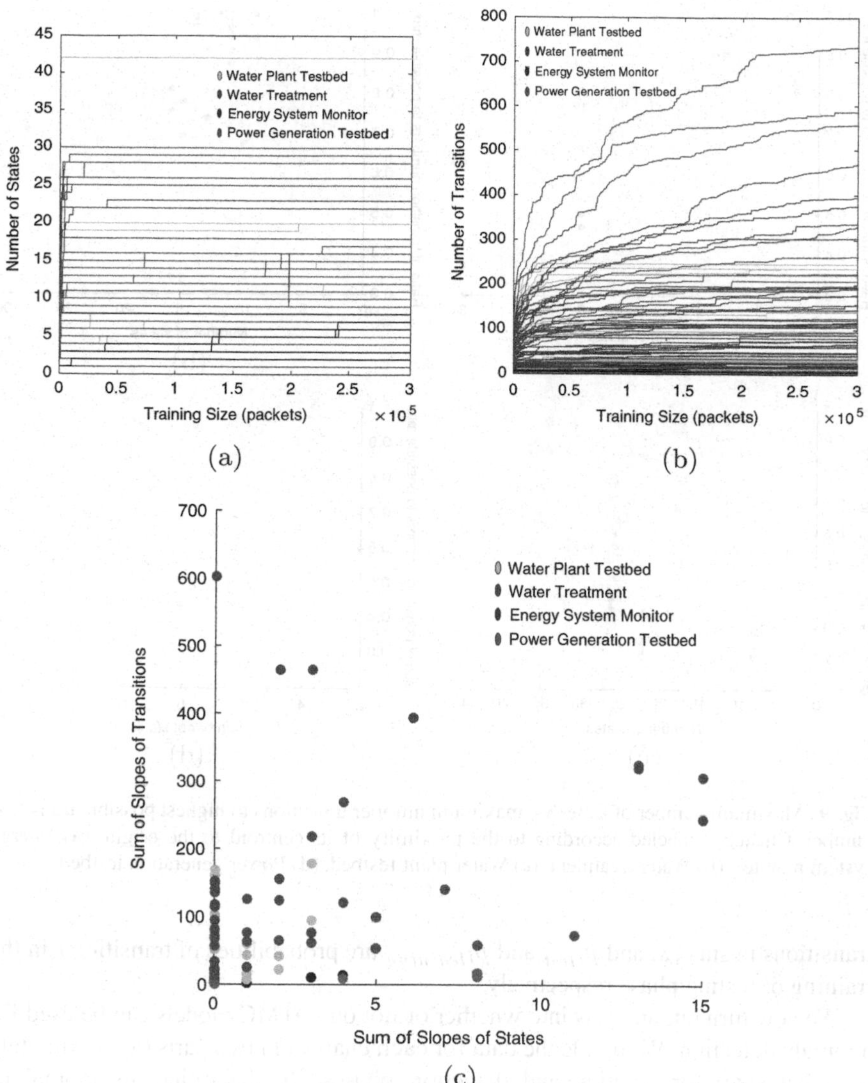

Fig. 8 States, transitions, and sum of slopes of them for all datasets. **(a)** States. **(b)** Transitions. **(c)** Sum of slopes of states vs. transitions

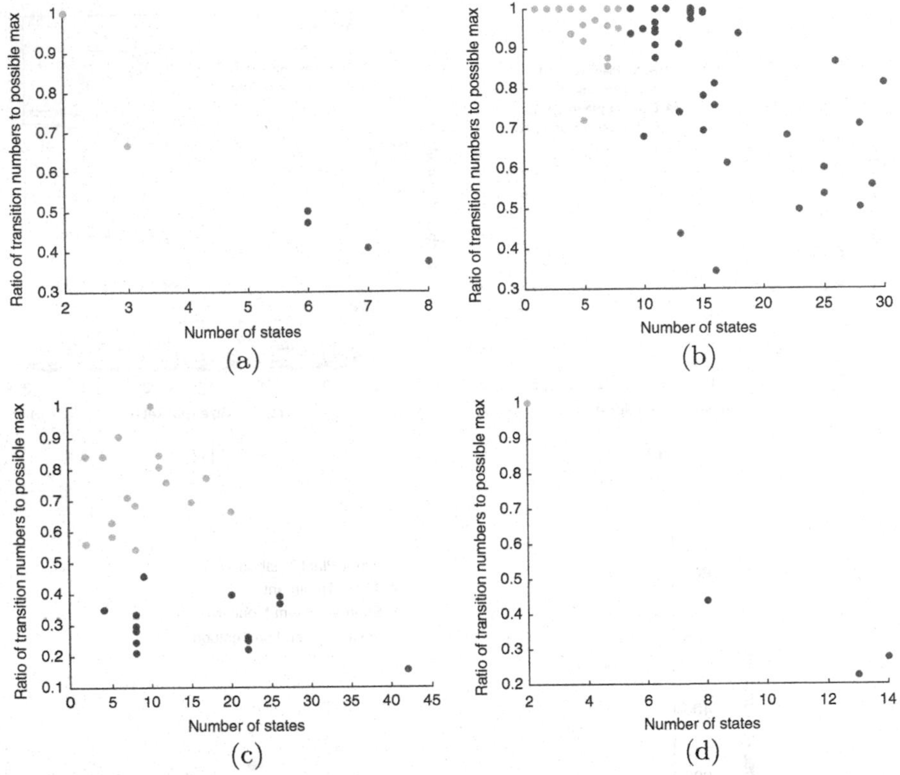

Fig. 9 Maximum number of states vs. maximum number transitions to highest possible transition number. Cluster is labeled according to the proximity of its centroid to the origin. (**a**) Energy system monitor. (**b**) Water treatment. (**c**) Water plant testbed. (**d**) Power generation testbed

transitions in state, s, and $p_{i_{test}}$ and $p_{i_{learning}}$ are probabilities of transition i in the training or testing phase respectively.

We now turn our analysis into whether or not our DTMC models can be used for anomaly detection. We divide the data for each channel in two parts (approximately of equal size) for learning and detection phases. To determine an anomalous probability (both anomalous states and transitions), we set a threshold value of $\theta = 0.1$ as previous work [2].

For the *ESM* dataset, for one channel (out of 6), no unknown (or new) state and transition are observed in the test phase. For the rest of the channels, the appearance of unknown states and transitions are frequent. A very high number of new states and transitions (248 and 479 respectively) occurred for one of the channels. We assume this could happen because of a configuration change, though we could not verify the reason due to our limited access to the configuration file. Anomalies related to unknown probabilities (related states and individual transition) are rare.

For example, in this dataset, we do not observe any unknown probabilities for 2 (out 6) channels and 1 or 2 (maximum) for the rest.

The *WT* dataset is the largest one among all of our datasets (94 channels). For 29 (≈30%) channels, we do not find any anomalies. The highest number of unknown states (12) was observed in one channel, however, for the most part, the number of these anomalies is zero or near to zero for the rest of the channels. For one channel, we received the highest number of unknown transitions (280) and unknown probabilities (115 and 18 for individual transitions and single state transitions respectively) and where the number of unknown states is 3. There are some cases where the number of unknown states is zero and where the number of other anomalies is non-zero. However, the reverse is not true that means that whenever an unknown state appears, it also introduced other types of anomalies.

We find that for the *WPT* dataset, no unknown states or unknown probability are observed. The result is intuitive as time duration for collected data is comparatively small (at most half an hour). Ten out of forty-one channels do not have any anomalies in the detection phase. Though 18 channels do not have any unknown transitions, the average unknown transition is 17.3 and the maximum one is 111 for which we also observe the highest number of unknown probabilities due to state.

Similar to the *WPT* dataset, for the *PGT* dataset, we do not observe unknown states and unknown probabilities due to single state's transitions. For this dataset, during data collection time, no configuration change happened. The highest number of unknown transitions (15) and unknown probabilities due to individual transitions (5) are observed for the channel between the SCADA master and generator 2. Both communications from the SCADA master to the relay at generator 2 and the relay of the transformer show a higher number (15 and 10 respectively) of unknown transitions. We observe some extra states due to invalid parameters and unreported class 2 events. For Modbus/TCP, which represent the communications between the SCADA master and generator 1, there were no anomalies observed. This is not surprising as this is one of the channels we pointed out in earlier sections, as having the smallest size (one a read request from the SCADA Master followed by its reply).

From our observation, the introduction of unknown state(s) in test phase also introduce unknown transition(s) which is intuitive. Unknown states and transitions are not necessary malicious. Unknown states can happen due to configuration changes. Network issues and the multi-threading nature of the controller or HMI may cause unknown transition.

7.1 Likelihood of Observed Sequences

In addition to the previously mentioned anomalies (new states, transitions not seen before, and probabilities between states that are different from the training dataset), we now turn to the question of how likely is it that a sequence of packets observed in the network comes from the same network configuration we observed in our training dataset, or whether the sequence of commands is suspiciously unlikely.

We now propose a new metric combining two anomalies (unknown state and unknown transition) by calculating the "information" of a sequence of packets by using a sliding window of length w: $\sum_{i=1}^{w} log(\frac{1}{p_i})$ where p_i is the probability of the i'th transition in the learning or training phases. The function $log(\frac{1}{p_i})$ is known as *self-information* or *information function* or *surprise* [14] and measures the information we receive for the i'th event with probability p_i. If the event happens all the time, we gain little information (as we already know that it happens certainly), but if it is an unusual event, then the amount of information we learn is high (as it is a surprising event and carries information for us). In our case, when an unknown state or transition (a "surprise" event) is observed in the testing phase, the corresponding probability is zero, which then forces the "information function" to be very high, i.e., infinity (∞). This means that when the transition probability in the training data is low or did not occurred, the "information function" will be high, and in case of a high transition probability, the "information function" will be low.

We now show some illustrative examples of this analysis. Figure 10a shows the "information" for different window sizes for a channel where we obtained a DTMC of size 2. For window size 3 (green) and 5 (black), we see a fluctuation as the channel has two transitions between two states and it is very periodic. For a window size which is multiple of 2 (here 4 and 6), the "information" remains constant (parallel to x-axis).

We see similar behavior for another channel in the *ESM* dataset (see Fig. 10b). In this case, for window size 6, the "information" becomes constant. For anomaly detection, a threshold value, τ can be set and if the "information" value exceeds the threshold, an alarm can be generated. As an example, Fig. 10c shows number of sliding windows vs. "information" for a channel between controller and generator 2 for window size 3. Here we can set the threshold as $\tau = 3$ and treat all "information" over 3 as anomalies.

For some channels with higher number of states and transitions, the "information" has larger value with frequent fluctuations. For example, Fig. 10a shows the "information" (for window size 3) of a channel in the *WT* dataset with 45 states and 255 transitions where higher "information" values and fluctuations are common.

As presented in previous work [2], DTMC models can be used for detecting anomalies. If an attacker attempts to send a malicious command (unknown state), invalid/harmful command sequence (unknown transitions) or sudden (high or low) interactions of a transition between two states (unknown probability), the model can capture such scenario and raise alarms. However, because all alarms are not malicious, a verification approach is necessary. In the training phase, statistics (mean, maximum, minimum, variance, etc.) of probabilities for a user-defined window size of each transition can be calculated. In the anomaly detection phase, statistics will be calculated for a sliding window for same size and compared with that of calculated in learning phase. And alarm will be raised in case of high dissimilarities.

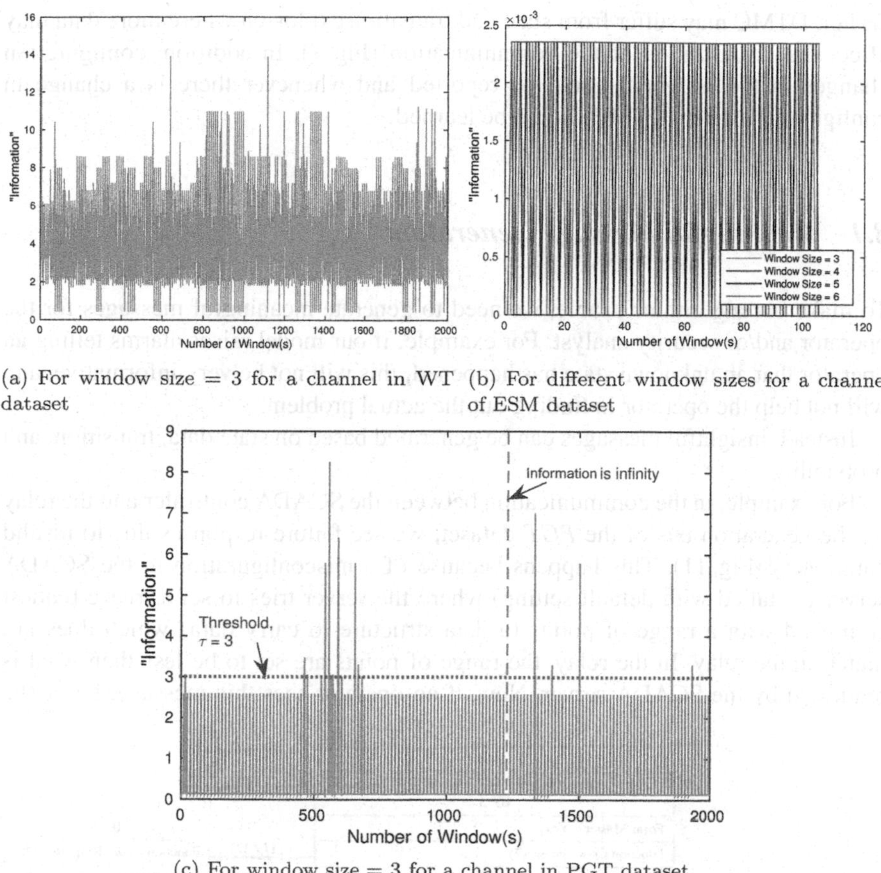

(a) For window size = 3 for a channel in WT dataset

(b) For different window sizes for a channel of ESM dataset

(c) For window size = 3 for a channel in PGT dataset

Fig. 10 "Information" for various datasets. (**a**) For window size = 3 for a channel in WT dataset. (**b**) For different window sizes for a channel of ESM dataset. (**c**) For window size = 3 for a channel in PGT dataset

8 Towards Understanding Industrial Networks

In addition to modeling the complexity of industrial networks and finding anomalies in sequences of commands, our DTMC models can be used to help operators understand their networks.

For example, DTMCs can be used as a visualization method for each channel, and a web interface or desktop based software can be developed to build a graphical representation of DTMCs learned from network data. Moreover, this can be performed in real-time as a new packets introduce new states and their corresponding transitions, and probabilities can be presented in a graph. This will allow an operator to monitor the recent condition of each channel. Thus, he or she will get a better understanding about the communications. However, as mentioned earlier, DTMC may suffer from state and transition explosion where more data may affect the visualization of the communication (Fig. 7). In addition, configuration changes in the network should be reported and whenever there is a change in configuration, a new DTMC should be learned.

8.1 Meaningful Message Generation

To make our algorithms usable, we need to generate meaningful messages for the operator and/or security analyst. For example, if our model raises alarms telling an operator that an unknown state has happened, this will not be very informative, and will not help the operator in finding out the actual problem.

Instead, insightful messages can be generated based on state data, transition, and probability.

For example, in the communication between the SCADA controller and the relay on the generation bus of the *PGT* dataset, we see failure responses due to invalid parameters (Fig. 11). This happens because of a misconfiguration in the SCADA server (installed with default setting) where the server tries to set a freeze request command with a range of points (a data structure to carry data) which does not match in the relay. In the relay, the range of points are set to be less than what is requested by the SCADA server. Now, if an operator sees this message, he or she

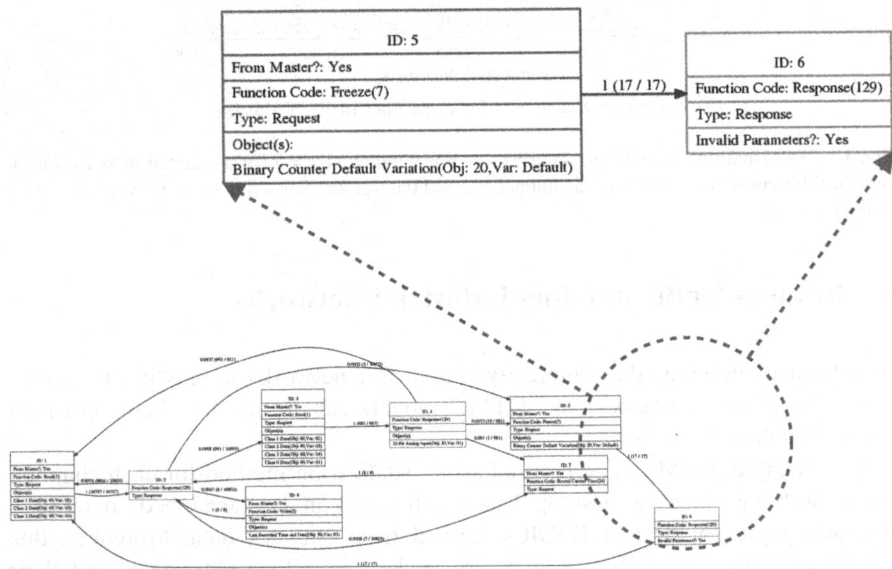

Fig. 11 We found a configuration error in the communication channel between the SCADA controller and the protection relay located at the generation bus of the PGT dataset

knows what an action should be taken (fixing configuration). Thus, DTMC can give us a detailed and meaningful way to generate messages for the operator.

9 Conclusions

In this chapter, we discussed a probabilistic modeling technique for learning the normal behavior of application layer ICS protocols connecting various IIoT elements, as well as its pros and cons.

We find that DTMCs are not free from shortcomings as they may suffer from state and transition explosion which will increase the variance of the model and computation cost. In future work we will explore ways to mitigate these problems (i.e., rebuilding DTMCs with new data, pruning the models based on policies or heuristics etc.)

From our observation, the number of states does not change much in most cases, which is good news for our goal of intrusion detection in industrial networks. However, the number of transitions may increase as we add new data to build the model. The appearance of new states or transitions are not necessarily malicious as they may be generated by configuration changes, operator interactions, networking issues, etc. Thus, a verification mechanism is required to validate unknown states, transitions and probabilities.

Though this kind of model (i.e., DTMC) is helpful for detecting an anomaly at the network protocol level, it cannot detect the manipulation of values from process control variables. For example, a DTMC may find that all states, transitions, and transition's probabilities are valid, while the attacker may send a higher current and voltage values for a protective relay at a generator (with write commands from the controller) potentially damaging the generator. Moreover, in an IIoT ecosystem, there could be a large number of channels, and maintaining a dedicated DTMC for each channel might be expensive. In addition, if the number of fields and their corresponding value ranges increase, the communication pattern using DTMC can be aperiodic. In future, we like to focus on these issues. In addition, we plan to explore the potential to model other IIoT protocols like MQTT (Message Queuing Telemetry Transport), 6LoWPAN (IPv6 over Low power Wireless Personal Area Networks), and AMQP (Advanced Message Queuing Protocol).

Acknowledgements This work was supported by a grant from the United States-Israel Binational Science Foundation (BSF), Jerusalem, Israel and the United States National Science Foundation (NSF) CNS-#1718848. This material was also supported by a grant from the Interdisciplinary Cyber-Research Center at TAU.

References

1. Astarloa A, Bidarte U, Jiménez J, Zuloaga A, Lázaro J (2016) Intelligent gateway for industry 4.0-compliant production. In: IECON 2016-42nd Annual Conference of the IEEE Industrial Electronics Society. IEEE, pp 4902–4907
2. Caselli M, Zambon E, Kargl F (2015) Sequence-aware intrusion detection in industrial control systems. In: Proceedings of the 1st ACM Workshop on Cyber-Physical System Security, pp 13–24
3. Fu Y, Yan Z, Cao J, Koné O, Cao X (2017) An automata based intrusion detection method for Internet of Things. Mob Inf Syst 2017:1–13
4. Goldenberg N, Wool A (2013) Accurate modeling of Modbus/TCP for intrusion detection in scada systems. Int J Crit Infrastruct Prot 6:63–75
5. IEEE: 1815–2012 – IEEE standard for electric power systems communications-distributed network protocol (DNP3) (2012). http://ieeexplore.ieee.org/document/6327578/. Online accessed 06 Mar 2017
6. Kleinman A, Wool A (2014) Accurate modeling of the Siemens S7 scada protocol for intrusion detection and digital forensics. J Digit Forensic Secur Law JDFSL 9(2):37
7. Kleinmann A, Wool A (2015) A statechart-based anomaly detection model for multi-threaded scada systems. In: International Conference on Critical Information Infrastructures Security, pp 132–144
8. Kleinmann A, Wool A (2016) Automatic construction of statechart-based anomaly detection models for multi-threaded scada via spectral analysis. In: Proceedings of the 2nd ACM Workshop on Cyber-Physical Systems Security and Privacy, pp 1–12
9. Modbus Organization (2012) Modbus application protocol specification V1.1b3. Online; Accessed 05 July 2016
10. Modbus Organization (2012) Modbus messaging on TCP/IP implementation guide V1.0b. Online; Accessed 05 July 2016
11. Open DeviceNet Vendor Association (2007) The CIP networks library volume 2: EtherNet/IP adaptation of CIP. http://www.tud.ttu.ee/im/Kristjan.Sillmann/ISP0051%20Rakenduslik%20Andmeside/CIP%20docs/CIP%20Vol2_1.4.pdf
12. Rubio JE, Alcaraz C, Roman R, Lopez J (2017) Analysis of intrusion detection systems in industrial ecosystems. In: 14th International Conference on Security and Cryptography (SECRYPT 2017)
13. Sadeghi AR, Wachsmann C, Waidner M (2015) Security and privacy challenges in industrial Internet of Things. In: 2015 52nd ACM/EDAC/IEEE Design Automation Conference (DAC). IEEE, pp 1–6
14. Tribus M (1961) Thermostatics and thermodynamics: an introduction to energy, information and states of matter, with engineering applications. van Nostrand, London
15. Zarpelão BB, Miani RS, Kawakani CT, de Alvarenga SC (2017) A survey of intrusion detection in Internet of Things. J Netw Comput Appl 84:25–37

Improving Security in Industrial Internet of Things: A Distributed Intrusion Detection Methodology

Giuseppe Bernieri and Federica Pascucci

Abstract The interaction among networking, sensing, and control in the modern industry results in a variety of new devices used in many sectors such as health, energy distribution, and transportation. The on-going tendency of exploiting automation and data exchange in manufacturing technologies leads to the Industry 4.0. The fourth industrial revolution deals with Cyber-Physical Systems, the Internet of Things, cloud computing, and cognitive computing converging towards the Industrial Internet of Things. To be successful, this new era requires innovative paradigms to ensure the security of provided services and connected systems. In the industrial field, the problem gets more complex due to the need of protecting a large attack surface while guaranteeing the availability of the systems and the real-time response to the presence of threats. In this chapter, we perform an analysis of the existing industrial threats and we present a distributed intrusion detection methodology to deal with attacks affecting the Industrial Internet of Things scenarios.

Keywords Industrial Internet of Things · Cyber-physical security · Intrusion detection system · Anomaly detection

1 Introduction

Industrial Control Systems (ICSs) play a vital role in several contexts that are frequently of primary importance for public safety. Therefore, the protection of ICSs is a relevant and challenging topic, both for academic and industrial parts. As can be

G. Bernieri (✉)
Department of Mathematics, University of Padua, Padua, Italy
e-mail: bernieri@math.unipd.it

F. Pascucci
Department of Engineering, Roma Tre University, Rome, Italy
e-mail: federica.pascucci@uniroma3.it

C. Alcaraz (ed.), *Security and Privacy Trends in the Industrial Internet of Things*, Advanced Sciences and Technologies for Security Applications,
https://doi.org/10.1007/978-3-030-12330-7_8

easily understood, the issues related to ICSs security are many and difficult to cope with.

ICSs show specific requirements due to the critical assets they manage [45]. Many differences can be noticed by comparing them with common Information and Communication Technology (ICT) systems. As for any ICT network, confidentiality, integrity and availability of data and systems should always be guaranteed [1]. However, in the ICSs scenario the security priorities can be set as: (1) Availability; (2) Integrity; and (3) Confidentiality, which turns to be the reverse order with respect to the ICT systems. The ICSs order grants that services are always available, given that the information are available. The integrity guarantees the robustness of the information and finally confidentiality protect it from un-authorized access. The scheme used for the priorities reflects the history of ICSs [2]: these systems were traditionally isolated and self-contained, as a consequence, confidentiality was not an issue and was placed bottom in the priority list.

Over the last decade, ICSs have becoming vulnerable to cyber threats due to the increasing number of interconnected distributed devices, sensors, and actuators widely spread in the field. Addressing the security challenges and ensuring ICSs reliability is a fundamental priority. Badly secured ICSs structures and services may be used as entry points for cyber-attacks and expose both data and systems to threats. In fact, as far as the operational and the information domains are becoming more interconnected, cyber-attacks to ICSs exploit weakness in ICT systems as entry points. Several cyber-attacks targeting ICSs, due to the lack of protection and the change in the design paradigm, have been performed. To the best of our knowledge, the first documented attempt to attack an ICS is dated back in 1982: a Trojan caused the explosion of a Trans-Siberian gas pipeline [3]. In 2000, an ex-employee exploited his knowledge to attack the ICS of the Maroochy water services [4]. One of the most complex and sophisticated attacks, discovered so far, was the Stuxnet [5] in 2010, a worm alleged developed to target and sabotage Iran nuclear plants. In 2011, a new malware, named Duqu [6], was reported. More recently, a variant of the malware BlackEnergy [7] shot down the power grid in Ukraine, affecting over 80,000 customers [8].

To protect ICSs, several security architectures have been designed. The National Institute of Standards and Technology (NIST) has published a reference document [9] that provides an overview of ICS threats and vulnerabilities, recommending adequate countermeasures and policies. The International Society for Automation (ISA) also published similar guidelines [10].

Today, the advent of the Industry 4.0 and the introduction of the Industrial Internet of Things (IIoT) are expanding the monolithic self-contained ICS model into geographically distributed systems requiring complex distributed supports and third party services, such as telecommunications and cloud infrastructures. The increasing convergence between the Information Technology (IT) and Operational Technology (OT) implies the transition from centralized control of production processes to decentralization [11]. This allows the development of Cyber-Physical Systems (CPSs), defining a new orientation for the design of the industrial automation and a new trend in the industrial control processes. CPSs, indeed, are composed of intelligent and autonomous local systems able to self-organize. The CPS realizes a paradigm shift in the factory, switching from one model computer integrated manufacturing pyramid [12] to cloud-based infrastructure with highly interconnected components foreseen by the IIoT. This change of perspective within industrial systems leads to several potential benefits: improves efficiency and productivity, speeds up the decision-making, fosters cooperation between the components, and increases scalability. Solutions to protect CPSs have been conceived [13, 14] but the rapid evolution of systems and their heterogeneity in the context of IIoT raises the need to devise ever more innovative distributed security mechanisms, exploiting also the physical behavior of the systems.

The evolution towards the distributed model brings also several challenges: it introduces new technologies to support the distributed deployment, requiring new patterns for protection. As ICSs become distributed, there is the need to prevent risks, both from operational errors and from cyber-attacks compromising their operation or resulting in disasters. The new security challenge can be adequately tackled by creating defense and reaction architectures built from the field level up to information level.

From the lessons learned during the recent years, a possibility to cope with attacks devoted to IIoT devices is to exploit the physical part of the system as probe for detecting anomalous behavior of the whole CPS. It is useful to underline that dealing with the vulnerabilities of a CPS can be quite challenging due to the impossibility of applying a software patch to the system or to the lack of knowledge about the existence of that vulnerability. In fact, what is known as zero-day vulnerability, is a software vulnerability that is unknown to the software designer, to the vendor, and to the user. Until the vulnerability is discovered and mitigated, it can be can exploited by the attacker for affecting the system. As it will be explained in the following sections, a possible detection strategy is based on the continuous monitoring of physical parameters for estimating the presence of cyber-attacks as well as physical failures.

In this chapter, the contribution addresses the challenges brought by the new IIoT paradigm by developing a methodology for distributed intrusion detection for IIoT scenarios. Specifically, in this work we investigate distributed anomaly detection based on fault-tolerant control for modern industrial networks. The main novelty of the approach states in exploiting both the cyber and the physical layer of the IIoT system to identify new vulnerabilities. A zero-day attack, indeed, cannot be easily identified by using the cyber layer, however, since the IIoT devices control physical systems, it can be identified by monitoring the effect of the attack on the plant.

The rest of the chapter is organized as follows. In Sect. 2, an overview of the existing approaches for ICSs security is proposed. In Sect. 3, the Distributed Intrusion Detection System methodology for cyber-physical threats is introduced. In Sect. 4 some preliminary results in emulated environment are discussed and in Sect. 5 some conclusive remarks are drawn.

2 Related Works

Several protection strategies have been presented in the literature for preventing cyber-attacks on ICSs in IIoT. The most promising cyber techniques are based on Intrusion Detection System (IDS) inspecting the network traffic to identify infection or compromised sensors/actuators. On the other hand, many solutions, based on fault tolerant control, exploit the physical layer of the IIoT devices to detect zero-day attacks. A brief review of these approaches is provided in the following.

2.1 IDS in ICS Protection

The IDS collects the network packets to extract information from header and/or payload. According to different features, different taxonomies can be drawn for classifying IDS.

Concerning the techniques proposed for the analysis of network packets, two different approaches are proposed in the literature [15]: passive and active measurements. The former exploits devices that passively observe the traffic network while the latter sends probe packets that will be analyzed for statical purposes. In the ICS framework, active measurements are not used since statistical information can be retrieved by considering the nominal operating condition. Furthermore, the active approach introduces additional traffic load, that hardly copes with the real-time constraints of the ICS.

Concerning the type of data analyzed, two main techniques are implemented. The first one computes statistics about packets (e.g., number of packets/byte, average length). The data can be gathered online by exploiting devices able to capture features contained in layers 2 and 3 of the Open Systems Interconnection (OSI) model [16]. The second one deeply inspects the packets and is generally performed off-line. The Deep Packet Inspection (DPI), also known as layer-7 filtering, processes the payload of the network packets. Since software-based algorithms for string matching are not fast enough for high speed networks, the DPI is usually implemented over specialized devices based on field-programmable gate Arrays, Content Addressable Memory, or Network Processors to reduce delays that may affect the overall system performances. The IDSes based on DPI apply pattern matching algorithms on the payload of a network packet. The main challenges of DPI [17] rely on the complexity of the search algorithm, the size of the pattern database, and the overlapping of multiple patterns.

Different modalities are adopted to collect data and inspect packets. The basic one is the port mirroring: all the network traffic is copied and forwarded to a specific port, thereafter a processor analyzes the traffic to identify malicious activities. The main disadvantage relies in the amount of data: for on-line detection, the IDS needs to be fast enough to process all the data, for off-line detection, the size of the storage devices becomes relevant. A different perspective can be adopted using statistical probes: according to this approach, only specific flows, i.e., a sequence of packets with common source and destination points, are considered. Malicious activities are detected by computing patterns and comparing it with the nominal behavior of the system. The IDS identifies an attack based on the deviation from the baseline pattern. Many vendors provide support for flow extraction and analysis: in 2003 the Internet Engineering Task Force (IETF) has accepted NetFlow, introduced by Cisco, as an industrial standard and proposed the Internet Protocol Flow Information eXport (IPFIX) [18]. Following this decision, other routers from other network equipment vendors support NetFlow: Huawei Technology introduced NetStream [19] and Juniper Networks J.Flow [20].

However, NetFlow represents a macro analytical tool for large volumes of traffic in real time, while an IDS needs to examine packet content at a micro-level. To this aim, different tools have been developed. Snort [21] is the most popular open source packet sniffer and logger used as IDS and is still under active development. The basic feature of Snort is to define rules which can be used to detect a large set of attacks (e.g., memory corruption bugs, stealth network port scans, web attacks). It also provides an integrated real-time logging system connected to email or syslog. It is available for all platforms: the list of pre-defined signatures is continuously updated by the community [21]. Another open source tool is Suricata [22]: it supports signature syntax sharing with Snort, however, the low level architecture of the tool is different.

The Bro network security monitor [23] is an open source security platform. It converts network traffic into series of events that can be analyzed using Bro-scripting to implement site-specific monitoring policies. Linux 2.4.x and later has a framework called Netfilter integrated into the kernel that provides packet filtering functionalities. An effective application layer classifier based on Linux is L7-filter [24]. The DPI library nDPI [25] provides cross-platform capabilities to inspect the application layer of network packets and it is designed to detect network protocols without relying on the port.

2.2 *Exploiting the Physical Layer for IIoT Security*

Concerning the L7-analysis, anomaly detection techniques can be applied to identify unexpected patterns (i.e., outliers, exceptions, faults, etc.). According to [26], most of the approaches exploited to detect anomalies are based on classification (i.e., nearest neighborhood, clustering, statistical inference, and information theoretic). A more interesting approach is derived from the context of the model based fault diagnosis. The ICS schema adopted is depicted in Fig. 1. It is composed of a plant, the controller, and the detection module (i.e, the observer and the detector).

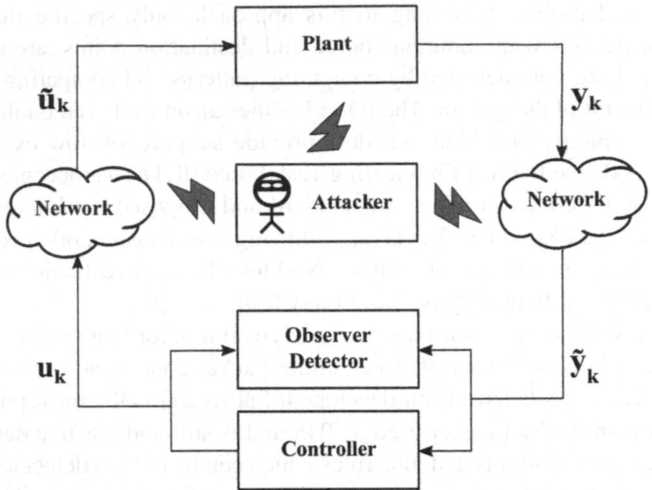

Fig. 1 ICS architecture using model based fault/attack diagnosis

The plant is regarded as a discrete time Linear Time-Invariant (LTI) system, i.e., according to [27]:

$$x_k = Ax_{k-1} + Bu_k, \\ y_k = Cx_k + Du_k, \tag{1}$$

where $x \in \mathbb{R}^n$ is the state vector, $u \in \mathbb{R}^p$ is the input vector, and $y \in \mathbb{R}^q$ is the output vector. Matrices A, B, C, D are real constant matrices and are defined as follows: $A \in \mathbb{R}^{n \times n}$ represents the state transition matrix, $B \in \mathbb{R}^{n \times p}$ is the input matrix, $C \in \mathbb{R}^{q \times n}$ is the output matrix, and $D \in \mathbb{R}^{q \times p}$ is the feedthrough matrix. The latter matrix is considered as $D = 0$ since generally physical systems are strictly causal. In presence of Gaussian noises, the systems described above become:

$$x_k = Ax_{k-1} + Bu_k + w_k, \\ y_k = Cx_k + v_k, \tag{2}$$

where $w(\cdot) \sim N(0, Q)$ and $v(\cdot) \sim N(0, R)$ are identical independent Gaussian noises having zero mean and variance Q and R, respectively. The controller implements the control law, and the detection module is composed of an observer (e.g., a Kalman filter) and a detector designed to identify randomly occurring faults. The detector compares the measurements of the plant with the expected output generated by the observer and set an alarm when they are not compliant according to a predetermined rule.

When the ICS is implemented in a distributed fashion, the integration of physical system devices with the advanced communication networks creates new interactions between these two domains. Consequently, the schema depicted in Fig. 1 becomes:

$$\begin{aligned} x_k &= Ax_{k-1} + B\tilde{u}_k + w_k, \\ y_k &= Cx_k + v_k, \end{aligned} \tag{3}$$

where $\tilde{u}_k$ is the control vector passed through the communication channel:

$$\tilde{u}_k = u_k + u_k^c + u_k^a, \tag{4}$$

where u_k^c and u_k^a are the quantization error and the attack vector on the input, respectively. Same definition can be done for the output vector passing through communication channels:

$$\tilde{y}_k = y_k + y_k^c + y_k^a, \tag{5}$$

where y_k^c and y_k^a are the quantization error and the attack vector on the output, respectively.

According to this approach, the detector can be used to discover cyber-attacks. However, if the anomaly is not random but properly designed, as in case of a cyber-attack, the detector may not be able to detect such an anomaly. Hence, it is worth investigating the ability of commonly used detectors to mitigate the impact of these undetectable attacks.

In the literature, three of the most interesting types of attack that have been addressed are: the Denial of Service (DoS), the deception attack, and the replay attack. Concerning the ICSs, the last two seems more interesting since a DoS is easily identifiable at cyber level.

Deception attack refers to the possibility of modifying the payload of a packet for attacking data integrity and availability while keeping stealthy the intrusion as in [28] and [29]. In [30], a false data injection attack is design for turning CPS unstable based on the knowledge of model eigenvalues. To detect this attack, a coding scheme, revealing the attack, is developed in [31]. In [32], Gaussian noise is injected into the controller for deceiving the alarm monitor and increasing the state estimation error. In [33] and [34], the attacks or induced phenomena in CPSs are described as stochastic behaviors. In more details, a security control scheme with quadratic cost criterion against a class of stochastic deception attacks, whose success probability obeys Bernoulli-distributed, is proposed in [33]. The desired controller gain is obtained based on the input-to-state stability in probability theory. For multi-sensors fusion procedures, different detection scenarios in presence of deception attacks are investigated in [35]. In particular, a novel type of detector is designed using both normal and compromised system information. It is worthy to underline that in order to implement an optimal deception attack, the attacker needs to collect information about system dynamics [36].

Replay attacks can have severe consequences on IIoT systems even without information about the external structure. Basically, by exploiting the replay attack, the attacker tries to influence the close-loop features of IIoT components and degrade the system performances. In [37], a recording-horizon control law is designed for dealing with the replay attack performed in the communication between controller and actuators. Under bandwidth constrains, a secure fusion estimation scheme is proposed for CPSs against replay attack in [38]. A noisy control scheme is proposed in [39], where an additional Gaussian signal is injected inside the actuators. The system performance is penalized for increasing the detection rate for the replay attack. Based on this noisy control countermeasure, a stochastic game approach is developed to decrease the loss in control performance [40].

3 Distributed IDSs for Cyber-Physical Threats in IIoT

Several tools have been developed for the security of legacy industrial networks. With the IIoT advent, these solutions can be considered still useful, however a re-design is mandatory to cope with a widely distributed and interconnected scenario. Innovative devices are connected using different connection layers: for example, cloud and fog computing paradigms are largely deployed in the new IIoT scenarios and it is necessary to address cyber security tools that foresee a distributed detection methodology [41, 42]. To this aim, we present a distributed IDS methodology: it represents an evolution of the anomaly-based IDS developed in [43, 44].

According to the IIoT paradigm, several IDS modules are deployed at a close range to the IIoT physical devices controllers in order to detect cyber or physical threats affecting the specific physical subsystem under consideration [44]. Each IDS module is designed to analyze network traffic and apply advanced detection methodologies that are based on innovative fusion between cyber security approaches and the control theory. In the following section, the IIoT cyber-physical network architecture is introduces to get insights of the distributed detection methodology conceived.

In this context, the Distributed-IDS (D-IDS) represents a behavioral-based anomaly detection system able to detect anomalies from both the cyber and the physical levels. The detection modules can be implemented in a developed fashion, directly coupled with the distributed controllers ($c_1, \ldots, c_n$) managing the IIoT processes, as sketched in Fig. 2.

As happens in legacy ICS scenarios, signature-based detection systems represent limited solution to secure IIoT implementations due to the analysis of the network traffic based on static rules. As already stated, zero-day attacks represent the worst cyber threat for the modern industry and take advance of vulnerabilities that have not been identified. Therefore, it is important to develop advanced and innovative detection tools.

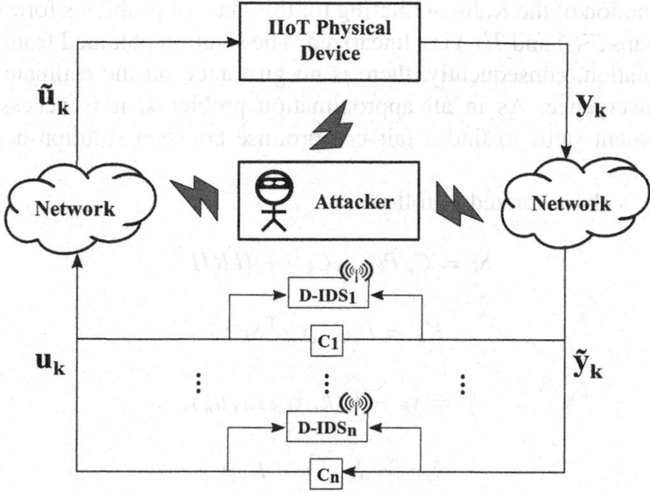

Fig. 2 IIoT schema with distributed IDS modules

To address cyber-physical threats affecting IIoT scenarios, we present two detection techniques that can be implemented in a distributed fashion in the D-IDS:

- An approach conceived using control theory knowledge and the Extended Kalman Filter (EKF) [46];
- An approach that exploits advanced behavioral-based anomaly detection on the cyber-physical data transmitted by IIoT devices in the control zone networks [10].

3.1 *Cyber-Physical Anomaly Detection Based on EKF*

The detection module implemented in this contribution proposes an enhancement with respect to the approaches in the literature: a cyber-physical approach using discrete time nonlinear system model has been adopted and paired with network traffic analysis.

In particular, being the majority of real dynamic systems governed by nonlinear functions, we consider an EKF to estimate the current state of the plant. Accordingly, the system equations become:

$$x_k = \mathcal{F}(k-1, x_{k-1}, u_k) + w_k,$$
$$y_k = \mathcal{H}(k, x_k, u_k) + v_k. \tag{6}$$

The adaptation of the Kalman filtering for this class of problems foresees that the nonlinear maps $\mathcal{F}(\cdot)$ and $\mathcal{H}(\cdot)$ are linearized. The solution obtained from the EKF is an approximation, consequently, there is no guarantee on the estimate optimality and the convergence. As in all approximation problems, it is necessary during the development steps to find a fair compromise between solution accuracy and efficiency.

The EKF is characterized as follows:

$$\begin{aligned} S_k &= C_k P_{k|k-1} {C_k}^{\mathrm{T}} + HRH^{\mathrm{T}}, \\ K_k &= P_{k|k-1} {C_k}^{\mathrm{T}} {S_k}^{-1}, \\ \gamma &= y_k - \mathcal{H}(k, \hat{x}_{k|k-1}, u_k), \\ \hat{x}_{k|k} &= \hat{x}_{k|k-1} + K_k \gamma, \\ P_{k|k} &= (I - K_k C_k) P_{k|k-1}, \\ \hat{x}_{k|k-1} &= \mathcal{F}(k, \hat{x}_{k-1|k-1}, u_k), \\ P_{k|k-1} &= A_k P_{k-1|k-1} {A^{\mathrm{T}}}_k + FQF^{\mathrm{T}}. \end{aligned} \tag{7}$$

where:

$$\begin{aligned} A_k &= \left[\frac{\partial \mathcal{F}(\cdot)}{\partial x}\right]_{x = \hat{x}_{k-1|k-1}}, \\ C_k &= \left[\frac{\partial \mathcal{H}(\cdot)}{\partial x}\right]_{x = \hat{x}_{k|k-1}}. \end{aligned} \tag{8}$$

where P_k is the covariance error matrix, K_k the kalman gain, and S_k the innovation covariance matrix.

The detector $\mathcal{D}_k$ is implemented as:

$$\mathcal{D}_k = \begin{cases} \mathbf{H}_0 & \text{if } r_k \leq \alpha \\ \mathbf{H}_1 & \text{if } r_k > \alpha \end{cases}, \tag{9}$$

where $r_k = y_k - \mathcal{H}(k, \hat{x}_{k|k}, u_k)$ is the residual, α is the threshold computed as the peak norm (i.e., the maximum value assumed by the residual norm under nominal conditions), $\mathbf{H}_0$ is the healthy, and $\mathbf{H}_1$ is the under attack hypothesis, respectively. When $\mathbf{H}_1$ is accepted, the attack detector system triggers an alarm.

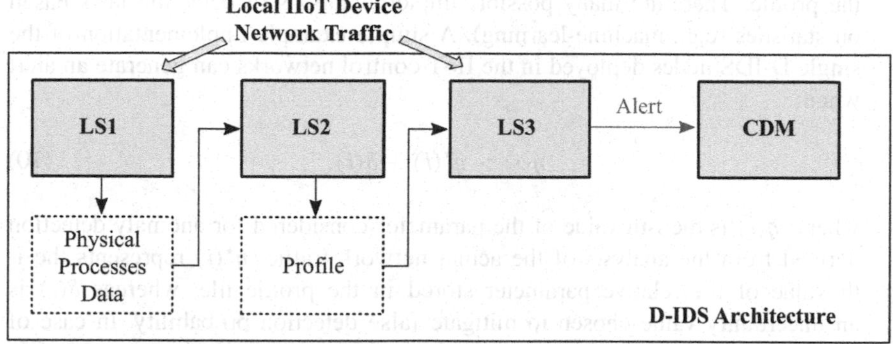

Fig. 3 D-IDS architecture schema

3.2 *Cyber-Physical Behavioral-Based Anomaly Detection*

Complex cyber-attacks can be detected by analyzing the network behavior, according to the processes that are carried out (e.g., read sensors, send commands to actuators). The activities related to the industrial physical processes can be associated to repetitive trends, useful for behavioral schema comparisons.

As sketched in Fig. 3, the D-IDS concept for the advanced behavioral-based detection presented in this work is conceived on three main phases:

- Local Stage 1 (LS1): in this phase the detection system analyses and filters the network traffic to extract specific data of interest from packets for a predefined training period. This stage is executed in a nominal condition without undergoing attacks or anomalous situations. Several iteration of this stage are performed to store physical processes data to better determine the normal behavior patterns. The time required depends on the period of the system. For example, if control operations of a production chain are repeated daily, it is necessary to save the 24-h network behavior;
- Local Stage 2 (LS2): during this phase, the system uses the physical processes data stored during LS1 to generate a profile of the normal network behavior. The methodology applied for the profile generation is an important step for the accuracy of the anomaly detection model application. Respect to the similar detection tools presented in the literature, the D-IDS allows to adapt the proposed distributed system to any network for analysis of the cyber-physical anomalies. Moreover, this module can extract specific cyber-physical data from the network traffic, selecting any traffic characteristic of the protocol under analysis in order to provide ad hoc anomaly detection solutions. This feature represents a valuable option for the security due to the rapidly evolving requirements of IIoT scenarios;
- Local Stage 3 (LS3): once the creation of the cyber-physical profile is completed, this is used for the behavioral-based anomaly detection. The traffic is analyzed and compared at regular intervals with the set of parameters generated within

the profile. There are many possible implementations of detection laws based on statistics (e.g., machine-learning). A simple example implementation of the single D-IDS nodes deployed in the IIoT control networks can generate an alert when:

$$\eta(i) > \eta^{\star}(i) + \delta(i), \tag{10}$$

where: $\eta(i)$ is the i-th value of the parameter considered for anomaly detection derived from the analysis of the actual network traffic, $\eta^{\star}(i)$ represents the i-th value of the relative parameter stored in the profile file, whereas $\delta(i)$ is an uncertainty value chosen to mitigate false detection probability. In case of anomalous behavior identified, alerts are sent to a Central Detection Module (CDM) able to correlate and analyze the alerts coming from the D-IDS nodes deployed in the distributed IIoT systems.

In this scenario, it is possible to address communication of the D-IDS modules exploiting the 5G telecommunication technology and exploiting publish/subscribe protocols (e.g., MQTT), largely used in the Internet of Things world. According to this, the alerts can be easily sent to the CDM for the alerts correlation.

4 Case Study: Water Distribution System

As a case study, we propose a water distribution system scenario. For the sake of simplicity and for better describing the detection methodology, in the following a single IIoT network zone is addressed. The system can be easily extended to the distributed configuration.

The methodology has been validated on the HYDRA testbed [47], a structure emulating a water distribution system. It is a low-cost testbed, designed and developed at the MCIPLab of the University of Roma Tre. The plant of the HYDRA testbed is composed of three tanks and a reservoir, as depicted in Fig. 4: the water in the system flows in the tanks by means of gravity and pumps. Specifically, the proportional Valve $v_{1,2}$ regulates the flow through the serial pipeline connecting Tank 1 to Tank 2. When the Valve $v_{1,2}$ is open, the water cascades due to gravity. The proportional Valve $v_{2,3}$ regulates the flow between Tanks 2 and 3. They are connected in a parallel configuration and the water moves according to Stevino's Law. The reservoir feeds the whole system by means of the centrifugal Pump P_1. It provides water to Tank 1, while the centrifugal Pump P_2 links Tanks 3 and 1. Three manual valves (the Valves v_1, v_2, and v_3) have been added to simulate leakages. Each tank is equipped with sensors to measure the level of the water.

The HYDRA physical system is monitored using a model-based approach, so an Observer/Detector has been implemented. The state transition model of the real

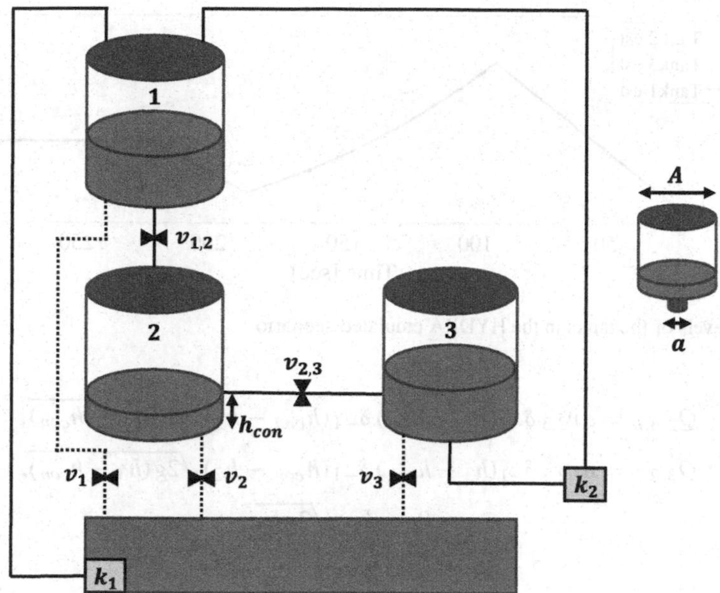

Fig. 4 The HYDRA plant

system behavior is replicated by using the following equations:

$$
\begin{aligned}
A\dot{h_1} &= P_1 + P_2 - Q_{1,2} - Q_{1,0},\\
A\dot{h_2} &= Q_{1,2} - Q_{2,0} - Q_{2,3} - Q_{2,3,h} + Q_{3,2,h},\\
A\dot{h_3} &= -Q_{3,0} + Q_{2,3} + Q_{2,3,h} - Q_{3,2,h} - P_2.
\end{aligned}
\tag{11}
$$

where:

$$
\begin{aligned}
Q_{1,2} &= av_{1,2}\sqrt{2gh_1},\\
Q_{1,0} &= av_1\sqrt{2gh_1},\\
Q_{2,0} &= av_2\sqrt{2gh_2},\\
Q_{3,0} &= av_3\sqrt{2gh_3},\\
Q_{2,3} &= av_{2,3}\,\delta_{-1}(h_2 - h_{con})\,\delta_{-1}(h_3 - h_{con}) \times\\
&\quad \times \mathrm{sign}(h_2 - h_3)\sqrt{2g|h_2 - h_3|},
\end{aligned}
\tag{12}
$$

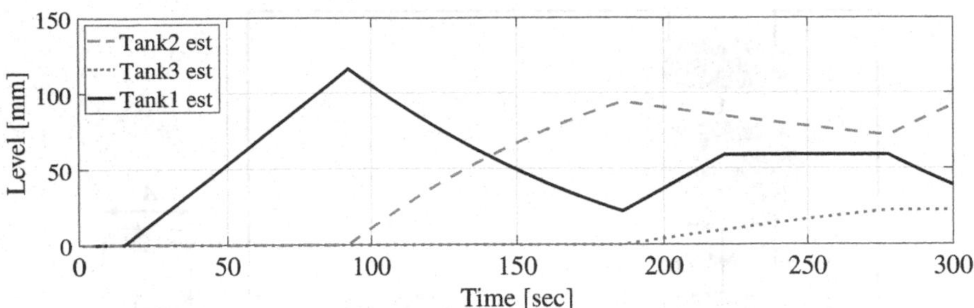

Fig. 5 Levels of the tanks in the HYDRA emulated scenario

$$
\begin{aligned}
Q_{2,3,h} &= av_{2,3}\, \delta_{-1}(h_2 - h_{con})\, \delta_{-1}(h_{con} - h_3)\sqrt{2g(h_2 - h_{con})}, \\
Q_{3,2,h} &= av_{2,3}\, \delta_{-1}(h_3 - h_{con})\, \delta_{-1}(h_{con} - h_2)\sqrt{2g(h_3 - h_{con})}, \\
P_2 &= k_2 a\sqrt{2gh_3}, \\
P_1 &= k_1.
\end{aligned}
$$

$Q_{i,j}$ represents the flow through the Tanks i and j, $\delta_{-1}(\cdot)$ is the step signal, and g the gravitational acceleration.

The sensors measure directly the state variable, thus the system is nonlinear concerning the state transition model and linear considering the observation model. The model based monitoring system is implemented as an Extended Kalman Filter. The residual is computed comparing the real measures with the estimated ones using a threshold.

The control system of the HYDRA testbed is implemented on an industrial Programmable Logic Controller (PLC) and exploits the Modbus/TCP protocol to communicate over the network. The controller emulates the scenario shown in Fig. 5. It starts considering all the tanks empty and lasts 300 s. Moreover, it can be divided into four steps:

- Step 1: Pump P_1 is activated to fill Tank 1 until the level of water in this Tank reaches 110 mm (time range [0 s,90 s]);
- Step 2: Pump P_1 is deactivated and Valve $v_{1,2}$ between Tank 1 and 2 is opened until the level of water in Tank 2 reaches about 90 mm (time range [90 s,180 s]);
- Step 3: Valve $v_{1,2}$ is closed and Valve $v_{2,3}$ between Tanks 2 and 3 is switched on. Pump P_1 is activated to fill Tank 1 until the level of water is above 50 mm (time range [180 s, 270 s]);
- Step 4: Valve $v_{2,3}$ is closed, Valve $v_{1,2}$ is opened (time range [270 s, 300 s]).

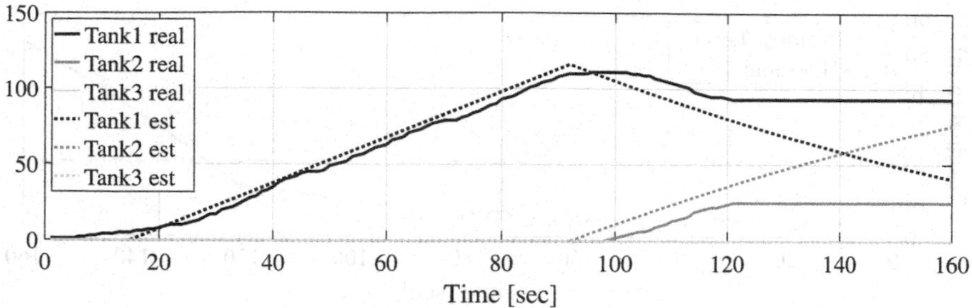

Fig. 6 Levels of the tanks during emulation under replay attack

4.1 Evaluation and Results

The performance of the D-IDS exploiting the approach conceived using control theory has been assessed in emulated scenario. In the following, preliminary results are presented. To this end, the HYDRA testbed has been attacked using Kali Linux [48] and Python scripting.

A simple static replay attack is performed during control routines of the proposed scenario. During Step 2, an ARP poisoning attack is set up so to corrupt the links between IP addresses and MAC addresses in the ARP table of the hosts. In this way, the attacker is able to realize a Man In The Middle attack between the controller and the plant. At time instant $t = 125$ s, the malicious agent starts to forward a constant value to the controller (i.e., the levels recorded when the attack starts), as shown in Fig. 6. On the other hand, the observer, considering the state of the actuators (i.e., valves and pumps), computes the expected measurements and compare them with the corrupted ones, also reported in Fig. 6.

Figure 7 shows the residual analysis for Tank 1 and Tank 2. As it can be seen, the detection system clearly identifies the anomalies few seconds after the beginning of the attack. The residuals for Tank 1 and Tank 2 are reported on the figure in solid blue line, while the red dotted lines represents the thresholds considered. It is worth noticing that during Step 2, Tank 3 is empty: the malicious measurement forwarded by the attacker is compliant with the Tank 3 status so the threshold is not violated.

5 Conclusion

The CPSs related to IIoT scenarios are gaining the attention of malicious organizations due to their fundamental role in several contexts that are essential for the functioning of a society and its economy. Examples of these scenarios include smart grids, water plants, autonomous automobile systems, medical monitoring, process control systems, robotics systems, etc.

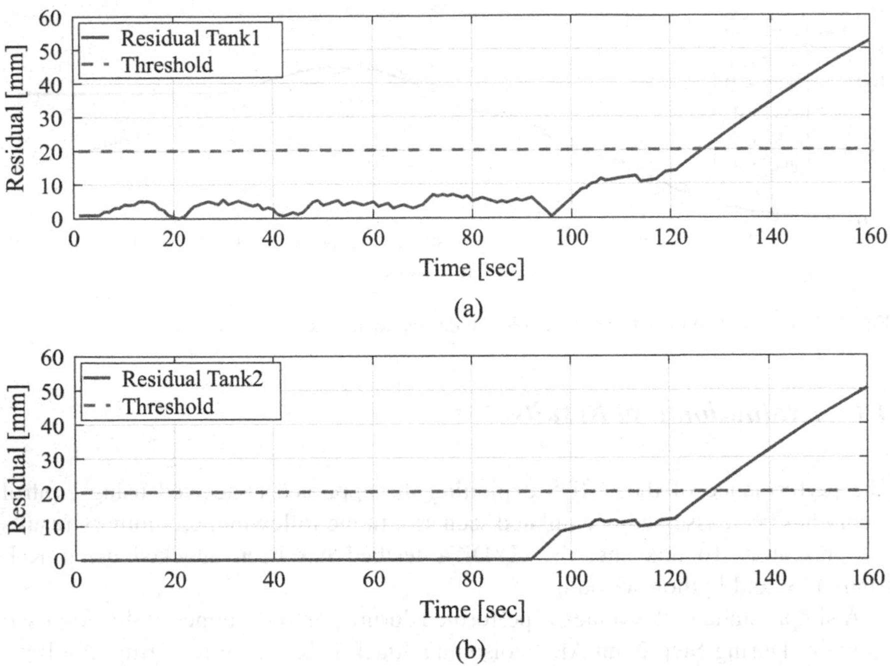

Fig. 7 Residuals of Tank 1 and Tank 2 during the replay attack. (**a**) Residuals of Tank 1 during the replay attack. (**b**) Residuals of Tank 2 during the replay attack

Modern ICS are based on the cooperation among several smart, autonomous, local systems that are able to self-organize. A central unit coordinating the whole system may be present. IIoT represents a strong innovation in the industrial framework, switching from the centralized model to the cloud-based infrastructure with highly interconnected components. This change of perspective leads to several potential benefits: improves efficiency and productivity, speeds up the decision-making, fosters cooperation between the components, and increases scalability. However, it leads also to severe security concerns. A distributed ICS in the IIoT scenario should be equipped with novel smart systems able to prevent as well as to mitigate risks resulting both from operational errors and from cyber-attacks.

In this chapter, we presented a brief analysis of the state of the art methodologies developed for increasing the security of IIoT. Moreover, we proposed a possible detection methodology for detecting cyber-physical threats in distributed IIoT scenarios. The basic idea of the D-IDS is to exploit the physical components for detecting the presence of cyber-attacks. It is based on a two-layers system (physical and cyber) for highlighting network anomalies affecting the physical plant.

The system is characterized by low computational complexity and it is easy to be implemented. It has been tested on a testbed emulating a water distribution plant. The achieved performances demonstrated the validity of the proposed architecture.

Ongoing work is devoted to the implementation of the whole system and to apply it to a distributed system. Furthermore, our goal will be the test of the system on a real infrastructure and to the understanding of the capacity of the detection methodology to highlight attack propagation in order to build a resilient control system.

Acknowledgements This work was funded by the Horizon 2020 ATENA Project (www.atena-h2020.eu). This project has received funding from the European Union's Horizon 2020 Research and Innovation Programme, under Grant Agreement no 700581. The authors would thank Matteo Sebasta for his support.

References

1. Bishop MA (2002) The art and science of computer security. Addison-Wesley Longman Publishing Co., Inc., Boston
2. Bailey D, Wright E (2003) Practical SCADA for industry. Elsevier Ltd., Amsterdam
3. Markoff J (2009) Old trick threatens the newest weapons. The New York Times. http://www.nytimes.com/2009/10/27/science/27trojan.html. Cited 22 Aug 2018
4. Slay J, Miller M (2008) Lessons learned from the maroochy water breach. In: Goetz E, Shenoi S (eds) Critical Infrastructure Protection, ICCIP 2007. IFIP International Federation for Information Processing, vol 253. Springer, Boston
5. Langner R (2013) To kill a centrifuge: a technical analysis of what Stuxnet's creators tried to achieve. The Langner Group https://www.langner.com/wp-content/uploads/2017/03/to-kill-a-centrifuge.pdf. Cited 22 Aug 2018
6. Symantec (2011) W32.Duqu – the precursor to the next Stuxnet (Version 1.4) http://www.symantec.com/content/en/us/enterprise/media/security_response/whitepapers/w32_duqu_the_precursor_to_the_next_stuxnet.pdf Cited 22 Aug 2018
7. F-Secure Labs (2014) BlackEnergy & Quedagh: the convergence of crimeware and APT attacks. https://www.f-secure.com/documents/996508/1030745/blackenergy_whitepaper.pdf. Cited 22 Aug 2018
8. E-ISAC (2016) Analysis of the cyber-attack on the Ukrainian power grid. https://ics.sans.org/media/E-ISAC_SANS_Ukraine_DUC_5.pdf. Cited 22 Aug 2018
9. Stouffer K, Lightman S, Pillitter V, Abrams M, Hahn A (2015) Guide to Industrial Control System (ICS) security, Revision 2. NIST Special publication 800-82 https://nvlpubs.nist.gov/nistpubs/SpecialPublications/NIST.SP.800-82r2.pdf. Cited 22 Aug 2018
10. ISA (2007) ANSI/ISA–62443-1-1 (99.01.01)–2007 security for industrial automation and control systems part 1: terminology, concepts, and models. Int Soc Autom
11. Soldatos JK, Gusmeroli S, Maló P, Di Orio G (2016) Internet of Things applications in future manufacturing. In: Vermesan O, Friess P (eds) Digitising Industry – Internet of Things Connecting the Physical, Digital and Virtual Worlds. River Publishers, Aalborg, pp 152–182
12. Groover MP (2001) Automation, production systems and computer integrated manufacturing. Prentice Hall, Englewood Cliffs
13. Mitchell R, Ing-Ray C (2014) A survey of intrusion detection techniques for cyber-physical systems. ACM Comput Surv (CSUR) 46(4). https://doi.org/10.1145/2542049
14. Han S, Xie M, Chen H, Ling Y (2014) Intrusion detection in cyber-physical systems: techniques and challenges. IEEE Syst J 8(4):1052–1062. https://doi.org/10.1109/JSYST.2013.2257594
15. Mahmood A, Leckie C, Hu J, Tari Z, Atiquzzaman M (2010) Network traffic analysis and SCADA security. In: Stavroulakis P, Stamp M (eds) Handbook of information and communication Security. Springer, Berlin/Heidelberg

16. Zimmermann B (1980) OSI reference model – the ISO model of architecture for open systems interconnection. IEEE Trans Commun 28(4):425–432. https://doi.org/10.1109/TCOM.1980.1094702
17. Rossi D, Sottile E (2009) Sherlock: a framework for P2P traffic analyis. In: IEEE Ninth International Conference on Peer-to-Peer Computing, Seattle, pp 321–330. https://doi.org/10.1109/P2P.2009.5284490
18. Boschi E, Mark L, Quittek J, Stiemerling M, Aitken P (2008) RFC 5153: IP flow information export (IPFIX) implementation guidelines. IETF. https://tools.ietf.org/html/rfc5153. Cited 22 Aug 2018
19. L. Huawei Technologies (2012) NetStream (integrated) technology white paper. http://e.huawei.com/au/marketing-material/onLineView?MaterialID=%7B7B805301-E381-4690-823A-4E73A0411E29%7D. Cited 22 Aug 2018
20. Juniper Networks, Flow monitoring feature guide. https://www.juniper.net/documentation/en_US/junos/information-products/pathway-pages/solutions/flow-monitoring/flow-monitoring.html. Cited 22 Aug 2018
21. Roesch M (1999) Snort – lightweight intrusion detection for networks. In: Proceedings of the 13th USENIX Conference on System Administration (LISA '99). USENIX Association, Berkeley, pp 229–238
22. Suricata (2018) Suricata user guide. https://suricata.readthedocs.io/en/suricata-4.0.5/#suricata-user-guide. Cited 22 Aug 2018
23. Bro, Bro manual. https://www.bro.org/sphinx/index.html. Cited 22 Aug 2018
24. Ayuso PN (2006) Netfilter's connection tracking system. USENIX magazine. http://people.netfilter.org/pablo/docs/login.pdf. Cited 22 Aug 2018
25. ntop (2013) nDPI – quick start guide. https://www.ntop.org/wp-content/uploads/2013/12/nDPI_QuickStartGuide.pdf. Cited 22 Aug 2018
26. Kohnen C, Uberall C, Adamsky F, Rakocevic V, Rajarajan M, Jager R (2010) Enhancements to Statistical Protocol IDentification (SPID) for self-organised QoS in LANs. In: International Conference on Computer Communications and Networks (ICCCN), Zurich
27. Ding SX (2013) Model-based fault diagnosis techniques. Advances in industrial control. Springer, London. https://doi.org/10.1007/978-1-4471-4799-2
28. An LW, Yang GH (2018) Data-driven coordinated attack policy design based on adaptive L2-gain optimal theory. IEEE Trans Autom Control 63(6):1850–1856. https://doi.org/10.1109/TAC.2017.2753466
29. Bai CZ, Gupta V, Pasqualetti F (2017) On Kalman filtering with compromised sensors: attack stealthiness and performance bounds. IEEE Trans Autom Control 62(12):6641–6648. https://doi.org/10.1109/TAC.2017.2714903
30. Mo Y, Sinopoli B (2010) False data injection attacks in control systems. In: First Workshop on Secure Control Systems, CPS Week
31. Miao F, Zhu Q, Pajic M, Pappas GJ (2017) Coding schemes for securing cyber-physical systems against stealthy data injection attacks. IEEE Trans Control Netw Syst 4(1):106–117. https://doi.org/10.1109/TCNS.2016.2573039
32. Bai CZ, Gupta V, Pasqualetti F (2017) Data-injection attacks in stochastic control systems: detectability and performance tradeoffs. Automatica 82:251–260. https://doi.org/10.1016/j.automatica.2017.04.047
33. Ding D, Wang Z, Han QL, Wei G (2018) Security control for discrete-time stochastic nonlinear systems subject to deception attacks. IEEE Trans Syst Man Cybern: Syst 48(5):779–789. https://doi.org/10.1109/TSMC.2016.2616544
34. Zhang D, Song H, Yu L (2017) Robust fuzzy-model-based filtering for nonlinear cyber-physical systems with multiple stochastic incomplete measurements. IEEE Trans Syst Man Cybern: Syst 47(8):1826–1838. https://doi.org/10.1109/TSMC.2016.2551200
35. Li Y, Shi L, Chen T (2017) Detection against linear deception attacks on multi-sensor remote state estimation. IEEE Trans Control Netw Syst. https://doi.org/10.1109/TCNS.2017.2648508
36. Teixeira A, Shames I, Sandbergand H, Johansson KH (2015) A secure control framework for resource-limited adversaries. Automatica 51:135–148. https://doi.org/10.1016/j.automatica.2014.10.067

37. Zhu M, Martinez S (2014) On the performance analysis of resilient networked control systems under replay attacks. IEEE Trans Autom Control 59(3):804–808. https://doi.org/10.1109/TAC.2013.2279896
38. Chen B, Ho DW, Hu G, Yu L (2018) Secure fusion estimation for bandwidth constrained cyber-physical systems under replay attacks. IEEE Trans Cybern 4(6):1862–1876. https://doi.org/10.1109/TCYB.2017.2716115
39. Mo Y, Chabukswar R, Sinopoli B (2014) Detecting integrity attacks on scada systems. IEEE Trans Control Syst Technol 22(4):1396–1407. https://doi.org/10.1109/TCST.2013.2280899
40. Miao F, Pajic M, Pappas GJ (2013) Stochastic game approach for replay attack detection. In: 52nd IEEE Conference on Decision and Control, Florence, pp 1854–1859. https://doi.org/10.1109/CDC.2013.6760152
41. Gazis V, Leonardi A, Mathioudakis K, Sasloglou K, Kikiras P, Sudhaakar R (2015) Components of fog computing in an industrial Internet of Things context. In: 12th Annual IEEE International Conference on Sensing, Communication, and Networking – Workshops (SECON Workshops), Seattle, WA, pp 1–6. https://doi.org/10.1109/SECONW.2015.7328144
42. O'Donovan P, Gallagher C, Bruton, K, O'Sullivan DTJ (2018) A fog computing industrial cyber-physical system for embedded low-latency machine learning Industry 4.0 applications. Manuf Lett 15(part B):139–142. https://doi.org/10.1016/j.mfglet.2018.01.005
43. Bernieri G, Pascucci F, Lopez J (2017) Network anomaly detection in critical infrastructure based on mininet network simulator. In: ITASEC
44. Etchevés Miciolino E, Setola R, Bernieri G, Panzieri S, Pascucci F, Polycarpou MM (2017) Fault diagnosis and network anomaly detection in water infrastructures. IEEE Des Test 34(4):44–51. https://doi.org/10.1109/MDAT.2017.2682223
45. Etchevés Miciolino E, Bernieri G, Pascucci F, Setola R (2017) Monitoring system reaction in cyber-physical testbed under cyber-attacks. Comput Electr Eng 59:86–98. https://doi.org/10.1016/j.compeleceng.2017.02.010
46. Ljung L (1979) Asymptotic behavior of the extended Kalman filter as a parameter estimator for linear systems. IEEE Trans Autom Control 24(1):36–50. https://doi.org/10.1109/TAC.1979.1101943
47. Bernieri G, Del Moro F, Faramondi L, Pascucci F (2016) A testbed for integrated fault diagnosis and cyber security investigation. In: International Conference on Control, Decision and Information Technologies (CoDIT), St. Julian's, pp 454–459. https://doi.org/10.1109/CoDIT.2016.7593605
48. Offensive Security (2018) Introduction to Kali Linux. https://docs.kali.org/category/introduction. Cited 22 Aug 2018

Who's There? Evaluating Data Source Integrity and Veracity in IIoT Using Multivariate Statistical Process Control

Iñaki Garitano, Mikel Iturbe, Enaitz Ezpeleta, and Urko Zurutuza

Abstract The security landscape in Industrial settings has completely changed in the last decades. From the initial primitive setups, industrial networks have evolved into massively interconnected environments, thus developing the Industrial Internet of Things (IIoT) paradigm. In IIoT, multiple, heterogeneous devices collaborate by collecting, sending and processing data. These data-driven environments have made possible to develop added-value services based on data that improve industrial process operation. However, it is necessary to audit incoming data to determine that the decisions are made based on correct data. In this chapter, we present an IIoT Anomaly Detection System (ADS), that audits the integrity and veracity of the data received from incoming connections. For this end, the ADS includes field data (physical qualities based on data) and connection metadata (interval between incoming connections and packet size) in the same anomaly detection model. The approach is based on multivariate statistical process Control and has been validated using data from a real water distribution plant.

Keywords Industrial internet of things · Anomaly detection · Source trust

1 Introduction

The field of industrial communications has evolved vastly in the past decades. From the basic and isolated first Programmable Logic Controllers (PLCs) created in the 1960s, to the totally interconnected Industrial Internet of Things (IIoT). In this new paradigm, industrial devices, sensors and servers collaborate to provide added-value services based on the gathered field data. This collaboration is generally based on communication over potentially insecure networks, such as the Internet. In

---

I. Garitano (✉) · M. Iturbe · E. Ezpeleta · U. Zurutuza
Mondragon Unibertsitatea, Mondragón, Gipuzkoa, Spain
e-mail: igaritano@mondragon.edu; miturbe@mondragon.edu; eezpeleta@mondragon.edu; uzurutuza@mondragon.edu

C. Alcaraz (ed.), *Security and Privacy Trends in the Industrial Internet of Things*, Advanced Sciences and Technologies for Security Applications, https://doi.org/10.1007/978-3-030-12330-7_9

addition of the inherent security risks of communicating over an insecure channel, environmental issues or other technical threats that compromise data sources such as sensors, can compromise the veracity and therefore, the validity of the received data.

Thus, it is necessary to secure the communications and to provide IIoT system operators information about the veracity of the data and the necessary operational awareness. Notable security incidents in the industrial setting such as Stuxnet [1] or Irongate [2] have demonstrated the danger of masking the real status of the industrial process to operators.

IIoT, as a new paradigm, is still developing in terms of security mechanisms. However, when designing security solutions for industrial applications, it is necessary to consider the particularities of these environments, that differ from the ones in traditional IT-based networks. But, at the same time, these particularities can be leveraged to build specialized security mechanisms.

In particular, as in industrial networks most of the traffic and data is generated by automated processes, Anomaly Detection Systems (ADSes) are particularly well suited for the task of attack detection [3]. ADSes are considered a subset of Intrusion Detection Systems (IDSes). In the most commonly deployed type of IDS, the signature-based ones, the system monitors the network to find known traces of malicious activity, known as signatures. On the contrary, ADSes are monitoring systems that focus on finding patterns that deviate from normal system behavior. In the case of signature-based IDSes, they are only effective against known threats whose traits are registered in the signature database. If an unknown attack is happening, signature-based IDSs will not be able to detect. In the case of the ADSes, however, it is possible to detect unknown attacks, as they are not looking for any known pattern, but just deviations from normality, ADSes are able to detect unknown attacks. However, this comes to a higher yield of false positives than signature-based IDSes. (IDSs).

In this chapter we present a two-fold contribution. First, we present a system where connection data is enriched by adding metadata to it using Big Data scalable framework suitable for real IIoT environments. And second, we present an ADS, based on Multivariate Statistical Process Control (MSPC) that monitors incoming IIoT connections and leverages the original and enriched connection data in order to detect and diagnose anomalies in it. These anomalies can range from an attacker performing Man-in-the-Middle (MitM) attacks, sensor malfunctioning or communication issues. The approach is validated using data from a real water distribution plant.

The rest of the chapter is organized as follows: Sect. 2 presents the related works in the field of IIoT security. Section 3 covers MSPC, the technique upon the ADS is based on. Section 4 describes the general approach while Sects. 5 and 6 discuss the experimental setup and the obtained results, respectively. Finally, Sect. 7 extracts some final remarks and concludes the chapter.

2 Related Works

Due to the mission critical role some industrial systems have, and after the security by obscurity practice has been rendered obsolete fading, research in the field of industrial security has received wide attention from the scientific community. Among different proposals, the field of anomaly detection is a particularly active one, both at a network level [4] and at the field level [5].

In the more recent field of IIoT, where many assets are connected to or through the Internet, it is necessary to provide security. Consequently, many works have been presented for this end [6].

For instance, Sajid et al. [7] analyzed the current state of the art and the future challenges of IoT based SCADA systems in a cloud environment. Furthermore, they also published a collection of recommendations and best practices to secure this kind of environments: Network segregation, monitoring and analysis of the device's activities, log analysis, file integrity monitoring, network traffic analysis, memory dump analysis, updating periodically, testing vulnerabilities regularly, proxy solutions, and use of malicious activity detector.

In the case of anomaly detection, there have been several proposals for (I)IoT: Rajasegarar et al. [8] presented a distributed algorithm for anomaly detection, based on data models created by a hyperspherical cluster. The system was implemented and tested in a real wireless network. Similarly, Thanigaivelan et al. [9] presented another distributed anomaly detection system. In this case, the proposal is based on a neighbor monitoring system to identify non-common behaviors. At the device level, the proposal of Summerville et al. [10] is centered in the lightweight monitoring of the embedded systems that lack proper computational power to deploy a full-fledged ADS. At the network level, Stiawan et al. [11] developed a monitoring solution for anomaly detection. The proposal was able to detect and diagnose the cause of the anomaly on the fly.

Node trust in IIoT remains an open challenge. As nodes are scattered and accessible to attackers it might be difficult to assess the amount of trust users put from data coming from a potentially compromised node. As such, trust for IoT environments is an active research field [12]. Bao and Chan [13] use honesty, cooperativeness and community-interest as reference parameters for evaluating node trust, assuming that most IoT devices are human-related. Mahalle et al. [14] proposed a trust-based access control system, focused on dynamic and decentralized IoT environments. In order to establish a relationship between devices, a trust score is calculated taking into account the experience, knowledge and recommendations, then the score is matched to a certain privilege and finally the access is granted with a set of credentials. Other proposals in the field include, trust evaluation based on a three-layer fuzzy approach [15]; a combination of location-aware and identity-aware information and authentication history [16]; hierarchical trust model [17]; a context-aware system based on past behavior of the device [18, 19]; or a dynamic mechanism for reputation selection [20].

Our approach aims to complement the previous works in the sense that provides a centralized anomaly detection system which fulfills the following objectives, which have not yet been addressed in previous proposals: (1) Detect network anomalies by monitoring incoming connection patterns on the data acquisition server side, (2) detect data anomalies by monitoring obtained data and, (3) reduce IIoT device computational requirements by placing all the data processing on the centralized server side.

3 Multivariate Statistical Process Control

This section presents the core technique used in our ADS, Multivariate Statistical Process Control (MSPC). MSPC [21] is a process monitoring methodology that relies on the use of multivariate control charts to detect unexpected changes in the monitored process. It is an extension of the univariate Statistical Process Control (SPC) approach. This approach has already been proposed as a viable solution for anomaly detection for IT systems [22] and for the detection and diagnosis of field-level anomalies in process control systems [23].

Stoumbos et al. [24] define SPC as a "set of statistical methods used extensively to monitor and improve the quality and productivity of manufacturing processes and service operations. SPC primarily involves the implementation of control charts, which are used to detect any change in a process that may affect the quality of the output".

Figure 1 shows an example of a control chart with 99% confidence level control limit. Under normal process operating conditions, 99% of all the points will fall under the control limit. In that case, we consider that the process is in a state

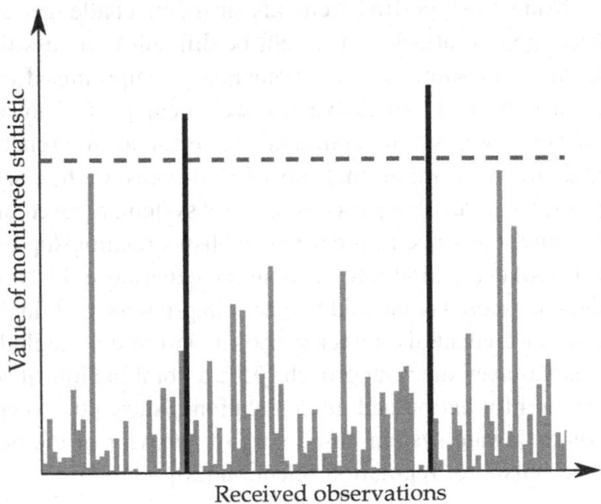

Fig. 1 Example of a control chart. The control limits is presented for a 99% (dashed line) confidence level. That is, 99% of the values should be registered below the control limit

of *statistical control*. It is important not to confuse this term with other similar expressions, such as control loop or automatic feedback control, as they refer to different concepts. Statistical control refers to the state of the process where only common causes of variation are present [21].

The existence of consistent observation series over the established control limit is likely to be attributed to a new special cause. In the case of a physical process, this variation source may be attributed to attacks or process disturbances, i.e., an anomaly.

The univariate nature of SPC means that only a single variable is monitored and visualized in a control chart. However, industrial processes and IIoT environments are multivariate by nature, as many process variables are observed in a plant (e.g., temperatures, pressures, volumes or distances). As monitoring all variables with SPC would be impractical, only a few of them are monitored, generally the ones related to product quality (e.g., purity of the produced chemicals).

Nevertheless, the monitoring of a few quality-related variables is impractical. The approach does not consider the information that other process variables give. For instance, the diagnosis of an anomalous event is complicated, as it relies on expert knowledge and a one-at-a-time inspection of process variables [25].

MSPC aims to solve these problems by providing tools to monitor all measured variables in an efficient manner. In that sense, MSPC does not only monitor the evolution of variable magnitude but also the evolution of the relationship it has to other variables. For this end, a main technique that MSPC uses is Principal Component Analysis (PCA).

3.1 PCA-Based MSPC

Let us consider process historical data as a $\mathbf{X} = N \times M$ two-dimensional dataset, where M variables are measured for N observations. PCA transforms the original M-dimensional variable space into a new subspace where variance is maximal. It converts the original variables into a new set of uncorrelated variables (generally fewer in number), called Principal Components (PCs) or Latent Variables.

For a mean-centered and auto-scaled[1] $\mathbf{X}$ and A PCs, PCA follows the next expression:

$$\mathbf{X} = \mathbf{T_A P_A^t} + \mathbf{E_A} \tag{1}$$

where $\mathbf{T_A}$ is the $N \times A$ score matrix, that is, the original observations represented according to the new subspace; $\mathbf{P_A^t}$ is the $M \times A$ loading matrix, representing the linear combination of the original variables that form each of the PCs; finally, $\mathbf{E_A}$ is the $N \times M$ matrix of residuals.

[1]Normalized to zero mean and unit variance.

In PCA-based MSPC, both the scores and the residuals are monitored, each in a separate control chart [26]. On the one hand, to comprise the scores, the D-statistic or Hotelling's T^2 [27] is monitored. On the other hand, in the case of the residuals, the chosen statistic is the Q-statistic or SPE [28].

For an n observation, both statistics are computed as follows:

$$D_n = \sum_{a=1}^{A} \left(\frac{t_{an} - \mu_{\mathbf{t}_a}}{\sigma_{\mathbf{t}_a}} \right)^2 ; \; Q_n = \sum_{a=1}^{A} (e_{nm})^2 \tag{2}$$

where t_{an} is the score of the observation in the a-th PC, $\mu_{\mathbf{t}_a}$ and $\sigma_{\mathbf{t}_a}$ represent the mean and standard deviation of the scores of the a-th PC in the training data respectively and e_{nm} stands for the residual value corresponding to the m-th variable.

D and Q statistics are computed for each of the observations in the anomaly-free training data, and control limits are set for each of the two charts. Training data is previously inspected through Exploratory Data Analysis to remove existing outliers that could change D and Q values. Later, these statistics are also computed for incoming data and plotted in the control chart. When an unexpected change occurs in one (or more) of the original measured M variables, one (or both) of these statistics will go beyond control limits. Thus, an M-dimensional monitoring scenario is effectively converted into a two-dimensional one.

An event is considered anomalous when three consecutive observations surpass the 99% confidence level control limit in either of the monitored statistics [29]. Leaving some of the observations out of bounds (1% of the observations with a control limit set on the 99% confidence level) improves the performance of the control charts in the monitoring phase [22, 29].

Once an anomaly has been detected, anomaly diagnosis in MSPC is generally carried out using contribution plots [25]. These plots show the contribution of the original measured variables to an anomalous event. Details of the calculation and analysis of contribution plots can be found in the work of Alcala and Qin [30].

In this work, we use oMEDA plots [31] to diagnose the anomaly causes by relating anomalous events to the original variables. In essence, oMEDA plots are bar plots where the highest or lowest values in a set of variables reflect their contribution to a group of observations. Therefore, when computed on a group of observations within an anomalous event, the most relevant variables related to that particular event will be the ones with the highest and lowest bars. Though similar, one of the main differences of oMEDA plots with traditional contribution plots is that the oMEDA plots are capable of comparing different sets of observations whereas traditional plots can only compute a single set of them. In that sense, oMEDA plots can be considered an extension to the contribution plots. In this case, to compute oMEDA we first define a dummy variable, $\mathbf{d}$, a vector of length N, in which the anomalous observations that are to be computed are marked with 1, leaving the rest as 0.

For a set of observations marked in **d**, oMEDA is computed as follows:

$$d^2_{A,(i)} = \frac{1}{N} \cdot \left(2 \cdot \sum_{(i)}^{d} - \sum_{A,(i)}^{d} \right) \cdot \left| \sum_{A,(i)}^{d} \right| \tag{3}$$

where $\sum_{(i)}^{A}$ and $\sum_{A,(i)}^{A}$ represent the weighted sum of elements for variable i in $\mathbf{X}$ and its projection $\mathbf{X}_A$ according to the weights in **d**, respectively. Larger absolute values of d^2 will indicate a larger contribution of that variable in causing the anomalous observation.

4 Proposed Approach

In this section, we describe the IIoT-oriented ADS based on MSPC. IIoT environments are, in essence, multivariate environments, where different physical qualities are constantly monitored. Generally, this monitoring is performed using site collectors, such as sensors, that are used to gather field data that is later sent through an edge device or IIoT gateway to the processing cloud. The gateways are the hardware devices that make possible sending the information towards the cloud. In the IIoT cloud backend, where the processing units are, data from different IIoT gateways is collected and processed.

With this our approach performs the anomaly detection in four main phases, depicted in Fig. 2:

Data enrichment When a field reading arrives to the IIoT backend cloud, some network-level statistics are computed, such as network packet size and time since the last reading was received. This newly created data is appended to the dataset, creating a cyber-physical hybrid dataset: Physical readings from the field and cyber processed network data. Later, the computed data can be used to detect network-level anomalies, such as abnormal latencies.

Creation of anomaly detection model Once the enriched dataset has been constructed, the ADS builds the MSPC model for detection with a dataset at rest. In order to discard outliers (i.e., anomalies registered during training), we perform a manual exploratory analysis for dataset cleaning. After, the D and Q statistic limits are calculated.

Anomaly detection Once the model has been built and the limits established, D and Q are calculated for each incoming reading, and checked whether the reading is out-of-bounds or not. If three consecutive out-of-bounds readings are registered, the event is flagged as anomalous.

Anomaly diagnosis After an anomaly has been flagged, the oMEDA vector is computed over the first out-of-bounds reading to examine the contribution of each of the variables to the anomalous event. Based on the oMEDA plot, the operator can check whether there has been some issue at the plant level

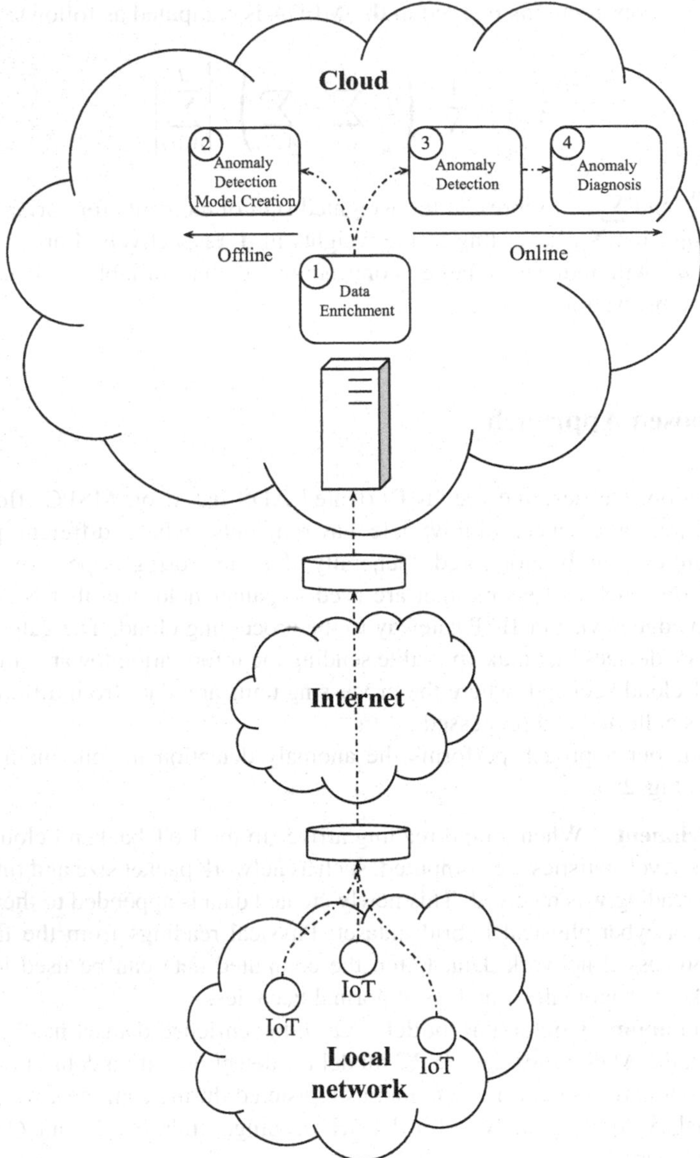

Fig. 2 Phases of the developed anomaly detection system

(anomalous readings) or with the network (unstable network connection). Repeatedly anomalous or unstable sources can be labeled as untrusted and compared to the ones that report few anomalies.

As IIoT is a scalable paradigm, where new sensor readings can be added or new functionalities deployed, it requires scalability for the increasing data complexity, the ADS has been developed using Big Data tools: Apache Kafka for the data enrichment phase, and Apache Spark [32] for the different phases of the anomaly detection and diagnosis.

5 Experimental Setup

This section describes the hardware and software resources, the network topology and the experimental setup considering anomaly and normality conditions along with the proposed experiments.

5.1 Architecture

The main objective is to emulate a common industrial setup where different locally installed sensors forward information to private/public cloud servers following a periodic pattern. Sensors can be either connected to Internet or not and thus, information can be forwarded through the Internet or by using private communication means.

Three nodes, each one emulating either an IoT device or a server and two different communication networks compose the necessary hardware equipment for this setup. While one of the servers captures and processes all the information gathered on a specific interface, the other server shapes the traffic under desired and controlled conditions. The emulating IoT device collects and sends some process variables which reflect the process status. Data collection and forwarding is performed on a regular basis, following a preestablished period.

The software side of the experimental setup is composed of four software tools: (1) A Python script which collects and further forwards the information on a regular basis, (2) a network traffic shaper tool, Traffic Control (TC) [33], which among other things allows to add a pre-established delay or to discard packets, (3) a modified version of Apache Kafka and (4) an Apache Hadoop instance.

The Python script used for emulating the data gathering and forwarding process is publicly available[2] for further testing and reconstructing the results shown within this work. The script basically gets a CSV file, a destination IP address and a packet sending frequency value as input, and as a result it sends the values of the variables of each row to the destination IP address in a given period. The file format and the protocol used for sending data are JSON and HTTP respectively. The Python script is installed on a host running GNU/Linux Debian distribution.

The TC network packets shaping tool allows to discard packets randomly or based on some other parameters. Furthermore, specific or random delays can also

[2]https://bitbucket.org/danzsecurity/dataforwarder

be introduced by the tool. Within this setup TC is installed on a separate server with two network interfaces, running GNU/Linux Debian distribution and with IP forwarding feature switched on.

Finally, both the modified Apache Kafka version and Apache Hadoop are installed on the same server. Again, running GNU/Linux Debian distribution. The modified Apache Kafka version, the Kafka REST Proxy, originally developed by Confluent [34], was modified in order to automatically evaluate some metrics related with captured packets. Those metrics include the time interval between two consecutive packets and each packet size. Together with the received packets, the evaluated metrics are further sent to Apache Hadoop both for storing and processing purposes. The modified Apache Kafka version is also publicly available.[3]

All these software tools allow the emulation of a real IoT to cloud data forwarding scenario/use case. Data forwarding not only mimics real delays and packet loss, but it also allows MitM attacks by modifying data values. Hence, both timing and value modification anomalies or attacks can be emulated.

Figure 3 shows both the emulated topology and the real one. As it is shown, the emulated topology is composed of three different networks: (1) A local network, (2) the Internet and (3) a cloud network. The local network is where different IoT devices are located. These devices basically measure the process and environmental variables, such as the temperature, following a preestablished periodicity; afterwards, they forward all the measurements to a cloud server. On the other hand, The Internet network could be a single public or private network or a combination of both of them; as in real networks, packets can be randomly delayed or dropped due to a network failure. Finally, there is the cloud network, which could be either a public, private or hybrid cloud infrastructure and managed by a cloud service provider, third party enterprise or internally. The cloud network hosts a server dedicated to acquiring and storing all data sent by the IoT device. Moreover, it evaluates the necessary metrics and stores them together with the acquired data.

The real network is composed of three servers connected directly through two different networks. Two out of three servers, the first and the third one, have a single network interface while the second server has two interfaces. The last one works as a transparent bridge, forwarding packets from one interface to the other and delaying or dropping packets.

During the experiment, two different datasets were created: (1) a normality dataset and (2) a manually altered or anomaly dataset. Figure 4 shows both setups and the servers where values were altered and network packets were either delayed or dropped. Both setups got the same CSV file as input; however, the output was stored in two different files. During the experiments, the first server read a row from the CSV file at a preestablished period. Then, some values were altered depending on the type of dataset we were creating, and were further sent to the second server. The same approach was applied in the second server. Under normal conditions, no packets were delayed neither dropped. However, under manually altered conditions,

[3]https://bitbucket.org/danzsecurity/modifiedkafkarest

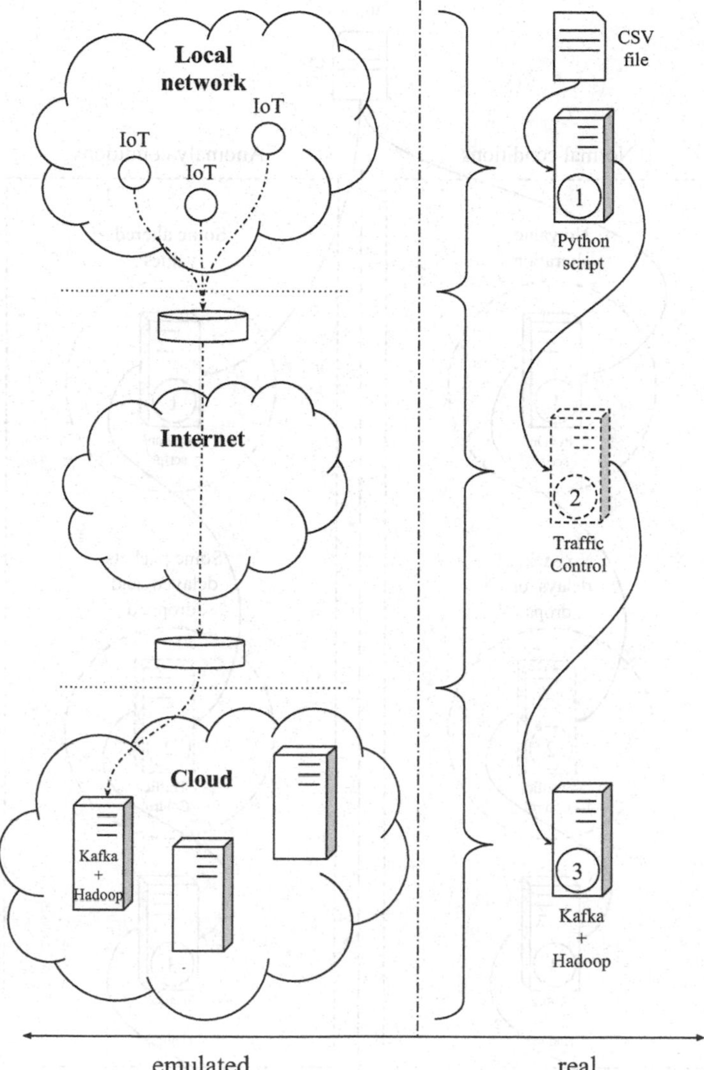

Fig. 3 Experiment topology

some of them were randomly delayed or dropped. Finally, the third server, evaluated a set of metrics and stored them, along with acquired data as a dataset for later analysis.

As a result, the experimental setup provides two different datasets given the same input, one of them, the normal dataset, created under normal conditions and the other one, the anomaly dataset, having altered some values and having delayed or dropped some packets.

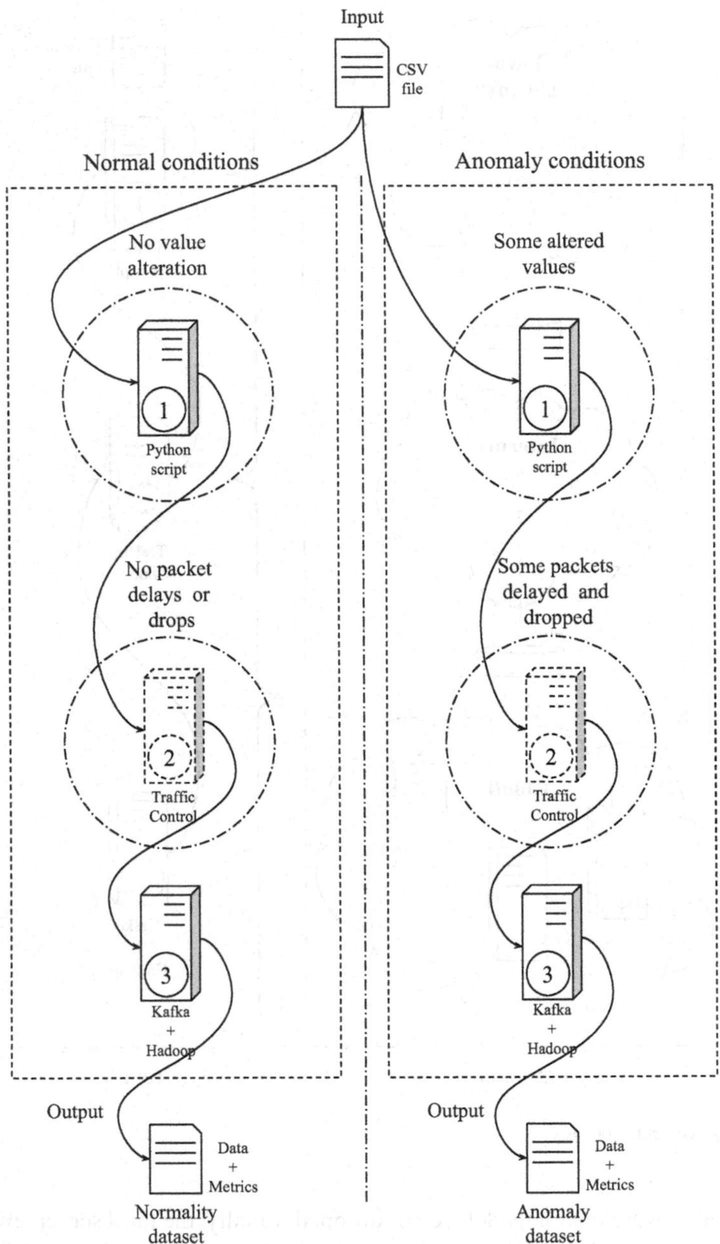

Fig. 4 Normal vs anomaly conditions

Table 1 Variables originally present in the dataset

Var. name	Units
Acidity	pH
Temperature	°C
Conductivity	$\mu S/cm$
Dissolved oxygen	mg/l
Reduction potential	mV
Organic matter	number of occurrences/m
Turbidity	NTU
Ammonia levels	mgN/l

Table 2 Scenarios for validation

Scenario	Description
Scenario 1	An attacker performs a Man-in-the-Middle attack and modifies packet size
Scenario 2	An attacker performs a Man-in-the-Middle attack drops half of the packets, that do not reach the backend cloud
Scenario 3	An attacker performs a Man-in-the-Middle attack and modifies the pH and temperature reading. The backend receives the following reading: $pH_{wat} = 9$ and $T_{wat} = 23$, both higher than the average
Scenario 4	An attacker performs a Man-in-the-Middle attack, drops half of the packets, and at the same time, injects the $pH = 5$ value, lower than usual

5.2 Dataset

The used dataset is a real dataset coming from a water distribution plant in Northern Spain. In there, several variables, shown in Table 1, are monitored to ensure water quality.

We enriched the dataset with the following variables, based on the received network data: Δt (time since the last reading was received, in ms) and network packet size in KB. Therefore, the final validation dataset consists of 10 variables, with a total of 22,000 readings.

5.3 Experiments

In order to validate our proposal, we have designed a set of experiments on top of the previously explained dataset. These experiments are shown in Table 2. All variations from the attack have been performed on the top of the dataset, where the middle node modifies the traffic before relaying it to the backend cloud.

6 Results

This sections shows the obtained results when applying our proposal in the setup described in Sect. 5. More specifically, this section shows the oMEDA plots of the detected anomalies. All four scenarios where identified as anomalous, and the oMEDA plot was computed over the first observation out-of-bounds.

Figure 5 shows the oMEDA plot for the scenario where an attacker modifies the packet size, doubling it in size. As we can see, the oMEDA plot shows that the variable regarding variable size is the most contributing factor to the anomaly, as it has a larger value than it should (large positive value).

In Scenario 2, the attacker drops half of the packets, so only one out of two packets reach the IIoT backend. As it is shown in Fig. 6, now the larger time between readings is the major contributing variable in the detected anomaly.

In the third scenario, the attacker does not drop packets nor alter their size significantly. In this case, it performs an integrity attack and sets the acidity and the water temperature to arbitrary values. As shown in the corresponding oMEDA plot (Fig. 7), it is noticeable how the pH level is higher than usual, as well as the temperature (albeit at a lower level). This is due to the fact that water temperature varies throughout the year, while pH levels are kept constant, so even small changes in pH can yield large variations in the oMEDA plot.

In the last scenario, as a combination of scenarios 2 and 3, the attacker drops half of the packets, while injects a lower-than-usual pH value to the packets that make it

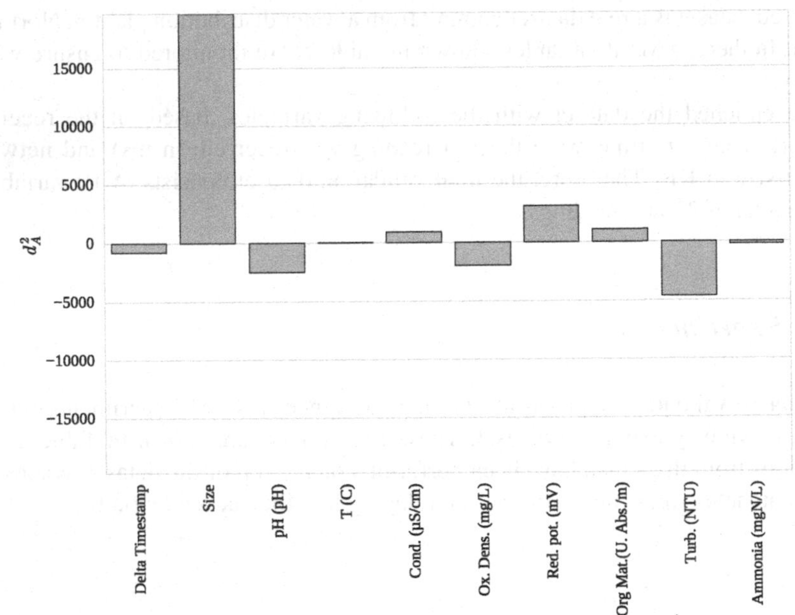

Fig. 5 oMEDA chart for the diagnosis of the anomaly in Scenario 1

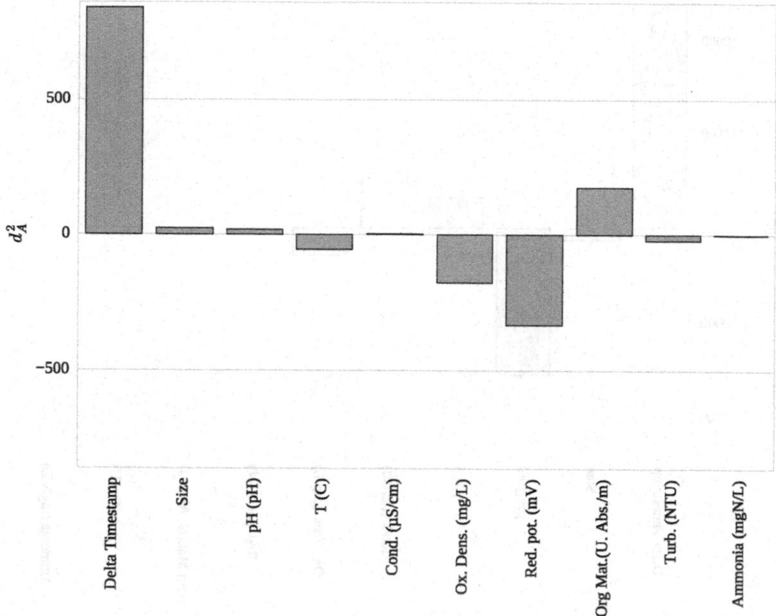

Fig. 6 oMEDA chart for the diagnosis of the anomaly in Scenario 2

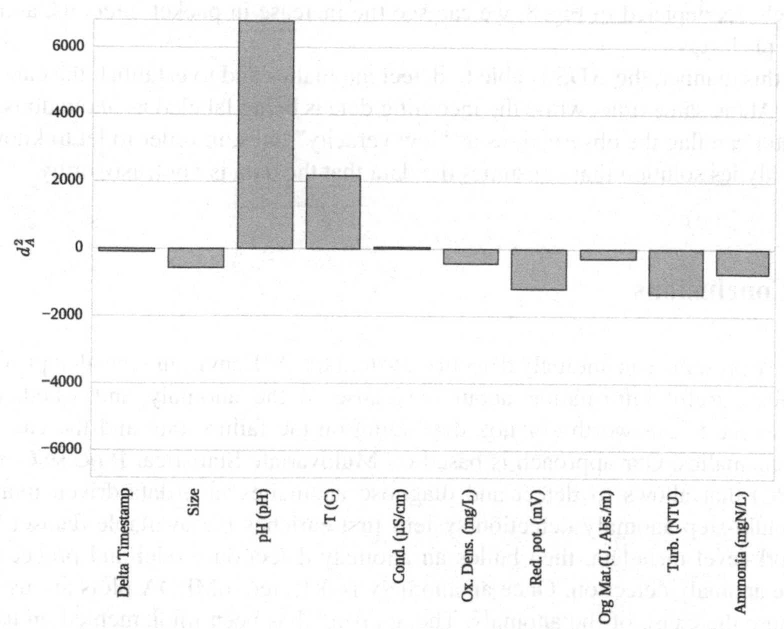

Fig. 7 oMEDA chart for the diagnosis of the anomaly in Scenario 3

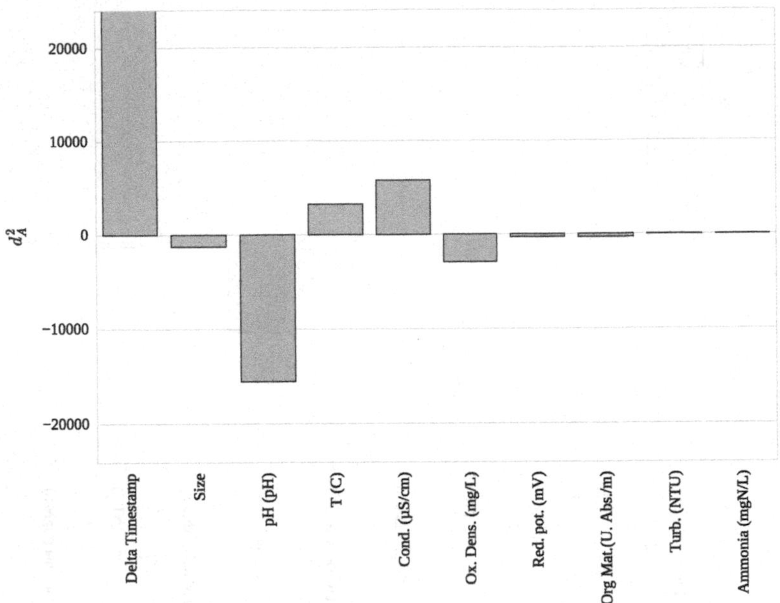

Fig. 8 oMEDA chart for the diagnosis of the anomaly in Scenario 4

through. As depicted in Fig. 8, we can see the increase in packet intervals, and the lower pH levels.

In this manner, the ADS is able to detect anomalies and to establish the cause of them. At the same time, while the incoming data is being labeled as anomalous, the operator can flag the observations as "low veracity" ones, in order to let to know to the analytics solution that consumes the data that the data is not trustworthy.

7 Conclusions

We have presented an anomaly detection system for IIoT environments that provides operators useful information about the cause of the anomaly and whether an IIoT source is trustworthy or not, depending on the failure rate and the cause of their anomalies. Our approach is based on Multivariate Statistical Process Control (MSPC) that allows to detect and diagnose anomalies in a data-driven manner. The multi-step anomaly detection system first enriches the available dataset with network-level metadata, then builds an anomaly detection model and proceeds to online anomaly detection. Once an anomaly is detected, oMEDA plots are used to diagnose the cause of the anomaly. The approach has been implemented on top of Big Data tools and has been validated using a real dataset from a water distribution plant. Results show that it is possible to detect and diagnose anomalies of different nature, even when appending data that was not present in the original dataset.

For further development of the solution, building network-related constructed variables can be designed for specific scenarios and can go well beyond packet size and interval. The choosing or building of relevant features for anomaly detection is a research field on its own, and approaches such as this one can benefit greatly from the results in this area.

Moreover, quantifying the result of anomaly detection and diagnosis into a continuous scale can help in determining a true "Trust score" for each of the sources, that will tell operators whether a data source yields more anomalies (and therefore, is less reliable), or on the contrary, is a source that produces few anomalies.

Acknowledgements This work has been developed by the intelligent systems for industrial systems group supported by the Department of Education, Language policy and Culture of the Basque Government. This work has been partially funded by the European Unions Horizon 2020 research and innovation programme project PROPHESY, under Grant Agreement no. 766994, and the Basque Government's Economic Development and Infrastructure departments Elkartek program project Cyberprest under agreement KK-2018/00076. Author Iñaki Garitano is partially supported by the INCIBE grant "INCIBEC-2015-02495" corresponding to the "Ayudas para la Excelencia de los Equipos de Investigación avanzada en ciberseguridad".

References

1. Falliere N, Murchu LO, Chien E (2011) W32.Stuxnet dossier. White paper, Symantec Corporation, Security Response
2. Homan J, McBride S, Caldwell R (2016) Irongate ICS Malware: nothing to see here... masking malicious activity on SCADA systems [Online]. Available: https://www.fireeye.com/blog/threat-research/2016/06/irongate_ics_malware.html (Retrieved: 2018-07-13)
3. Cheminod M, Durante L, Valenzano A (2013) Review of security issues in industrial networks. IEEE Trans Ind Inf 9(1):277–293
4. Ding D, Han QL, Xiang Y, Ge X, Zhang XM (2018) A survey on security control and attack detection for industrial cyber-physical systems. Neurocomputing 275:1674–1683
5. Urbina DI, Giraldo J, Cardenas AA, Valente J, Faisal M, Tippenhauer NO, Ruths J, Candell R, Sandberg H (2016) Survey and new directions for physics-based attack detection in control systems. NIST GCR 16–010. Technical report, National Institute of Standards and Technology
6. Sadeghi AR, Wachsmann C, Waidner M (2015) Security and privacy challenges in industrial internet of things. In: 2015 52nd ACM/EDAC/IEEE Design Automation Conference (DAC), pp 1–6
7. Sajid A, Abbas H, Saleem K (2016) Cloud-assisted IoT-based SCADA systems security: a review of the state of the art and future challenges. IEEE Access 4:1375–1384
8. Rajasegarar S, Leckie C, Palaniswami M (2014) Hyperspherical cluster based distributed anomaly detection in wireless sensor networks. J Parall Distrib Comput 74(1):1833–1847
9. Thanigaivelan NK, Nigussie E, Kanth RK, Virtanen S, Isoaho J (2016) Distributed internal anomaly detection system for internet-of-things. In: 2016 13th IEEE annual consumer communications networking conference (CCNC), pp 319–320
10. Summerville DH, Zach KM, Chen Y (2015) Ultra-lightweight deep packet anomaly detection for internet of things devices. In: 2015 IEEE 34th International Performance Computing and Communications Conference (IPCCC), pp 1–8
11. Stiawan D, Idris MY, Malik RF, Nurmaini S, Budiarto R (2016) Anomaly detection and monitoring in internet of things communication. In: 2016 8th International Conference on Information Technology and Electrical Engineering (ICITEE), pp 1–4

12. Sicari S, Rizzardi A, Grieco L, Coen-Porisini A (2015) Security, privacy and trust in internet of things: the road ahead. Comput Netw 76:146–164
13. Bao F, Chen IR (2012) Dynamic trust management for internet of things applications. In: Proceedings of the 2012 International Workshop on Self-Aware Internet of Things. Self-IoT'12, New York. ACM, pp 1–6
14. Mahalle PN, Thakre PA, Prasad NR, Prasad R (2013) A fuzzy approach to trust based access control in internet of things. In: Wireless VITAE 2013, pp 1–5
15. Wang JP, Bin S, Yu Y, Niu XX (2013) Distributed trust management mechanism for the internet of things. Appl Mech Mater 347:2463–2467
16. Liu Y, Chen Z, Xia F, Lv X, Bu F (2012) An integrated scheme based on service classification in pervasive mobile services. Int J Commun Syst 25(9):1178–1188
17. Liu WM, Yin LH, Fang B, Zhang HL (2012) A hierarchical trust model for the internet of things. Chin J Comput Phys 35(5):846–855
18. Saied YB, Olivereau A, Zeghlache D, Laurent M (2013) Trust management system design for the internet of things: a context-aware and multi-service approach. Comput Secur 39:351–365
19. Liu Y, Gong X, Feng Y (2014) Trust system based on node behavior detection in internet of things. J Commun 35:8–15
20. Tormo GD, Mármol FG, Pérez GM (2015) Dynamic and flexible selection of a reputation mechanism for heterogeneous environments. Futur Gener Comput Syst 49:113–124
21. MacGregor JF, Kourti T (1995) Statistical process control of multivariate processes. Control Eng Pract 3(3):403–414
22. Camacho J, Pérez Villegas A, García Teodoro P, Maciá Fernández G (2016) PCA-based multivariate statistical network monitoring for anomaly detection. Comput Secur 59:118–137
23. Iturbe M, Camacho J, Garitano I, Zurutuza U, Uribeetxeberria R (2016) On the feasibility of distinguishing between process disturbances and intrusions in process control systems using multivariate statistical process control. In: 2016 46th Annual IEEE/IFIP International Conference on Dependable Systems and Networks Workshop (DSN-W), Toulouse, pp 155–160
24. Stoumbos ZG, Reynolds MR Jr, Ryan TP, Woodall WH (2000) The state of statistical process control as we proceed into the twenty-first century. J Am Stat Assoc 95(451):992–998
25. Kourti T (2002) Process analysis and abnormal situation detection: from theory to practice. Control Syst IEEE 22(5):10–25
26. Camacho J, Pérez Villegas A, Rodríguez Gómez RA, Jiménez Mañas E (2015) Multivariate exploratory data analysis (MEDA) toolbox for matlab. Chemometrics Intell Lab Syst 143:49–57
27. Hotelling H (1947) Multivariate quality control. In: Eisenhart C, Hastay MW, Wallis WA (eds) Techniques of statistical analysis. McGraw-Hill, New York
28. Jackson JE, Mudholkar GS (1979) Control procedures for residuals associated with principal component analysis. Technometrics 21(3):341–349
29. Ramaker HJ, Van Sprang EN, Westerhuis JA, Gurden SP, Smilde AK, Van Der Meulen FH (2006) Performance assessment and improvement of control charts for statistical batch process monitoring. Statistica Neerlandica 60(3):339–360
30. Alcala CF, Qin SJ (2011) Analysis and generalization of fault diagnosis methods for process monitoring. J Process Control 21(3):322–330
31. Camacho J (2011) Observation-based missing data methods for exploratory data analysis to unveil the connection between observations and variables in latent subspace models. J Chemometrics 25(11):592–600
32. Zaharia M, Xin RS, Wendell P, Das T, Armbrust M, Dave A, Meng X, Rosen J, Venkataraman S, Franklin MJ et al (2016) Apache spark: a unified engine for big data processing. Commun ACM 59(11):56–65
33. The Linux Foundation: Iproute2. https://wiki.linuxfoundation.org/networking/iproute2. Accessed 18 Aug 2018
34. Confluent: Confluent rest proxy. https://docs.confluent.io/current/kafka-rest/docs. Accessed 18 Aug 2018

Secure Machine to Machine Communication in Industrial Internet of Things

Mauro Conti, Pallavi Kaliyar, and Chhagan Lal

Abstract In todays world, *Internet of Things* (IoT), is an emerging technology, where many smart devices are connected with each other. The rapidly growing deployment of IoT in real-world applications and the advancement in technology has attracted the concept of *Industrial Internet of Things* (IIoTs). The large number of applications such as smart oil and gas industry, smart transportation, smart grid, smart health-care and smart metering are the few examples of the use of smart devices in IIoTs. These intelligent devices have the capabilities of sensing, actuating, storing, and processing of the data, and it causes challenge-able problems (e.g., communication security and reliability) in the network. In this book chapter, first we present the key benefits and challenges of the use of IoT technologies in today's industries. Second, to address few of the identify challenges, we propose *SCOUT*, which is a secure machine to machine communication technique for IIoTs. In particular, *SCOUT* makes efficient use of the Routing Protocol for Low Power and Lossy Networks (RPL), the de facto routing protocol for IoT and an optimized remote software attestation algorithm to improve the communication security and scalability in large scale heterogeneous IIoT network scenarios. Finally, to show the deployment feasibility and working efficiency of SCOUT, we explain it with a real-world industrial use case.

Keywords IIoT · Security · Machine-to-machine communication · RPL · Remote attestation

M. Conti · P. Kaliyar
Department of Mathematics, University of Padova, Padua, Italy
e-mail: conti@math.unipd.it; pallavi@math.unipd.it

C. Lal (✉)
Department of Mathematics, University of Padova, Padua, Italy

Manipal University Jaipur, Jaipur, Rajasthan
e-mail: chhagan@math.unipd.it

C. Alcaraz (ed.), *Security and Privacy Trends in the Industrial Internet of Things*, Advanced Sciences and Technologies for Security Applications,
https://doi.org/10.1007/978-3-030-12330-7_10

1 Introduction

Internet of things (IoT) [1] facilitates billions of devices to be enabled with network connectivity to collect and exchange real-time information for providing intelligent services. Thus, IoT allows connected devices to be controlled and accessed remotely in the presence of adequate network infrastructure [2]. Due to this huge advancement in IoT, Industries are shifting their modules towards the application based services, where smart connecting devices have control environment for storing and processing their data [3]. This new ubiquitous environment where data sensing, collection, analyzing, and processing is done by connecting and collaborative automatic machines considered to be a very promising environment for the Industrial Internet of Things (IIoT) [4].

More recently, Industries has shifted its paradigm towards the new Industry Revolution (IR) 4.0 from the traditional industrial communication system. In IR 4.0, the people, machines, and products will stay connected throughout the whole logistics process, i.e., the point of origin of the product to the point of its consumption. It is believed that IR 4.0 will boost the productivity of the whole system, and also it will provide benefits regarding the economies as it will allow customized and flexible production. The "SMART Factory 4.0" is an example of this revolution, which is also considered as "Cyber-Physical Systems (CPS)". CPS [5] is seen as a system which performs the integration of Computational, Network and Physical processes. However, the transition from the third industrial revolution to IR 4.0 raises an array of new security and privacy threats. It is because unlike the traditional industrial communication system, in IR 4.0, the factories will increasingly rely on diverse communication technologies including wireless standards to ensure connectivity, interoperability, and remote operation and control of production processes through the Internet. This provides an unprecedented attack surface for the attackers. Additionally, a large number of smart devices, which are being used in various industrial applications produce massive amount of real-time data while communicating with each other. Hence, any security and privacy compromise can have catastrophic impact on business. Hence, security along with data privacy, and the data communication reliability are crucial for smooth functioning of Industrial Systems [6–8].

In this book chapter, we propose SCOUT, a secure and reliable data communication technique for machine-to-machine (M2M) interaction in IIoTs networks. SCOUT uses the advantages of secure remote attestation process to check the integrity and confidentiality of device in large-scale IoT network. In SCOUT, the attestation process runs in parallel with the routing process, which is done by Routing Protocol for Low-Power and Lossy Networks (RPL) [9]. The use of software remote attestation of smart devices along with the RPL improves the reliability in data communication process of RPL as it makes RPL robust against different malicious activities in the network. For example, the routing attacks such as rank [10] and sybil [11] could be avoided or detected by removing nodes that are identified as malicious during the attestation process. The main goal of our approach is to ensure the integrity of the smart IoT devices as well as the data packets they exchange with each other. However, communication security and reliability are significant challenges in today's large-scale IoT networks. In particular, this chapter has the following contributions.

- We survey the state-of-the-art regarding the secure frameworks and protocols for various IoT applications (such as smart production-logistics systems) that are based on CPS and IIoT. We also present the benefits and challenges of the use of IoT technologies in the industrial domain.
- We propose SCOUT, which is a secure machine to machine communication technique for IIoTs applications. In SCOUT, we use smart device attestation approach to check the integrity of the devices. Our attestation approach is secure and lightweight and it performs the software remote attestation after a specific time-interval while doing routing of data packets through RPL routing protocol.
- Finally, we discuss the feasibility and effectiveness of our proposal on existing IIoT applications through a use case (i.e., smart metering).

The rest of this book chapter is organized as follows. In Sect. 2, we provide background and related work on IIoT networks. It includes the details about the standards and networking protocols used for communication in IIoT, and the working methodology of IIoT routing protocol called RPL. Section 2.3, provides description of state-of-art solutions related to communication security in IIoT.In Sect. 3, we discuss the benefits and challenges of IoT in industrial domain. Section 4 includes our proposed approach along with its working methodology, design considerations, and a use-case study. Finally, in Sect. 5, we conclude the work done in this chapter along with possible future directions.

2 IIoT: Background and Related Work

In this section, Firstly we present the IIoT architecture along with the standards and networking protocols that are currently in use. Later, we present the related work which includes generic works that are done to improve the security in IIoT, and works that are specific to the use of RPL routing protocol for enhancing the security of M2M communication in IIoT networks.

IoT architecture and its protocols [12] are the hugely funded topics in today's research fields of industry and academia. IoT is currently a highly invested industry, already billions of dollars are spent, and more are expected in the upcoming years [1]. To keep the network homogeneous in IoT ecosystem, standardized architectures and networking protocols are used for the hardware manufacturing, micro-computing, and M2M communication. In order to use IoT services in industries, every application tries to follow these standards.

In IIoT applications, the smart devices lacks in computational power and storage. These devices have to act smart and make collaborative decisions according to the requirements of various IoT applications. The devices have to see, hear (sensing), and think (processing) very carefully to make decisions and also act smartly in order to communicate with other. Sometimes in real time environment making the dynamic decision can be very critical. In this situation, the standardization of architecture and protocols provides significant benefits by making communication easier [13]. In the next section, we will discuss the standard architecture and protocols at various layers that are being used for deploying IoT networks in IIoT applications.

2.1 Networking Architecture

Several survey papers have already been published on the different aspects of standardization in IoT. For instance, in [14] authors discuss various layer standardization's and their protocols. The OSI/TCP layered protocol stack is used as a standard protocol stack for the wired and wireless networks, but the versatile nature, interoperability, and heterogeneity of IoT devices push towards a new networking structure for IoT networks [15]. For this purpose, various groups such as Internet Engineering Task Force (IETF) [2, 16], World Wide Web Consortium (W3C), EPC global, Institute of Electrical and Electronics Engineers (IEEE), and the European Telecommunications Standards Institute (ETSI), are working towards the development of a standard protocol stack which is more suitable for IoT networks.

Figure 1 shows a six layered model of IoT architecture, and all layers show the names of their standard protocols. All the devices in IoT ecosystem follows this architecture and the structure of their protocol to send and receive data from each other. The internet uses Internet Protocol Version 6 (IPv6) for a device to device communication in IoT networks [16]. RPL [9] has been recently standardized as a routing protocol for IoT networks. The main reason underlying the development of RPL is to provide an efficient routing scheme for low powered IoT devices.

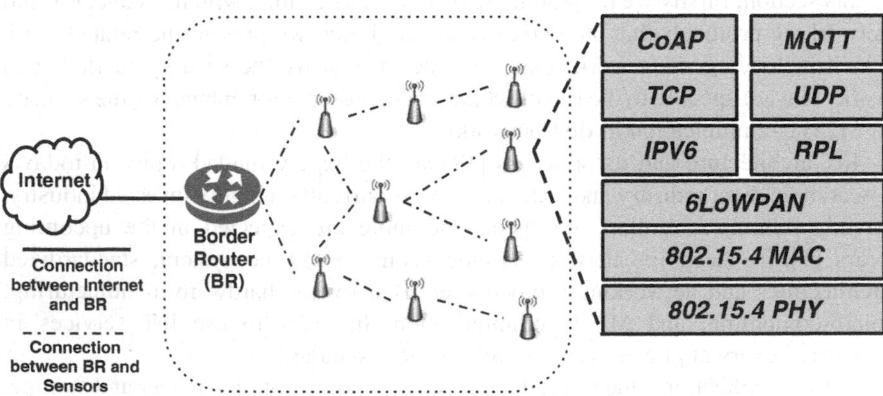

Fig. 1 Standards and networking protocols for IoT developed by IETF/IEEE

The adaptation layer protocol, i.e., IPv6 over Low-powered Wireless Personal Area Networks (6LoWPAN) [17] is used to provide support from IPv4 to IPv6. In the next section, we briefly explain the methodology and working of RPL communication protocol for IoT systems [18].

2.2 RPL: Routing Protocol for Low Power and Lossy Networks

For communication purpose, IoT systems usually use RPL routing protocol. The RPL is a proactive routing protocol, which follows a distance vector routing technique. It establishes a spanning tree topology, and it organizes network devices into Directed Acyclic Graphs (DAGs). A DAG represents a network where all the nodes are connected in a way there are no round-trip paths, and the traffic is routed to reach one or more root nodes. In the DAG resides one or more Destination-Oriented DAG (DODAG), where every node has only one route from each leaf node to the root node. The DODAG is always routed at a single destination, which is called the root node, where all the data will sink. The graph is constructed by the use of an Objective Functions (OFs) which defines important configurations such as routing metrics used, optimization objectives, how to calculate the rank, and how to select parents in the DODAG. The OF specifies how routing constraints and other functions are taken into account during topology construction [19].

Nodes in RPL can be stateless (non-storing) or stateful (storing) mode. A node in non-storing mode can only keep the details of its parents due to memory limitations. While in the storing mode, a node can able to store more information like details of the children and parent and routing information. In network, only root has the complete details of all the nodes in all the DODAGs connected to it. Hence, all communications of the complete network will go through the root node in every case. In RPL, there are four different types of ICMPv6 control messages [9] for DODAG formation, maintenance, and graph information exchange. The brief details of these control messages are given below. Later by using Fig. 2, we also explains how the network topology is created and maintained using RPL protocol.

1. DODAG Information Object (DIO): This message is advertised by each node (starting from the root node) during the DODAG creation phase, and also later to maintain the DODAG. DIO is the main source of routing control information and it contains information such as DODAG ID, DODAG version number, RPL instance ID, RPL mode of operation, the rank of sending node, DODAG configuration (including the OF adopted), and other information and options. DIO is sent when a DODAG Information Solicitation (DIS) message is received [19].
2. DODAG Information Solicitation (DIS): This message is sent by nodes when they want to join a DODAG, but did not received any DIO messages for a period of time from a reachable neighbor.

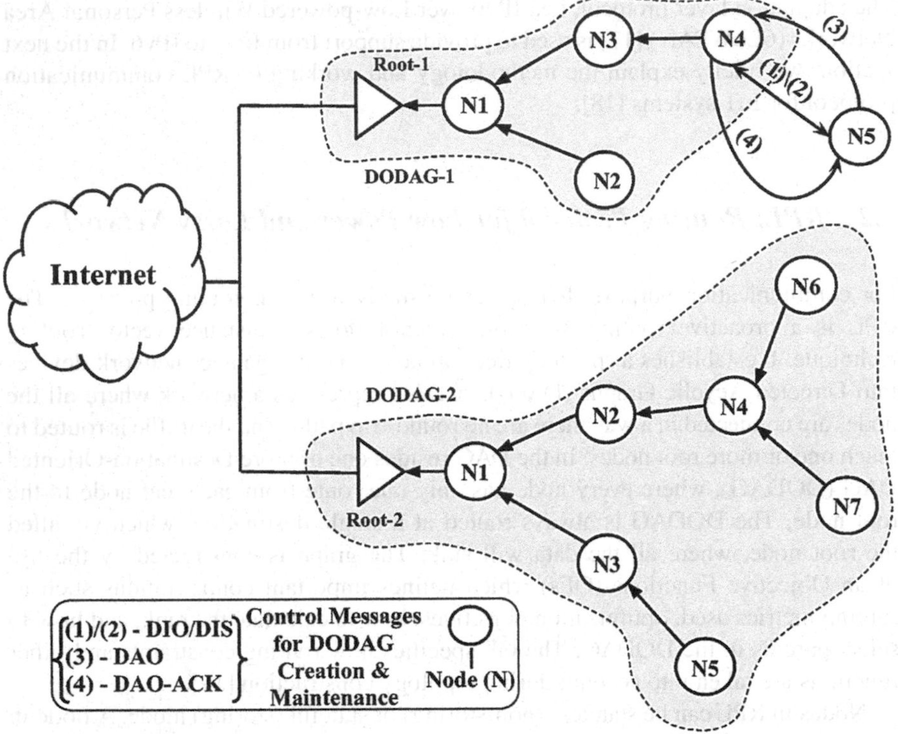

Fig. 2 DODAG creation and maintenance in RPL

3. Destination Advertisement Object (DAO): While RPL uses DIO and DIS messages to create and maintain the downward routes in the DODAG, DAO enables the support for upward traffic, it is used to propagate destination information upwards along the DODAG. DAO also works as beacon message, which helps maintaining the DODAG topology through out the network. The transmission of DAO messages depends on RPL's mode of operation, in the storing mode the DAO messages are send as unicast to the node's preferred parent(s), while in the non-storing mode it will be unicast toward DODAG's root node.
4. Destination Advertisement Object Acknowledgment (DAO-ACK): After receiving the DAO message, according to the information in the flag fields of the DAO message, the recipient may respond by sending an acknowledgment message (DAO-ACK).

Figure 2 shows the working methodology of RPL protocol in an IoT network where various number of nodes are creating 2 DODAG's by connecting to their root node and then to the internet. In Fig. 2, we show, a new node's joining process, which means when a node from outside joins the network, for instance the Node 5 ($N5$) joins DODAG-1. To join the DODAG-1, first $N5$ waits to receive DIO message from the neighboring nodes. If $N5$ did not receive any DIO message for a certain time interval, it sends DIS message to all the neighboring nodes and receives the DIO message in response. After this, $N5$ sends DAO message to ask to join the network. If the neighboring node that is $N4$ wants $N5$ to join the DODAG-1, it sends DAO-ACK, i.e., the permission for $N5$ to join $N4$ as a child node. This process shows the topology creation and management of RPL, and how the nodes add themselves to the DODAG they choose in the IoT networks.

2.3 Related Work

In this section, we discuss state-of-the-art works on RPL with IIoT scenarios such as smart grid and smart meter technology. RPL provides energy efficiency, adaptive to work in various environments and scalability, which makes it best suited routing protocol for resource-constrained large scale IoT networks [20]. Being the only stranded routing protocol for lossy and low power networks (LLN) such as IoT and IIoT, various research works on IoT applications (such as smart grid technology [21]) uses the RPL protocol [22] for routing tasks. Most of these works has focused on ensuring secure and robust communication among devices (i.e., smart meters) during data routing process, but these research works have not properly addressed the problem of device authenticity.

Authors in [23] proposes a RPL based routing for advanced metering infrastructure in smart grid. In particular, the concept of routing through IPv6 RPL is implemented in a smart grid network. The concept of DODAG is used to maintain node states of the grid in a tree form. The DIO and DAO messages are used to maintain inward and outward traffic respectively. The Expected Transmission Time (ETX) is used as link matrix, which is used for fast rerouting computational technique. Further, the outward traffic management is done by every node by analyzing the inward data with minimal overheads. However, the method for rerouting relays on unreliable links whose less knowledge is conveyed to the nodes. Also, the method of computation rely on ETX whose sudden change can cause increase in memory utilization. Authors in [24] proposes an analysis of the unreliability issues in advanced metering infrastructures. It shows how RPL is still in early stages and is unreliable in terms of links utilization in tree formation. However, as compare to other similar protocols like LOAD and AODV, RPL is proven to perform better.

The use of M2M communication techniques is a key for building a secure and reliable IIoT environment where smart devices are able to exchange the required information in an autonomous way with minimal or no human intervention [25]. To this end, authors in [7] proposes a lightweight authentication technique for M2M communications in IIoT ecosystem. The technique is based on hash and XOR operations, and it is characterized by low computational cost, communication and storage overhead, while achieving mutual authentication, session key agreement, devices identity confidentiality, and resistance against the following attacks: Replay attack, Man-in-the-middle attack, Impersonation attack, and Modification attack.

To address the challenges such as excessively long waiting time and a serious waste of energy in Smart Production Logistics (SPL) systems that are based on CPS and IIoT, authors in [26] proposes a framework depicting the design and working methodology of SPL systems to implement intelligent modeling of key manufacturing resources and investigate self-organizing configuration mechanisms. Additionally, to implement the self-organizing configuration, a data driven model is developed using analytical target cascading techniques.

Authors in [27] presents a self-organizing mesh networking solution based on RPL extension for smart metering communications. The paper proposes the concept of self-configuring wireless radio network is introduced. Each node maintains a best-Rank and Channel-id. Traditional mechanism is used to calculate rank and an array is maintained, so when connection loss takes place, the direct connection to next best rank will trigger automatically. Also the connected channel is removed from the channel array for node. Separate algorithms are proposed for channel selection and connection to a node. Also when there a node is connected ideally, it does a channel evaluation on random channels and store their information for further purposes. In the same line, authors in [28] analyzes the role of RPL routing protocol for smart grid communication. The basic aim is to analyze the key components of SGC. A smart grid needs guaranteed data acquisition and timely deployment of commands, and the paper shows that RPL stands out to be the perfect candidate due to its minimal memory consumption, low complexity routing, reduce overheads for constrained link layer frames.

3 Benefits and Challenges of IIoT

In this section, we briefly explain few of the benefits of IIoT in industrial growth and development, and we also list few of the challenges of IIoT in today's network.

3.1 Benefits

1. Use of IIoT in today's environment vastly improved the operational efficiency, for example it improves the up-time, facilitate the remote management, and helps in asset utilization through predictive maintenance.
2. IIoT increases the development of an outcome economy in the market by software-driven services and innovations in hybrid (Hardware-Software co-design) architecture.
3. IIoT improves and increases the visibility into processes, products, customers, and stakeholders.
4. IIoT creates a connected and improved ecosystem using software platform, which helps in collaboration between humans and machines.
5. IIoT also helps to substantial improvement in productivity.

3.2 Challenges

The primary goal of IIoT is to improve the productivity, efficiency, and safety of the operations with a focus on return on investment. The IIoT is all about making machines more efficient, simpler to monitor, and to get more comfortable work experience. All these functionalities and benefits also have few challenges, which we briefly explain below.

1. Precision: In IIoT, automotive machines work instead of humans intervention. In mission critical systems where run-time decision making is required, the precision is a challenge because expecting right decisions from machines can be critical. The system without human interaction and supervision can lead to failure.
2. Adaptability and Scalability: Adaptability and scalability are two challenges which are interdependent. Firstly, adaptability means adjusting with a large number of heterogeneous devices and different protocols. Applications in IoT predominantly faces this challenge due to a large number of different platforms used in IoT systems. The different platforms exist due to the existence of billions of smart devices in IoT environment, which leads to another challenge that is scalability. In IoT network, these billions of devices use different IoT applications with different hardware and software configurations.
3. Security and Reliability: Security and reliability are also interdependent. As we previously mentioned IoT network are very vast and heterogeneous, so security is a big challenge. If the IIoT systems are not secure, it can make the system unreliable, which leads to system failure.
4. Maintenance: Management of large number of IoT devices is a challenge in itself. Managing the device means providing required services on time, keeping the track of failures, their performance, and configuration. In IIoT, where for a single application we have thousands of user, managing their needs on time and all other things is a challenge.
5. Updates Flexibility: As we already discussed that in IIoT a large number of users exist for a single application. Due to different types of services and heterogeneity of devices, and security features it is a challenge to make updates flexible.

4 Proposed Approach: SCOUT

We previously mentioned in Sect. 1 about the recent buzz of the IoT that has engulfed the industrial sectors as well. The advancement in IoT and its technology has its sheer effect on industries, thus IIoTs has been adopted by the industry for smooth and low-cost operations. Due to the easy deploy-ability and low-cost, the use of smart devices is perfect to achieve work automation. For the industrial purposes, smart devices often employed as a group to perform a specific task. These group of devices is called “Swarms”. Swarm’s are intelligent and capable of performing tasks without constant human intervention, which has the direct impact on the deployment and operation cost. Thus, swarms are being adopted in different fields (e.g., smart meters, oil and gas exploration, smart factory, military applications, smart cities). As these low-constrain devices perform critical operations by communicating with each other, it is indeed crucial to check the legitimacy of these devices as they are prone to attacks [29]. Recently, researchers have started working on the security aspects of swarms like [30–33], by proposing the concept of remote attestation to check the integrity and confidentiality of the devices. However, the proposed solutions are not employable for communication purpose in IIoT applications. Hence, the various issues of communication security of devices in IIoT are still need to be addressed properly.

In this section, we first present the details of the network and adversary models in which we envision the use of SCOUT. Later, we discuss our proposed approach and its working methodology. Finally, to show the validity of SCOUT, we show how it can be efficiently deployed to improve the security and communication reliability using a case study on smart grid applications such as smart metering.

4.1 System Model

- As shown in Fig. 3, the IIoT networks consist of a set $N = \{N_1, N_2, \ldots N_x\}$ of size n resource constraint IoT devices (i.e., sensors and actuators). These devices can be static or mobile, depending upon the services of the IIoT applications running on top of the underlying network. The devices can also be inter-operable according to their functionalities (different underlying software or hardware). The various sub-networks can also be homogeneous or heterogeneous concerning the resources of their devices.
- The topology of RPL creates a virtual DODAG on top of the physical network topology. The DODAG is a spanning tree consist of all the nodes. The root node plays a critical role in creating and maintaining the DODAG in the existing network but it also plays the role of the Verifier (V) during the attestation process. Figure 3, depicts the RPL topology in IoT network in an Industrial application scenario.
- From the previously proposed research literature on the attestation, we knew that the verifier is assumed to be secure from all types of attacks. Therefore, we are also assuming that our verifier, i.e., the root node (V), is trusted and cannot be compromised. All the other devices in the network have trusted execution environment [34, 35]. The brief introduction about trusted execution environment is provided in Sect. 4.3.

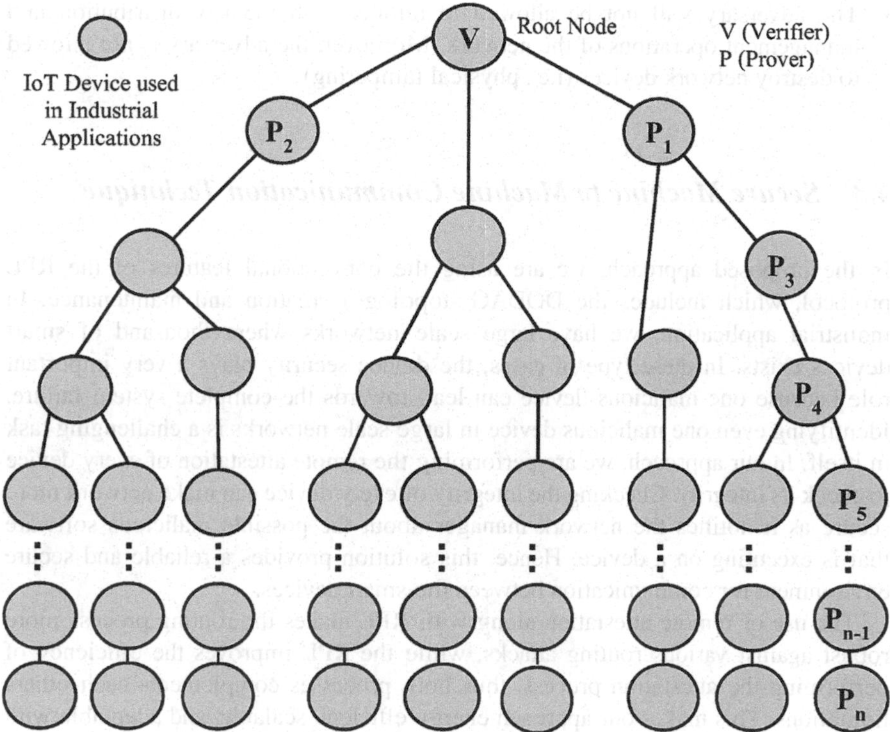

Fig. 3 IIoT system architecture

4.2 *Adversary Model*

The use of IoT devices in a large array of user-centric industrial applications make these kind of networks a profitable target for the adversaries. Hence, the adversaries in this type of networks try their best to perform malicious activities and to harm the integrity of the device(s). By doing these efforts, an adversary can genuinely make the network "question about its reliability and confidentiality". For our industrial environment network, we are considering an attacker who is capable of performing software-only attacks, and the physical tempering is out of this chapters scope.

In our targeted IIoT network scenarios, the adversary are assumed to have the following characteristics:

- As we previously discussed that the adversary is resourceful, and it could perform the clone, rank, sybil, jamming, blackhole, eavesdropping, and wormhole attacks in the network. To perform these attacks, the adversary can compromise an existing node in the network or it can join the network as a new node.

- The adversary will not be allowed to tamper with the key distribution and management operations of the network. Moreover, the adversary is not allowed to destroy network devices (i.e., physical tampering).

4.3 Secure Machine to Machine Communication Technique

In the proposed approach, we are using the conventional features of the RPL protocol, which includes the DODAG (topology) creation and maintenance. In industrial application, we have large scale networks where thousand of smart devices exists. In these type of cases, the device security plays a very important role because one malicious device can lead towards the complete system failure. Identifying even one malicious device in large scale networks is a challenging task in itself. In our approach, we are performing the remote attestation of every device to check its integrity. Checking the integrity of every device can make network more secure as it notifies the network manager about the possible malicious software that is executing on a device. Hence, this solution provides a reliable and secure environment for communication between the smart devices.

The use of remote attestation along with RPL makes the routing process more robust against various routing attacks, while the RPL improves the efficiency of performing the attestation process, thus both processes complements each others limitations. This makes our approach energy efficient, scalable, and adaptable with respect to different industrial applications. In IIoT, our approach is the first one who introduces the concept of device attestation process through RPL routing. Next, we will explain the design and working methodology of SCOUT.

4.3.1 Design Considerations

- Our approach works in both modes of RPL that is storing and non storing mode (i.e., MOP1 and MOP2). It is beneficial because the non storing mode is best suited for resource constrained smart devices due to its support for minimal memory and computational requirements. Furthermore, in both of these modes every device in the network is allowed to send the messages directly to the root node, which means no intermediate device is allowed to drop these messages. If any intermediate device tries to a drop message that is scheduled to deliver to the root node, it can easily be detected by pre-installed Intrusion Detection System (IDS).
- For our approach we define one trickle-timer [36] and two internal timers in RPL's DODAG topology. The trickle-timer is basically used to control the generation of beacon messages in the RPL IoT networks, here it is used to perform the attestation process by triggering the internal-timers that are even and odd, i.e., one after another in a cyclic way as shown in Fig. 4. We can also change the value of the trickle-timer according to the security requirements of an

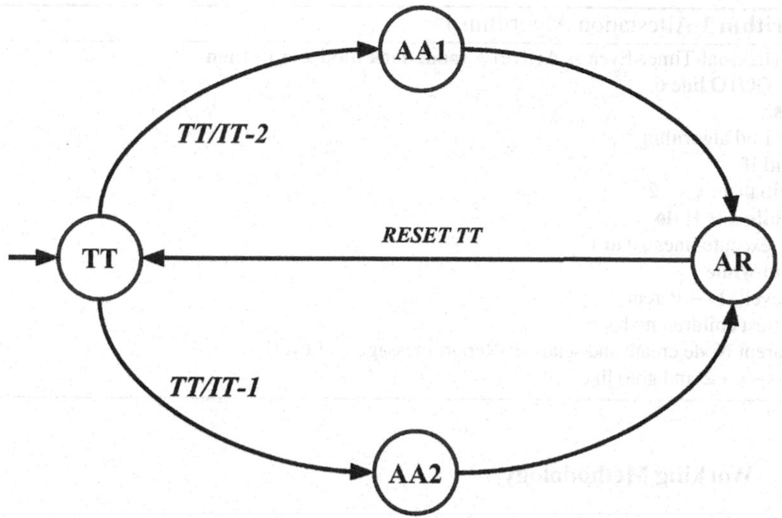

Fig. 4 Model of proposed approach

industrial application. For both the internal timers (IT-1 and IT-2), it is fixed to run the attestation process of all the devices in the RPL topology network.

- Except the existing control messages in RPL, in our approach, we create one more type of control message namel "Att-Report Message". This message is only used for the attestation process in the network.

The proposed attestation process for smart devices that are deployed in an industrial application will be performed by executing a series of steps, which are as follow:

- **Step 1:** Perform the initial network setup by deploying a set of nodes. All these nodes have the attestation details and device-ID, and initially we set the PS-Trickle-Timer to IT-2.
- **Step 2:** The RPL algorithm is executed which completes the DODAG formation in the network.
- **Step 3:** Initialize the following:
  Root → Level 1
  IT-1 → Internal-Timer-Even
  IT-2 → Internal-Timer-Odd
- **Step 4:**: For all active states of Trickle-Timer execute the steps 5 – 7.
- **Step 5:** IF (PS-Trickle-Timer == IT2) then execute Step 6 else Step 7
- **Step 6:** Execute attestation Algorithm 1 and set $PS - Trickle - Timer \rightarrow IT1$
- **Step 7:** Execute attestation Algorithm 1 and set $PS - Trickle - Timer \rightarrow IT2$

Algorithm 1 Attestation Algorithm 1

```
if (Internal-Timer-Even ≡ Active) ∧ (node-rank mod 2 ≡ 0) then
    GOTO line 6
else
    End algorithm
end if
Initialize: x ← 2
while x ≠ H do
    execute lines 10 to 13
end while
Level(x) ← Parent
Attest children nodes
Parent Node create and send Att-Report message to Level1
x ← x + 2 and goto line 7
```

4.3.2 Working Methodology

- As we previously explained in Sect. 2.2 that the node-rank is an important feature in the RPL DODAG topology. The rank of a node depends on the number of hops between the node and the root, which means "how far it is from root node". A child node always choose the node with the smaller rank out of his neighboring node as a parent node. In DODAG, multiple nodes can have the same rank, which also means that they are on the same level. We start from the root node by assuming its rank is 1, and it is on level 1. Accordingly, all other nodes which have rank 2 are considered on level 2 and so on.
- In our approach, the two internal timers, i.e., IT-1 and IT-2, are used to perform the attestation process in the RPL topology. Whenever the trickle-timer initiates the attestation process in the DODAG, it checks for the node rank. If the rank of the node is even, then IT-1 gets activated, which performs the attestation Algorithm 1 else it activates the IT-2, which performs attestation Algorithm 2.
- In attestation Algorithm 1, all the nodes which are on even levels, for example level 2, 4, 6, 8, and so on up to the maximum height (H) of the tree will attest their child nodes, i.e., the nodes that are on the levels 3, 5, 7, and so on. In attestation Algorithm 2, all the nodes which are on odd levels, for example level 1, 3, 5, 7 up to the maximum limit of the tree will attest their child nodes which are on level 2, 4, 6, 8 and so on as shown in Fig. 5.
- Whenever the attestation process got finished, the node who performs the attestation process creates the Att-Report message (Fig. 7), which include details like "device-id, attestation report of the attested device with the time-stamp" and send it directly to the root node (i.e., Verifier). In Att-Report message the 'device-id" field contains the id of the node,, and the "attestation report" field contain the hash value of the key of the software running on the attested device and the "time-stamp" contains the time of the attestation to prove its time-bound freshness. The Att-Report message is encrypted using root node's public key which allows only the root node to decrypt it.

Algorithm 2 Attestation Algorithm 2

```
1: if (Internal-Timer-Odd ≡ Active) ∧ (node-rank mod 2 ≠ 0) then
2:     GOTO line 6
3: else
4:     End algorithm
5: end if
6: Initialize: x ← 1
7: while x ≠ H do
8:     execute lines 10 to 13
9: end while
10: Level(x) ← Parent
11: Attest children nodes
12: Parent Node create and send Att-Report message to Level1
13: x ← x + 2 and goto line 7
```

- Later, depending upon the attestation report, the Verifier (i.e., root node) decides either the device is malicious or not. Using this approach, we are successfully able to attest all the devices in the network to check their integrity.
- Any event of malicious activity will be notified at the time of remote-attestation process because only the owner of the device/node will have the read and write access to the hardware executing the attestation software. Previous [30, 31] results of remote attestation on IoT devices show that it is a very well formulated, optimum, faultless, efficient for low power devices and best security solution for today's IoT devices used in IIoT applications.

Previously, a number of research articles [37] have proved that the secure hardware-software co-design is more suitable for advance computing platforms as a low-cost security solution. To apply hardware security features in the overall network, we assume that the IoT devices in our IIoT application scenarios are equipped with minimal secure hardware protection unit [34, 35]. For our approach we will consider the SMART [34] based implementation. In Fig. 6, we show the inclusion of our optimized SMART based security module in SCOUT. For more detailed information on attestation process, we refer interested readers to more comprehensive works given in [34] research article.

4.4 Case Study

In this section, we present a discussion on the validation and effectiveness of SCOUT, with respect to the security and communication reliability of the networking infrastructure in a large scale IIoT scenario.

Smart grid communication system has multiple interconnections of a large number of small-scale networks organized into a hierarchical architecture covering huge geographic areas. They use different network topologies, communication technologies, and functionalities according to the application requirements. One of

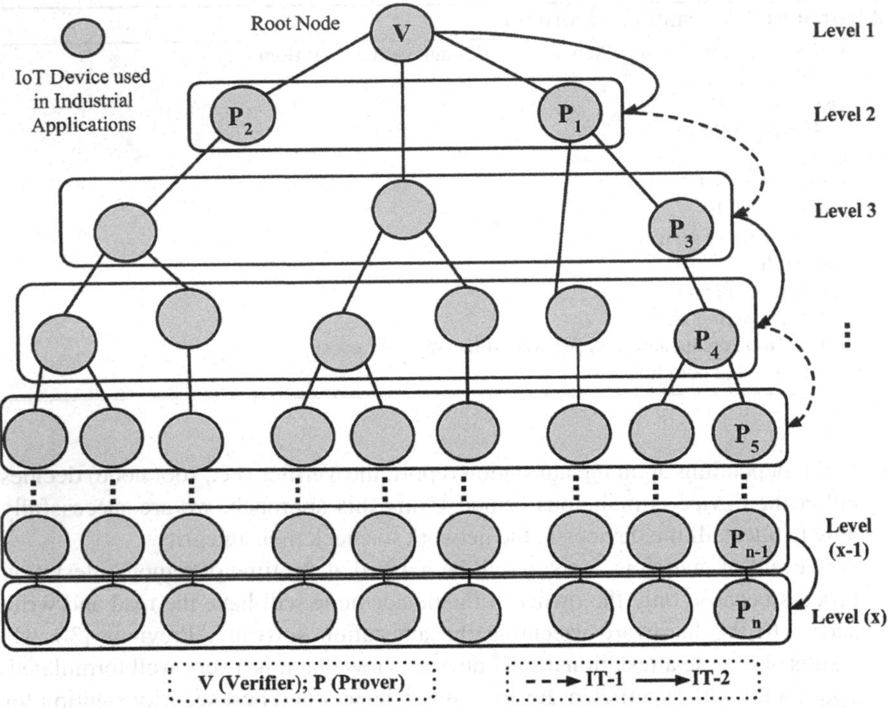

Fig. 5 Proposed approach

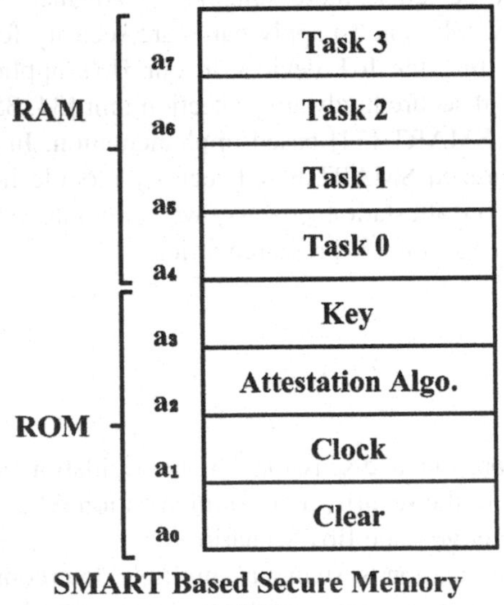

Fig. 6 SCOUT hardware security module

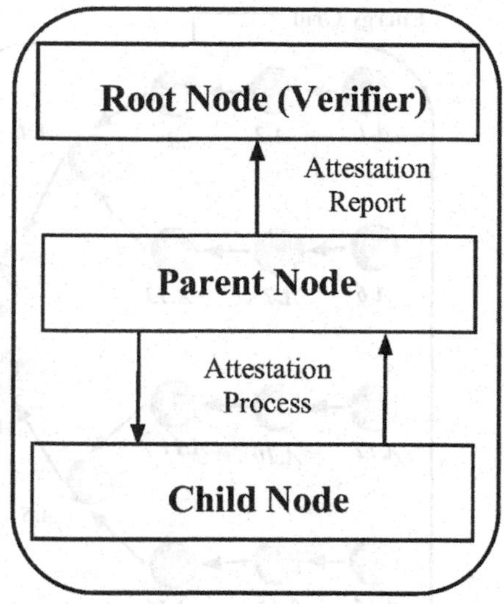

Fig. 7 SCOUT's attestation phase in RPL DODAG

the best suitable examples of smart grid applications is smart metering [40, 41]. A smart meter is an internet-connected device that calculates the energy, water, and gas consumption of your house or building. The traditional meters only measure the total consumption of resources whereas the smart meter monitors the consumption as well as it adjusts the price and usage according to the situations (i.e., time, day and season). Smart meters performs various tasks such as reducing energy usage, monitoring and tracing renewable powers, minimizing the power consumption, performing manual operations remotely, proving customer-oriented services, to name a few [38, 39].

For analysis purposes, we consider a smart metering use-case-study as depicted in Fig. 8. The security support in this case study is the attestation process which we perform to check the integrity and confidentiality of the devices, e.g., smart meters . The devices $A.1, A.2,\ldots, A.16$ are the smart-meters that we use at homes for measuring electricity and the gateway (V) which is a border router (i.e., root node) of these devices. The smart meters and gateway embeds with the "SMART based secured memory structure" and communicate through this smart environment. The security module shown in Fig. 6 is installed at both the communication ends to ensure that the devices are secure from various external and internal attacks. To ensure that an adversary does not temper the smart meter's sensing and transmitting modules, we run periodic attestation process that is well suited for such IIoT networks.

In Fig. 8, their is a smart building (say, A) connected with the energy grid, which is used to provide energy to the flats within the building. The building A consists of sixteen smart meters termed as $A.1, A.2,\ldots,A.16$ each connected to one flat in the building. Whenever a smart meter $A.1, A.2,\ldots,A.16$ sense usage of energy (for example electricity, water, or gas) consuming activity in the flat, it start calculating and recording the consumption units and according to the time, the

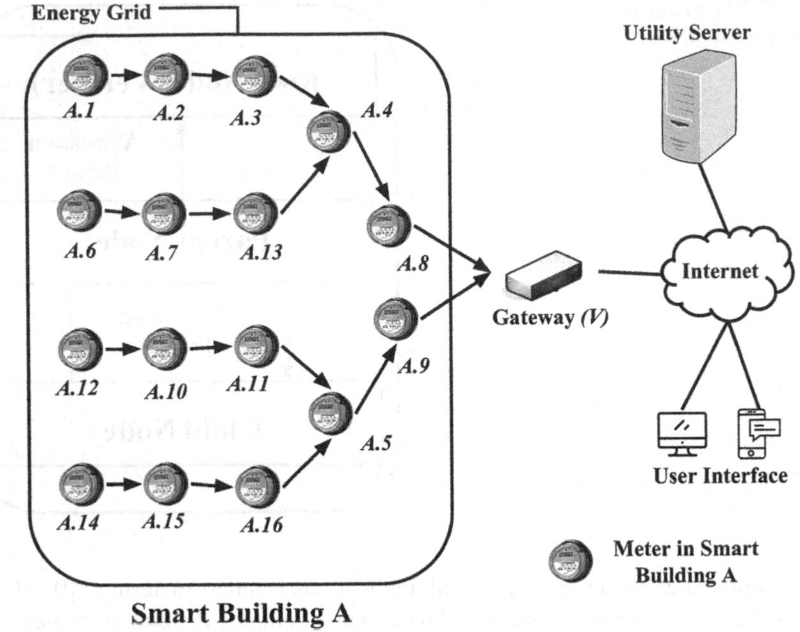

Fig. 8 Use case study of smart metering in IIoT

smart meter send the data to the gateway. Later, the gateway aggregates the data and send it to the nearby utility server at the Authenticated Control Center (ACC). These ACC's prepare bills and charge the user according to their usage. As per our security approach SCOUT, these smart meters calculate their rank according to their hop distance from the root node (gateway). For example, $A.8$ and $A.9$ has rank 2, $A.4$ and $A.5$ has rank 3, $A.3$, $A.13$, $A.11$, and $A.16$ has rank 4, and so on. Than whenever the trickle timer is active, the internal timers i.e., IT-1 and IT-2, will perform the attestation process in the RPL topology of smart meters. As per the attestation Algorithm 1, the smart meter $A.8$ performs the attestation of $A.4$, smart meter $A.9$ performs the attestation of $A.5$, smart meter $A.3$ performs the attestation of $A.2$, smart meter $A.13$ performs the attestation of $A.7$, and so on. While as per the attestation Algorithm 2, the root node (V) performs the attestation of smart meter $A.8$ and $A.9$, smart meter $A.4$ performs the attestation of $A.3$, smart meter $A.4$ performs the attestation of $A.13$, smart meter $A.5$ performs the attestation of $A.11$, and so on. These two attestation phases will rotate one after the other, in this way all the devices in the DODAG gets attested in a efficient way. Whenever the attestation process got finished, the smart meters, who performs the attestation process creates Att-Report message, which includes details like "device-id, attestation report of the attested device with the time-stamp" and send it directly to the root node (i.e., gateway). With attestation report, if the root node find any smart meter with malicious or adversarial behaviour, than gateway can report this to the user and to the service provider for further actions.

In this use case study, we are only taking an example of intra-network smart metering IIoT application to provide intra-network security to the IoT devices. However, our approach can also be applied to large-scale inter-network IIoT applications which might be using heterogeneous devices and communication technologies within their networks. As it can be concluded from the example mentioned above that with the help of SCOUT approach, we can achieve not only device security, but the security with respect to communication in between devices and also the communication reliability concerning scalability with ease. Additionally, the use SCOUT makes Industrial applications more secure to use in heterogeneous and large scale IoT network, which is one of the main concerns for the Industrial applications.

5 Conclusion and Future Work

In this book chapter, we present an approach to improve data communication security in between devices for the Industrial application. In our approach, we perform an RPL-based software remote attestation of devices while they communicate with each other. In support of the proposal, we argue that the remote attestation can be used efficiently to check the integrity of a large number of IoT devices that could be used in various IIoT applications. The applications in IIoT mainly consist of IoT devices from various heterogeneous networks, and our proposed approach is helpful to make the M2M communication reliable and secure. The resource constrained nature of IoT devices with the limitation in communication abilities are significant factors that will determine the future of IoT in Industries. As per se, our approach is a step towards the improvement in communication security in IIoT. As a future work, we will implement our RPL based attestation approach and evaluate its performance over the Industrial networks regarding the security as well as for scalability. We will also try to minimize the hardware assumptions of the devices due to their constraint nature.

References

1. Sheng Z, Yang S, Yu Y, Vasilakos A, McCann J, Leung K (2013) A survey on the IETF protocol suite for the internet of things: standards, challenges, and opportunities. IEEE Wirel Commun 20:91–98
2. Abera T, Asokan N, Davi L, Koushanfar F, Paverd A, Sadeghi A-R, Tsudik G (2016) Invited things, trouble, trust: on building trust in IoT systems. In: 2016 53nd ACM/EDAC/IEEE Design Automation Conference (DAC), Austin, pp 1–6. https://doi.org/10.1145/2897937.2905020
3. Xu LD, He W, Li S (2014) Internet of things in industries: a survey. IEEE Trans Ind Inf 10:2233–2243

4. Perera C, Liu CH, Jayawardena S (2015) The emerging internet of things marketplace from an industrial perspective: a survey. IEEE Trans Emerg Top Comput 3:585–598
5. Wollschlaeger M, Sauter T, Jasperneite J (2017) The future of industrial communication: automation networks in the era of the internet of things and industry 4.0. IEEE Ind Electron Mag 11:17–27
6. Wurm J, Hoang K, Arias O, Sadeghi AR, Jin Y (2016) Security analysis on consumer and industrial IoT devices. In: Proceedings of the 21st Asia and South Pacific Design Automation Conference (ASP-DAC), pp 519–524
7. Esfahani A et al (2019) A lightweight authentication mechanism for M2M communications in industrial IoT environment. IEEE Internet Things J 6(1):288–296
8. Conti M, Kaliyar P, Lal C (2017) REMI: a reliable and secure multicast routing protocol for IoT networks. In: Proceedings of the 12th International Conference on Availability, Reliability and Security (ARES'17). ACM, New York. Article 84, pp 8. https://doi.org/10.1145/3098954.3106070
9. Winter T, Thubert P, Brandt A, Hui J, Kelsey R, Levis P, Pister K, Struik R, Vasseur J, Alexander R (2012) RPL: IPv6 routing protocol for low-power and lossy networks (RFC 6550). https://tools.ietf.org/html/rfc6550.2012
10. Dvir A, Holczer T, Buttyan L (2011) VeRA – version number and rank authentication in RPL. 2011 In: IEEE Eighth International Conference on Mobile Ad-Hoc and Sensor Systems, pp 709–714
11. Wallgren L, Raza S, Voigt T (2013) Routing attacks and countermeasures in the RPL-based internet of things. Int J Distrib Sensor Netw. http://dx.doi.org/10.1155/2013/794326.2013
12. Granjal J, Monteiro E, Sa Silva J (2015) Security for the internet of things: a survey of existing protocols and open research issues. IEEE Commun Surv Tutorials 17:1294–1312
13. Yang W, Wan Y, Wang Q (2017) Enhanced secure time synchronisation protocol for IEEE802.15.4e-based industrial internet of things. IET Inf Secur 11:369–376
14. Karagiannis V, Chatzimisios P, Vázquez-Gallego F, Alonso-Zarate J (2015) A survey on application layer protocols for the internet of things. Trans IoT Cloud Comput 3:11–17
15. Conti M, Droms R, Hassan M, Lal C (2018) Fair-RTT-DAS: a robust and efficient dynamic adaptive streaming over ICN. Comput Commun. https://doi.org/10.1016/j.comcom.2018.07.033
16. Al-Fuqaha A, Guizani M, Mohammadi M, Aledhari M, Ayyash M (2015) Internet of things: a survey on enabling technologies, protocols, and applications. IEEE Commun Surv Tutorials 17:2347–2376
17. Kushalnagar N, Montenegro G, Schumacher C (2007) IPv6 over low-power wireless personal area networks (6LoWPANs): overview, assumptions, problem statement, and goals, (RFC 4919). https://tools.ietf.org/html/rfc4919.2007
18. Kim HS, Ko J, Culler DE, Paek J (2017) Challenging the IPv6 routing protocol for low-power and lossy networks (RPL): a survey. IEEE Commun Surv Tutorials 19:2502–2525
19. Shelby E, Zach S (2010) Chakrabarti and E. Nordmark.: Neighbor Discovery Optimization for IPv6 over Low-Power Wireless Personal Area Networks (6LoWPANs) (RFC 6775). https://tools.ietf.org/html/rfc6775.2010
20. Conti M, Kaliyar P, Lal C (2017) REMI: a reliable and secure multicast routing protocol for IoT networks. In: Proceedings of the 12th International Conference on Availability, Reliability and Security (ARES'17), pp 1–8
21. Meng W, Ma R, Chen HH (2014) Smart grid neighborhood area networks: a survey. IEEE Netw 28:24–32
22. Sabbah AI, El-Mougy A, Ibnkahla M (2014) A survey of networking challenges and routing protocols in smart grids. IEEE Trans Ind Inf 10:210–221

23. Wang D, Tao Z, Zhang J, Abouzeid AA (2010) RPL based routing for advanced metering infrastructure in smart grid. In: 2010 IEEE International Conference on Communications Workshops, pp 1–6
24. Ancillotti E, Bruno R, Conti M (2012) RPL routing protocol in advanced metering infrastructures: an analysis of the unreliability problems. In: 2012 Sustainable Internet and ICT for Sustainability, pp 1–10
25. Barki A, Bouabdallah A, Gharout S, Traore J M2M security: challenges and solutions. IEEE Commun Surv Tutorials 18(2):1241–1254, Secondquarter
26. Zhang Y, Guo Z, Lv J, Liu Y (2018) A framework for smart production-logistics systems based on CPS and industrial IoT. IEEE Trans Ind Inf 14:4019–4032
27. Kulkarni P, Gormus S, Fan Z, Motz B (2011) A self-organising mesh networking solution based on enhanced RPL for smart metering communications. In 2011 IEEE International Symposium on a World of Wireless, Mobile and Multimedia Networks, pp 1–6
28. Ancillotti B, Bruno R, Conti M (2013) The role of the RPL routing protocol for smart grid communications. IEEE Commun Mag 51:75–83
29. https://www.wired.com/2014/11/countdown-to-zero-day-stuxnet/
30. Asokan N, Brasser F, Ibrahim A, Sadeghi A-R, Schunter M, Tsudik G, Wachsmann C (2015) SEDA: scalable embedded device attestation. In: Proceedings of the 22nd ACM SIGSAC Conference on Computer and Communications Security (CCS'15), pp 964–975
31. Ambrosin M, Conti M, Ibrahim A, Neven G, Sadeghi A-R, Schunter M (2016) SANA: secure and scalable aggregate network attestation. In: Proceedings of the 2016 ACM SIGSAC Conference on Computer and Communications Security (CCS'16), pp 731–742
32. Ibrahim A, Sadeghi A-R, Tsudik G, Zeitouni S (2016) DARPA: device attestation resilient to physical attacks. In: Proceedings of the 9th ACM Conference on Security & Privacy in Wireless and Mobile Networks (WiSec'16), pp 171–182
33. Ambrosin M, Conti M, Lazzeretti R, Rabbani MdM, Ranise S (2018, in press) PADS: practical attestation for highly dynamic swarm topologies. In: Proceedings of the 7th International Workshop on Secure Internet of Things (ESORICS 2018 workshop: SIoT 2018), 3–7 Sept 2018, Barcelona
34. Eldefrawy K, Tsudik G, Francillon A, Perito D (2012) SMART: secure and minimal architecture for (establishing dynamic) root of trust. In: Proceedings of the 19th Annual Network & Distributed System Security Symposium (NDSS'12), pp 1–15
35. Brasser F, El Mahjoub B, Sadeghi A-R, Wachsmann C, Koeberl P (2015) TyTAN: tiny trust anchor for tiny devices. In: 52nd ACM/EDAC/IEEE Design Automation Conference (DAC), pp 1–6
36. Levis P, Clausen T, Hui J, Gnawali O, Ko J (2011) The trickle algorithm (RFC 6206). https://tools.ietf.org/html/rfc6206.2011
37. Koeberl P, Schulz S, Sadeghi A-R, Varadharajan V (2014) TrustLite: a security architecture for tiny embedded devices. In: EuroSys'14, pp 1–10
38. Ambrosin M, Hosseini H, Mandal K, Conti M, Poovendran R (2016) Despicable me (ter): anonymous and fine-grained metering data reporting with dishonest meters. In: Proceedings of the 2016 IEEE Conference on Communications and Network Security, (CNS'16), pp 163–171
39. Illera AG, Vidal JV (2014) Lights off! The darkness of the smart meters. In: BlackHat Europe
40. Gungor VC, Sahin D, Kocak T, Ergut S, Buccella C, Cecati C, Hancke GP (2011) Smart grid technologies: communication technologies and standards. IEEE Trans Ind Inf 7:529–539
41. Zheng J, Gao DW, Lin L (2013) Smart meters in smart grid: an overview. In: IEEE Green Technologies Conference (GreenTech), pp 57–64

Part IV
Privacy Issues in Industrial Connected Networks

Modelling the Privacy Impact of External Knowledge for Sensor Data in the Industrial Internet of Things

Salaheddin Darwish, Ilia Nouretdinov, and Stephen Wolthusen

Abstract Some type of privacy-preserving transformation must be applied to any data record from Industrial Internet of Things (IIoT) before it is disclosed to the researchers or analysts. Based on the existing privacy models such as Differential Privacy (DP) and k-anonymity, we extend the DP model to explicitly incorporate feature dependencies, and to produce guarantees of privacy in a probabilistic form that generalize k-anonymity. We assume that additional (external) knowledge of these relations and models can be represented in the form of joint probability distributions, such as Mutual Information (MI). We propose an enhanced definition of DP in conjunction with a realisation for non-randomizing anonymizing strategies such as binning, reducing the extent of binning required and preserving more valuable information for researchers. This allows the formulation of privacy conditions over the evolving set of features such that each feature can be associated its own allowance for privacy budget. As a case study, we consider an example from the Industrial Medical Internet of Things (IMIoT). We have identified some challenges that are not completely addressed by existing privacy models. Unlike physiological measurements in conventional medical environments, IMIoT is likely to result in duplicate and overlapping measurements, which can be associated with different

S. Darwish
School of Mathematics and Information Security, Royal Holloway University of London, Egham, Surrey, UK
e-mail: salaheddin.darwish@rhul.ac.uk

I. Nouretdinov (✉)
Computer Learning Research Centre and School of Mathematics and Information Security, Royal Holloway, University of London, London, UK
e-mail: i.r.nouretdinov@rhul.ac.uk

S. Wolthusen
School of Mathematics and Information Security, Royal Holloway, University of London, London, UK

Department of Information Security and Communication Technology, Norwegian University of Science, Gjøvik, Norway
e-mail: stephen.wolthusen@rhul.ac.uk

C. Alcaraz (ed.), *Security and Privacy Trends in the Industrial Internet of Things*, Advanced Sciences and Technologies for Security Applications,
https://doi.org/10.1007/978-3-030-12330-7_11

personally identifiable items of information. As an example, we present a model of sequential feature collection.

Keywords Industrial internet of things · Industrial medical internet of things · Anonymizing · Differential privacy

1 Introduction

The principal challenge for this work is how to perceive the dependence of different features within the task of data privacy protection. Taking into account the dependencies between the features is necessary to make the privacy mechanism correct, avoiding indirect leakage of the information. Therefore, this work attempts to address two connected goals. First, we cater for the privacy protection mechanism to be safe under highly dependent features. At the same time, if existing knowledge about the form of their dependence is deep enough, we aim to use it for saving the privacy budget and to make more information available for disclosure.

The source of such information about feature dependencies may step from knowledge of experts as well as analysis of existing data. Unlike machine learning methods of data analysis, we do not rely on the data examples as a direct source of the knowledge.

This work was initially motivated with Technology Integrated Health Management (TIHM) project on multi-device system of IMIoT from several different producers.

General security requirements for this system were observed in [1]. However, that review only mentioned data-related aspects of privacy. The anonymising scheme has to prevent an identification of a patient by an adversary by means of data analysis, but make the data available for research aims.

It appeared from the analysis of the system that some new challenges have to be taken into account. Each of IMIoT sensors produces a series of possibly dependent data measurements. The information of distribution and dependence of measurements is initially provided from Meta-Data sources, in the form of ready conclusions from the experts' information and data analysed before. Serial measurements from different sensors must be processed as they arrive in some (partial) time order. We also assume that the exact research goal may be unknown, therefore we avoid data transformations which work by adding noise.

The contribution of this work is creating a new version of the privacy framework, that is applicable to a record on the background of prior probabilistic knowledge. Our assumption is that some prior knowledge is available before having the measurements in the form of restrictions on their joint distribution. This makes existing models from the literature (k-anonymity [2] and DP [3]) inapplicable to the problem directly. The principal privacy requirement is making two individual data records indistinguishable with high probability. This approach combines some properties of DP and k-anonymity but it is not identical to any of them. Some ideas of this work were briefly presented in the conference paper [4].

2 Related Work

To protect the privacy of the system, we have to answer the following questions: to select the way of data protection (privacy transformation), to take into account the prior knowledge (information from experts), to quantify the level of protection (by a privacy constraint). Our principal assumption is that the information of dependence is initially provided by an expert, and their opinions may be derived from the other dataset that are not available for the analysis.

2.1 Anonymizing Strategies

The principal challenge in the anonymizing is to keep a balance between disclosure risk and data utility. It is assumed that the data set is anonymized by some transformation before it becomes open for researchers.

A radical way of data mapping is **summarizing**, replacing data sets with statistical summaries. An example is On-line Compression Model (OCM) [5] drawing a boundary between the useful research information and the noise with respect to a concrete statistical model. The drawback of such methods is that they are more focused on suppressing the noise in data, and this may be insufficient for privacy defence.

A close approach is **data shuffling** i.e., creating a synthetic data set with artificial data records that can not be matched to the original ones [6]. In terms of summarizing, shuffling can often be understood as creating another data set with the same summary, It contains risk that an interesting research property is lost in the data image.

Data swapping [7] can be considered as a moderate form of synthetic data creation. It has the same principal drawback (some undiscovered dependence may be lost from the permutation) but can be applied in a compromised form (by putting a restriction on the number of permutations). However, it is not very suitable for continuous data values.

Adding random noise to data records is a popular way of obscuring continuous data (e.g., used in [3]). Its disadvantage is intentionally feeding the data records with imprecise information. This is specifically harmful if the aims of research directions are not known in advance and one does not wish to restrict them.

Suppression includes such strategies as imputing gaps or missing values, nulling, deletion of some records [8]. It has advantages of being fair (not imputing wrong information) and (if performed in a random way) not affecting statistical conclusions principally (they just may have smaller support). The shortcoming of this approach is inconvenience of processing data with missing values, that usually requires special methods. Also, it may make too many examples useless, if the number of features is small.

Binning (generalisation or discretisation) [9] i.e., replacing exact values with intervals or rounded numbers, can be understood as 'partial' suppression of the values. The advantage of binning over complete suppression is that it can be applied to all the examples in a uniform way, with preserving the overall structure of the data set as it is. It also covers some useful effects achievable by adding random noise to the continuous variables. An example of using binning in medical area can be found in [10].

Therefore, we consider binning as the preferable type of anonymizing strategy for the case of continuous (real-valued) features which is being modelled in this work. It does not create any intentionally wrong or imprecise records, it is moderate in information reduction. In addition, it can be applied to different features and examples in a more or less 'fair' (uniform and systematic) way.

2.2 *Prior Knowledge*

The information about feature dependencies may be obtained by data analysis (e.g., see [11]) or come from experts. The strongly correlated information (redundant features) is known to be important for medical decisions: "in attempting to synthesize the predictive value of multiple considerations... if some of those considerations are strongly correlated with one another, this redundancy should greatly affect how those facts are used" [12, p. 654].

However, involving elements of experts in data analysis is very desirable. This is discussed e.g., in the work [13] where a Bayesian causal network for diagnostic is created, with elements of human feedback. The role of experts in that work was to identify relevant symptoms for the diseases, limited to those suitable for a self-diagnosis by a patient. Expert knowledge may also include some prior knowledge collected from earlier research on different data sets, such as connection between pulse pressure and coronary heart disease detected in [14]. Therefore, we assume that prior knowledge comes in the form of elements of probabilistic model rather than in the form of data examples.

Our task is expanding such or similar way of processing knowledge for the aims of privacy defence. However, there is a difference in the way of its usage. Exclusion of redundant variables is recommended in some of the prediction models [12], but in the privacy context they have to be addressed in a different way: features duplicating each other have to be disclosed in a similar way in order to prevent leakage of one of them through another. Also, in IMIoT area the diagnosis (disease) come from different sources than the symptoms (features), that are more subjective, therefore we prefer not to give it a principal role in data analysis.

2.3 Privacy Models

In order to put one of the anonymizing approaches into practice, the level of achieved privacy protection should be quantified. The known criteria include **k-anonimity** [2]: a release of data is said to have the k-anonymity property if the information for each person contained in the release cannot be distinguished from at least $k-1$ other individuals from the release. Similar variations are l-diversity and t-closeness [15].

These criteria are convenient and straightforward for using with deterministic anonymizing strategies. The disadvantage of k-anonymity model for us is that it relies on having the background knowledge in the form of a data collection, although, as we have just discussed above, the expert knowledge may come in a more generalized form. Instead of estimating the probability that a data record becomes indistinguishable from other data records, we have to refer to probability of two different data records being identical after the privacy transformation,

Another criterion known as **DP** [3] is motivated by the threat of linking several databases by an adversary. Its advantage is a probabilistic nature. As a criterion it requires that two data sets (collections of data records) originally different only in one feature of record become indistinguishable after the anonymizing transformation with high probability. This method can be applied to the observations of a medical sensor in such works as [16] where time series of measurements come from a fixed sensor.

DP ensures anonymity in the sense that a feature related to personal identification is not predictable from other features of the same patient as they are presented to the researchers. However, it is discussed in [17] that when the features are dependent, this kind of difference (only in one feature before applying a transformation) may be a too rare event. Therefore, the standard DP may be irrelevant for the actual privacy aims. The solution suggested in [17] (**ε-dependent DP** given by Definitions 3,4) is to extend the notion of similarity. Its disadvantage in the context of medical problems is that the form of their dependence is assumed to be deterministic.

There also developed some criteria related to the potential quality of **machine learning** results made an adversary (used in such works as [18]). They are much less relevant for our type of knowledge.

3 Theory and Methodology

In our statement, the privacy goal is to prevent distinguishing a record from other possible records with the similar schedule of measurements. We consider a record separately on the background of general distribution of comparable ones, so their number is not known exactly.

Both k-anonymity and DP put an individual record into a set of other individual records, while we consider a record in the context of the distribution. Therefore, we have to formulate another constraint.

3.1 Suggested Privacy Constraint and its Properties

The principal meaning of the constraint is that privacy transformation has to make two individual data records indistinguishable from each other with high probability. This is formalized in the following way.

Definition 1 Denote an individual data record as $D = (d_1, \ldots, d_m) \in R^m$, let $\mathbf{A}$ be a privacy algorithm on R^m (data transformation function), $\mathbf{P}$ be the set of possible density functions P on R^m. The $(\varepsilon, \mathbf{P})$**-DP** constraint for $\mathbf{A}$ is

$$\forall P \in \mathbf{P} : Prob_P \left\{ \mathbf{A}(D) = \mathbf{A}(D') \right\} \geq e^{-\varepsilon},$$

or, formulated in terms of density functions,

$$\int_{D \in R^m s.t. \mathbf{A}(D) = \mathbf{A}(D')} P(dD) \geq e^{-\varepsilon}$$

where P generates m-dimensional data records D and D' independently of each other, $\mathbf{P}$ is the class of possible density functions of distribution P on R^m according to the prior knowledge, and the privacy budget parameter ε quantifies strength of the constraint.

Small values of the parameter ε make the condition stronger. The larger ε is, the easier the constraint is to satisfy.

The proposed constraint is different from the standard version of DP in the following. The data transformation is applied to each of the records individually, instead of the whole data set. The criterion is applicable to deterministic privacy transformations such as rounding (binning), which look much more preferable than obscuring strategies adding the noise for the data. The probabilistic mechanism refers to the data generation algorithm (completely or partially known from the prior knowledge), instead of a noising strategy.

The principal goal of privacy protection is similar to k-anonymity. The constraint of k-anonymity requires a record to be indistinguishable from $k - 1$ other records in a concrete database. If there are n records in the whole data set, this means that two records have the same image with probability about k/n. Having the distribution knowledge instead of data examples, we have to formulate a condition directly in a probabilistic form, referring to probability that two records become indistinguishable after the data transformation. In a nutshell, it is important to stress that we can say that the suggested constraint is closer to k-anonymity in the sort of privacy guarantee, but has more similarity with DP in the mathematical model and the way of calculation.

The advantage of incorporating the prior knowledge is that having any sort of dependence (MI, covariance) for the same values of variance typically makes the ε-privacy condition easier to satisfy. It also allows to gain some economy of the privacy budget for further measurements.

A practically useful property of Definition 1 is its decomposability. For ordered features, one of the ways to satisfy the privacy constraint is decomposing it into a sequence of conditional constraints:

$$Prob_P\{\mathbf{A}(d_1,\ldots,d_j)=\mathbf{A}(d'_1,\ldots,d'_j)$$

$$|d_1=d'_1,\ldots,d_{j-1}=d'_{j-1}\}$$

$$\geq e^{-(\varepsilon_j-\varepsilon_{j-1})}$$

where ε_k is the part of privacy budget allocated for first k measurements.

3.2 Application to Binning Strategy

The strategy should satisfy the privacy constraint for selected ε, at the same time being as non-aggressive (keeping as much data for research) as possible. In this example, we consider a strict order $\{i : i \Leftarrow j\} = \{1,\ldots,j-1\}$, for the general case it may be defined by analogy.

As justified in the review section, we consider deterministic binning as the preferable approach of anonymizing. For simplicity, we exploit uniform binning (rounding) strategies that use bins of equal size when working with a concrete feature. They are given by formulas of type $\mathbf{A}(D) = B$ where $D = (d_1,\ldots,d_m)$, $B = (b_1,\ldots,b_m)$ and

$$b_j = \left[\frac{d_j}{r_j}\right] r_j \pm \frac{r_j}{2}$$

where square brackets mean rounding off (replacing a number with the closest integer) and the values r_j is the resolution level for the j-th feature. For example, if $r_j = 10^{-3}$, this means leaving only 3 significant digits after the point. Let us also use the notation $\bar{b}_j$ for the average $\left[\frac{d_j}{r_j}\right] r_j$.

In time scale, r_j can depend on previous measurements $b_1,\ldots,b_{j-1}$ as variables. If the dependence is essential, we call this dependent binning unlike independent binning with constant values of $r_1,\ldots,r_m$. The main challenge is to find a balanced value for r_j on each step. It has to be as small as possible (in order to save informativeness of the data) but large enough to satisfy the privacy requirement.

In order to define r_j in a unique way, some details of the model has to be chosen. They include the original parameters (such as ε, **P**) and some other decisions of the realisation. The most important degree of freedom is the balance between different features: there may be several equally good solutions, different in the priority given to the features. In the case of dependent binning, there is also a choice of strategy in how the load is distributed by different cells: some of them may be divided with better resolution at cost of the others.

We normally expect that the constraint weakens as the number of features increases: ε is an increasing function of the number of dimensions (features). Otherwise, as far as the constraints have been satisfied earlier in a tight manner, there will be no room for non-trivial representation of the new features at all. Therefore, for on-line modelling an increasing sequence

$$\varepsilon_1 < \varepsilon_2 < \cdots < \varepsilon_{m-1} < \varepsilon_m = \varepsilon$$

where the increment $(\varepsilon_j - \varepsilon_{j-1})$ means the part of privacy budget allocated to the feature j.

4 Modelling Example

In this paper, we will present some artificial samples following the key challenges related to the problem. Therefore, we discuss modelling the data structure and the structure of possible extra information related to the data on a synthetic example motivated by the problems appearing in IMIoT.

Let the data instance for a patient have the following form:

$$D = (d_1, \ldots, d_m)$$

where d_j is j-th measurement (momental observation of one of the sensors), m is the overall number of measurements. In general, j-th feature is a measurement of a sensor s_j taken at the time moment t_j. The schedule for measurements of a sensor may be regular (e.g., each sensor produces a measurement daily at a fixed time) and known in advance or partially unpredictable (stochastic). When a new measurement arrives, the vector becomes larger by one feature. The same happens if a new device is installed and used for the first time.

Let $i \Leftarrow j$ roughly mean measurement i is registered in the system before j. This implies logical meaning rather than directly time-related, saying more exactly: the value of measurement i and the knowledge about joint distribution of features i and j can be used while processing the feature j. More generally, this may be a non-strict and/or a partial order. We will use notation $i \leftarrow j$ meaning: "$i \Leftarrow j$ and not $j \Leftarrow i$". This means j that the measurement i is taken strictly before j, so i can be used while processing j but not the other way. This is the most typical situation.

It may happen that some measurements may have to be processed independently on each other: when one of them is processed, no information about the second one can be used. In this case neither $i \Leftarrow j$ nor $j \Leftarrow i$, and any mutual knowledge about i and j can be used only later for such k that both $i \Leftarrow k$ and $j \Leftarrow k$. In a more fortunate case, some measurements may appear simultaneously and can be processed together, then both $i \Leftarrow j$ and $j \Leftarrow i$. However, we assume that if the measurements are marked with integers i, j then it never happens that $i \leftarrow j$ but $j < i$.

In this description, the measurements have to be transformed for keeping privacy immediately after their addition to the data. Alternatively, it may be assumed that they are kept as they are due to limited disclosure on the queries. In this condition, there is no need to state the problem in a principally different way, but the order of the measurements would refer to the time of query rather than to the time of actual measurement.

4.1 Modelling External Knowledge

External knowledge can be understood as some system of restrictions applied to the joint distribution of data features. Therefore, we discuss in in the form of statistical model (a class of density functions) satisfying these restrictions.

Initially, we concentrate on a measurement having the form of continuous variables. For clarity, we assume that all the distributions have densities.

Anonymizing of records from a patient is based on the assumption that there may be patients with the same (or similar) schedule of measurements, so the feature vector D is generated by a relevant distribution. We assume that feature vectors D are generated independently of each other but their features are dependent and some information about their joint distribution is available. This information stems from dependence between the parameters of a patient's health state, and the noise (imprecision) level of observed measurements that we actually have to work with.

We use the notation $\mathbf{P}$ for the class of density functions on the feature vectors space compatible with knowledge about the joint distribution of observed measurements. In the best case, the density function of the distribution of D is known exactly, so $\mathbf{P}$ is a one-element set. Otherwise, it may be known to belong to a class $\mathbf{P}$ of size larger than 1.

Typically, it is a parametric model with some constraints on its parameters. The constraints reflect the knowledge of independence and correlations between the features.

The presentation of the knowledge as a class $\mathbf{P}$ is a general approach. Usually, it can be divided into:

1. The knowledge of the distribution of individual features.
2. The knowledge of the character of their dependence or independence.

The first type of information may be such as a parametric model for data distribution. Generally saying, it is desirable to present the second kind of knowledge (dependence) in a quantitative way such as MI mentioned e.g., in [19].

MI is a form of generalisation of correlation. In [20], MI concept is exploited in the area of DP in another form: MI between the sets of data instances (not features) was taken into account. For example, **P** may be the set of all density functions P from some parametric class such that MI for two features (i, j) is above a known threshold I_{ij}, and the correlation between these two features is known to be either positive or negative for each of the pairs. At the same time some pairs of features are known to be completely independent i.e., $I_{ij} = 0$.

In particular, in the Gaussian context, the correlation is known to be connected to MI ρ by the following relation:

$$I_{ij} = -\frac{1}{2}\log(1-\rho_{ij})^2.$$

While working with multi-dimensional normal distributions as example, the language of covariance/correlation might be preferable because it also includes the sign information (whether the dependence is positive or negative).

4.2 Modelling a Sequence of Measurements

Our task is to convert knowledge about the data distribution into a binning strategy satisfying the property of DP for a given ε. In the computational schemes, we concentrate on presenting a scheme of sequential updates with dynamical distribution of privacy budget. Therefore, we mainly address one-distribution model and a strict totally defined order.

Our practical aim is to deal with multi-sensor systems where each sensor produces multiple measurements at different time points. The schedule itself is actually a sequence of pairs (s_j, t_j) where s_j refers to a sensor's number in the system, t_j is the time moment. Although the schedule may be regular, it is more interesting to assume for the aims of modelling that the sequence of time moments is random (stochastic) and not completely known in advance.

Looking realistically, we cannot rely on a complete synchronisation between different patients. However, for a particular patient it may be considered as a reasonable approximation, assuming that many others patients have a similar schedule.

According to principles formulated in Sect. 4.3, when a new feature is added, is required to satisfy the property:

$$\begin{aligned} &Prob_P\{\mathbf{A}(d_1,\ldots,d_k) \\ &= \mathbf{A}(d'_1,\ldots,d'_k)|\mathbf{A}(d_1,\ldots,d_{k-1}) = \mathbf{A}(d'_1,\ldots,d'_{k-1}) \\ &\geq e^{-(\varepsilon_k-\varepsilon_{k-1})} \end{aligned}$$

for any density function $P \in \mathbf{P}$. Indeed, this ensures a privacy requirement for j dimensions if it is already satisfied for $j-1$ dimensions.

When a new feature is added to the model, it follows the sequence of already binned ones. A new version of binning can include both re-binning (further compression) of previous features (especially too old ones) and binning of the new features. We start with models where only the second type of binning is done.

On the other hand, we allow the binning to be dependent: r_j can be a function of previous feature values. Normally, in that case there are many possible ways to satisfy the constraint for j dimensions given it is satisfied for $j-1$ dimensions. One of them is to find a constant function that in its pure form would reduce the case to independent binning. The second approach is the opposite: to ensure some uniformity of conditional binning. Within each of the cells $(b_1, \dots, b_{j-1})$, the conditional probability of two instances being indistinguishable in the 'new' binning is about the same. In this work, we prefer modelling the second option purely although in the future work, some desirable compromise may be found.

4.3 Dynamic Distribution of Privacy Budget

Initially, we can have some knowledge useful for selection of the privacy budget distribution strategy. In may include the variance or precision of each sensor's measurement, relative importance of sensors (that may be not completely reducible to the variance) and third, about a deterministic or (as in our example) a stochastic mechanism generating the schedule of measurements for each of the sensors.

We quantify two value related to a sensor. The first one is a weight factor assigned to each sensor i. It may be increased for the sensors that produce more important information: for example, medical measurements compared to environmental ones. The other parameter is relative expected frequency f_i of measurements coming from the sensor i. We also assume that the privacy budget each feature has its natural limit, so the j-th feature is initially rounded at level $\hat{r}_i$ called sufficient (desirable) level of resolution.

This leads to a possible solution: the privacy budget is shared by measurements proportionally to a sensor's weighting factor, and normalized in such way that expected sum of ε_j over $j = 1, \dots, m$ is equal to ε, i.e.,

$$\varepsilon_j^A - \varepsilon_{j-1} = \frac{w_{s_j}}{m \sum_{i=1}^{q} \left(w_i \frac{f_i}{\sum_i f_i} \right)} = \frac{w_{s_j} \sum_{i=1}^{q} f_i}{m \sum_{i=1}^{q} w_i f_i}$$

where q is the total number of sensors. This formula gives a preliminary allocation of the privacy budget: it is constructed for the worst case when the measurements are completely independent. Having any sort of dependence (covariance) for the same values of variance allows to make savings of the budget for the future steps.

4.4 Computational Scheme and Complexity

The whole computation scheme with selection of ε_j is presented in Algorithm 1. Each step j related to a specific measurement from one of the sensors. The number of the sensor taking the measurement j is denoted as $i = s_j$. Making a step includes allocation of the privacy budget (selection of ε_j). When a feature arrives, we first allocate it the maximal limit of ε_j calculated from the sensor's weight and frequency and then try to decrease it by utilizing information about the limit resolution of the sensor.

Algorithm 1 Computational scheme

1: INPUT m, number of measurements
2: INPUT ε, total privacy budget parameter
3: INPUT q, number of sensors
4: INPUT $(f_1, \ldots, f_q)$, expected frequency
5: INPUT $(w_1, \ldots, w_q)$, weight factors
6: INPUT $(\hat{r}_1, \ldots, \hat{r}_q)$, sufficient resolution
7: set remaining funds $\varepsilon^r := \varepsilon$
8: set $\varepsilon_0 := 0$
9: **for** $j := 1, \ldots, m$ **do**
10: INPUT the observation d_j of the sensor $s_j \in \{1, \ldots, q\}$, with the time label t_j
11: make corresponding update of the statistical model
12: make the preliminary binning $b_j := \left[\frac{d_j}{\hat{r}_{s_j}} \right] \times \hat{r}_i$
13: allocate $\varepsilon_j := \varepsilon_{j-1} + \frac{w_{s_j} \sum_{i=1}^{q} f_i}{(m-j+1) \sum_{i=1}^{q} f_i w_i} \varepsilon^r$
14: **if** actual ε_j is already smaller than required **then**
15: replace ε_j with its actual value
16: **else**
17: make a step of sequential binning s.t.

$$\begin{aligned} Prob\{(\delta_1, \ldots, \delta_j) : \mathbf{A}(\delta_j) = \mathbf{A}(\delta'_j) \\ | \mathbf{A}(\delta_1) = \mathbf{A}(\delta'_1) = b_1, \ldots, \mathbf{A}(\delta_{j-1}) = \mathbf{A}(\delta'_{j-1}) = b_{j-1}\} \\ \geq e^{-(\hat{\varepsilon}_j - \varepsilon_{j-1})} \end{aligned}$$

for any distribution $Prob$ on the first j measurements which is compatible with the statistical model. (To reach the equality is desirable, to keep the inequality true is required.)

18: **end if**
19: cut remaining funds $\varepsilon^r := \varepsilon - \varepsilon_j$
20: **end for**

Computational complexity of a concrete step actually depends on the method of performing the followig lines of the Algorithm 1, that are the most time-consuming and have some freedom in the form of their computational realisation:

- 14–15: IF actual ε_j is smaller than required even without binning, THEN replace ε_j with its actual value;

- 16–17: ELSE make a step of sequential binning s.t. the (conditional) privacy constraint for j and preceding features is satisfied for ε_j;

Let us start with case of 'exact' knowledge $\mathbf{P} = \{P\}$.

For the second (more typical) case, the required result of binning, acceptable for a given privacy constraint, may sometimes be calculated analytically. In a more general case, this can be done by Monte-Carlo simulation of a large enough number K of artificial data examples, following the model. At a step j, the conditional distribution to be simulated is:

$$P\left\{d_j|\mathbf{A}(d_1) = b_1, \ldots, \mathbf{A}(d_{j-1}) = b_{j-1}\right\}$$

that may be approximated by

$$P\{d_j|d_1 = \overline{b}_1, \ldots, d_{j-1} = \overline{b}_{j-1}\}$$

if cutting a proper segment from the space becomes problematic for a large number of dimensions. Then is possible to make scanning of possible resolution levels up to some precision R. Note that making r_j larger than it is does not affect validity of the model. Therefore, by selection of K and R the calculation can be fitted into a time limit given for a step j.

The first case appears if the initial version of binning (up to $\hat{r}_j$) already satisfies the privacy constraint with some surplus, i.e., for a smaller version of ε_j than required. This may be done e.g., by measuring the empirical conditional probability P that two randomly generated sequences D_j and D'_j become equal after the binning, and reverting it backwards to

$$\hat{\varepsilon}_j = -\log Prob\left\{\mathbf{A}(d_k) = \mathbf{A}(d'_k)\right\}.$$

Again, the time can be regulated by the number of artificially generated examples.

The same approach can be used in the case of 'inexact' knowledge when $\mathbf{P}$ consists of several distributions, the privacy constraint can be checked for each of them. If $\mathbf{P}$ has such form as parametric distribution with some range of parameters, then a scanning can be made over a grid within the allowed parameter range.

4.5 Creating Synthetic Data

In this section, we present some artificial examples presenting the principal phenomena which we expect to appear in real data. We assume that each of the individual measurements taken separately is normally distributed. This is not the only possible case but the simplest of typical cases. However, we model the case where different measurements taken from the same patient are dependent. For measurements taken by the same sensor the correlation is positive, and decreases as the time between them increases. Dependence between measurements from different sensors also becomes smaller with the time, but it may be negative as well, or negligibly small. Our model tries to include these possibilities.

When dependence between the features is actually presented, it may be presented in the knowledge completely or partially. We model this by offering the versions of 'exact' and 'approximate' knowledge. As a simple example of approximate knowledge, we take the case when the distribution is known to be one of two.

We start with basic two-feature tasks (independent and dependent binning) that may be visualized. The principle of visualisation of a two-dimensional distribution is generation of its random sample and showing in two dimensions together with binning cells. Then, we model a multi-feature case with 10 measurements from 2 sensors.

4.6 Basic Two-Feature Examples

In the examples of this section, we presume that $\mathbf{P}$ consists of 1 or 2 distributions from the following list:

- P_0 is the density of a two-dimensional normal distribution with independent features and diagonal covariance matrix $\begin{pmatrix} 1 & 0 \\ 0 & 2 \end{pmatrix}$
- P_1 and P_2 are the densities of skewed normal distributions with covariance matrices $\begin{pmatrix} 1 & 1 \\ 1 & 2 \end{pmatrix}$ and $\begin{pmatrix} 1 & 0.5 \\ 0.5 & 2 \end{pmatrix}$

Three possible distribution classes are used as examples:

1. Independence known: $\mathbf{P}_I = \{P_0\}$
2. A dependence known exactly: $\mathbf{P}_D = \{P_1\}$
3. Partially agnostic (the dependence known approximately): $\mathbf{P}_A = \{P_1, P_2\}$

If a $(\mathbf{P}, \varepsilon)$-privacy constraint is satisfied, we need some criterion for the evaluation of the results to compare different versions of binning. Clearly, a binning is more efficient if its resolution is higher i.e., resolution coefficients r_i are lower, and the binning cells are smaller. One possible criterion is using a small cell volume as evaluation criterion, but for better uniformity by features, it might be preferable to assess the cell's diameter, which we use here.

It can be seen from Table 1 how these examples illustrate principal features of the suggested privacy criterion. If the binning was done according to the model which does not include the real density of the data distribution, then it may be invalid i.e., break the required privacy constraint. If the model behind the binning covers the data distribution, then the results are valid anyway, but typically they are more efficient (in terms of an evaluation criterion) if the class $\mathbf{P}$ of possible distributions is smaller. For example, it is easier to fit the binning to one possible density function than to two or more functions.

Table 1 Applicability of binning models to data records generated by P_0 or P_1

Model	Applied to P_0	Applied to P_1
Independence $\mathbf{P}_I = \{P_0\}$	(Sect. 4.6.1)	INVALID
Exact kn. $\mathbf{P}_D = \{P_1\}$	INVALID	(Sect. 4.6.2)
Approx. kn. $\mathbf{P}_A = \{P_1, P_2\}$	INVALID	(Sect. 4.6.3)

The plots for two-dimensional example will be included into Fig. 1. The points a random sample of the size $K = 10^4$ of the 'real' data distribution behind the data record. The cells are areas of possible records that become identical after applying the privacy transformation. More details will be explained further in the corresponding paragraphs.

4.6.1 Independent Case

We start with a model that does not include any update: two features does appear synchronically (both $1 \Leftarrow 2$ and $2 \Leftarrow 1$ are true). The task is to select resolution levels for two dimensions at the same time.

Assume that P is a normal distribution with independent dimensions, i.e., the vectors d and d' are generated independently from $N(0, \sigma_1) \times N(0, \sigma_2)$. Let us consider the function

$$Q(r_1, r_2) = Prob\left\{\mathbf{A}(D) = \mathbf{A}(D')\right\}.$$

Due to independence, $Q(r_1, r_2) = Q_1(r_1)Q_2(r_2)$ where

$$Q_l(r_l) = Prob\left\{\left[\frac{d_l}{r_l}\right] = \left[\frac{d'_l}{r_l}\right]\right\} = Q_0\left(\frac{r_l}{\sigma_l}\right);$$

$Q_0(r)$ is the same as $Q_1(r_1)$ in assumption that $\sigma_1 = 1$.

A pair (r_1, r_2) is acceptable for ϵ if $Q_0\left(\frac{r_1}{\sigma_1}\right) Q_0\left(\frac{r_2}{\sigma_2}\right) \geq e^{-\epsilon}$. From different acceptable pairs (r_1, r_2) we prefer one with the smallest cell diameter $\sqrt{r_1^2 + r_2^2}$. This can be also found by simulation experiments. For the case $\sigma_1^2 = 1, \sigma_2^2 = 2$, the resulting split of the feature space is shown in Fig. 1. Each of the records has two features, and the records in the same cell are indistinguishable after the binning.

4.6.2 Introducing Feature Dependence

Assume that P is known to be a two-dimensional skewed normal distribution with covariance matrix $\begin{pmatrix} 1 & 1 \\ 1 & 2 \end{pmatrix}$.

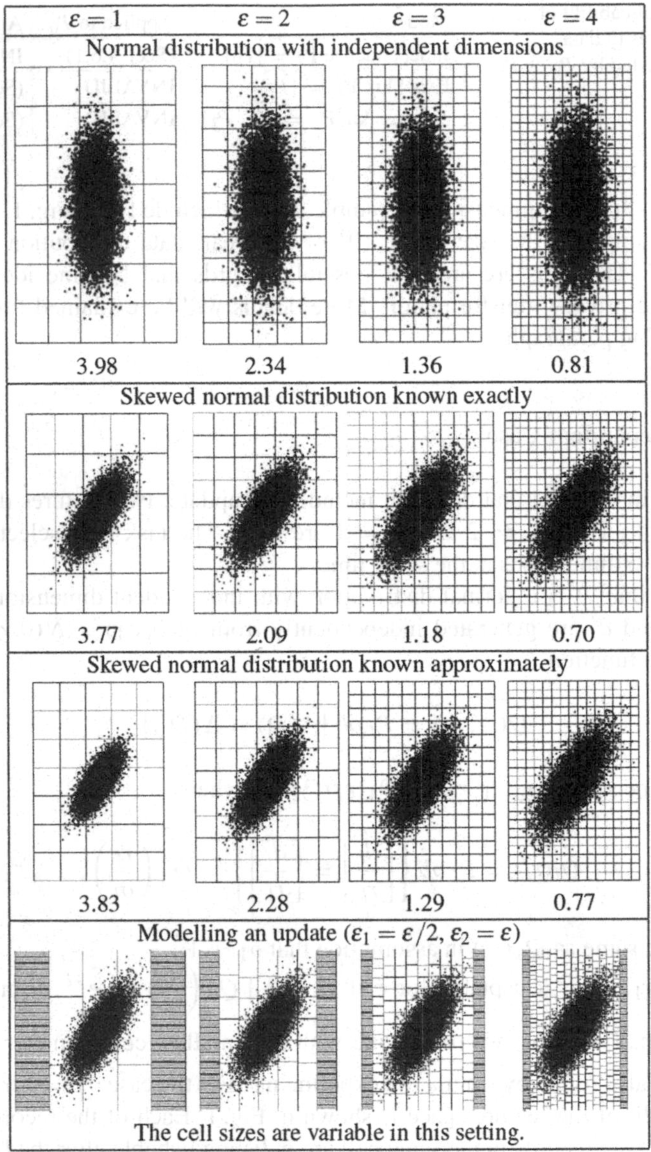

Fig. 1 Optimal resolutions and corresponding binning cells.

We are looking for resolution level pairs (r_1, r_2) acceptable for ε. In this example, we suggest that the features are not ordered by time, as in the case of simultaneous measurements, so these coefficients may be selected together.

In this case we also have $Q(r_1, r_2)$ as a function of the resolution levels r_1 and r_2. The pairs (r_1, r_2) such that $Q(r_1, r_2) \geq e^{-\varepsilon}$ are called acceptable. From different acceptable pairs (r_1, r_2) we would like to choose the optimal one with the smallest cell diameter $\sqrt{r_1^2 + r_2^2}$.

The resulting split of the feature space is shown in Table 1. Typically, the cells are smaller for the same values of ε than analogous cells for independent case. It illustrates the economy gained by using extra information about the feature dependence.

4.6.3 Partially Agnostic Model

Let us now consider an example of inexact knowledge of the distribution. Assume that the density of the distribution behind the data is known to be either P_1 or P_2 but not known which of them. This means that we can choose resolution from the pair (r_1, r_2) which are acceptable for both distributions. By analogy, this can be extended to a larger family of distributions.

In the example, we assume that the true density is P_1. The results are shown on figures in Table 1. As expected, incomplete knowledge about the distribution leads to some loss of efficiency (lower resolution).

4.6.4 Update by Adding a New Measurement

Let us now use the same skewed example as in Sect. 4.6.2 for modelling an update and dependent binning. Use $\varepsilon = \varepsilon_2 = 1, 2, 3, 4$ for two features and $\varepsilon_1 = \frac{1}{2}\varepsilon_2$ for feature 1. According to our approach of modelling update discussed in Sect. 4.3, the constraint is split into two steps:

$$Prob\{\mathbf{A}(d_1) = \mathbf{A}(d_1')\} \geq e^{-\varepsilon_1}$$

$$Prob\left\{\mathbf{A}(d_1, d_2) = \mathbf{A}(d_1', d_2') | \mathbf{A}(d_1) = \mathbf{A}(d_1') = b_1\right\}$$

$$\geq e^{-(\varepsilon_2 - \varepsilon_1)}$$

Here, we force the second condition to be uniform by cells. It has to be satisfied separately for any possible value of b_1. i.e., for each of the first dimension bins. This actually makes the second dimension resolution r_2 dependent on the first dimension value b_1. The results are shown on figures in Table 1.

4.7 *Multi-dimensional Sequential Example*

Now, we generate sequences of 10 measurements from 2–3 sensors. In our model, we set the parameters according to Table 2. Time between the measurements of the

Table 2 System parameters

Sensor. no i		s1	s2
Time between measurements		Exponentially distributed $\lambda = 1$	$\lambda = 0.5$
Frequency f_i		1	2
Variance		1	1.5
Covariance as a	s1	e^{-t}	$-e^{-4t}$
function of time t	s2	$-e^{-4t}$	$1.5e^{-2t}$
Privacy budget		$\varepsilon = 10$ for $m = 10$ measurements	
Weight w_i		1.5	0.75
Sufficient resolution $\hat{r}_i$		0.1	0.1

Table 3 Random schedule and covariance matrix (2-sensor version)

No.	Sensor	Time	Covariance matrix									
1	1	0.05	1	−0.32	−0.16	−0.13	−0.06	0.23	−0.00	0.14	−0.00	−0.00
2	2	0.33	−0.32	1.5	1.27	1.20	0.97	−0.01	0.29	−0.00	0.15	0.15
3	2	0.50	−0.16	1.27	1.5	1.42	1.15	−0.02	0.34	−0.00	0.18	0.17
4	2	0.56	−0.13	1.20	1.42	1.5	1.21	−0.02	0.36	−0.00	0.19	0.18
5	2	0.77	−0.06	0.97	1.15	1.21	1.5	−0.05	0.45	−0.01	0.24	0.23
6	1	1.51	0.23	−0.01	−0.02	−0.02	−0.05	1	−0.15	0.61	−0.01	−0.01
7	2	1.98	−0.00	0.29	0.34	0.36	0.45	−0.15	1.5	−0.89	0.80	0.76
8	1	2.01	0.14	−0.00	−0.00	−0.00	−0.01	0.61	−0.89	1	−0.09	−0.07
9	2	2.61	−0.00	0.15	0.18	0.19	0.24	−0.01	0.80	−0.09	1.5	1.43
10	2	2.66	−0.00	0.15	0.17	0.18	0.23	−0.01	0.76	−0.07	1.43	1.5

same sensor is assumed to be random (exponential) with a known mean value. A random schedule for measurements created according to this setting is shown in Table 3. The experiments are conducted using Monte-Carlo simulation with $K = 10^7$.

The whole computation cycle is performed according to Algorithm 1. The input values of f_i, w_i and $\hat{r}_i$ (where i refers to a sensors' number) are taken from Table 2.

The results are reflected in Table 4. For comparison, the 'static' version of allocated privacy budget is shown in brackets. It is calculated as if the knowledge of the mutual dependence were not used. The difference between 'allocated' and 'used' actually reflects the economy.

5 Conclusions

In this work, we modify a privacy model for the sequential measurement data supplied with external information about feature dependencies. For this aim, we formulate a modified version of differential privacy constraint (that is correct in the case of dependencies) and provided a binning strategy that satisfies it. Our work

Table 4 Dynamic allocation of the privacy budget (2 sensors)

No.	Sensor	Time	('Static')	Allocated	Used	Remains
–	–	0	–	–	–	10/10
1	1 (1.5)	0.05	(1.50)	1.50	1.40	8.60/9
2	2 (0.75)	0.33	(0.71)	0.72	0.70	7.90/8
3	2 (0.75)	0.50	(0.73)	0.74	0.67	7.23/7
4	2 (0.75)	0.56	(0.76)	0.77	0.75	6.48/6
5	2 (0.75)	0.77	(0.79)	0.81	0.70	5.78/5
6	1 (1.5)	1.51	(1.65)	1.73	1.62	4.16/4
7	2 (0.75)	1.98	(0.72)	0.78	0.76	3.40/3
8	1 (1.5)	2.01	(1.57)	1.70	1.36	2.04/2
9	2 (0.75)	2.61	(0.59)	0.77	0.71	1.33/1
10	2 (0.75)	2.66	(0.74)	1.00	0.91	0.42/0

was intentionally focused on incorporating external information about the data. Motivated by MIoT challenges, the proposed model is applicable to the cases where external information can be extracted in the form of knowledge of dependence between the features.

During the modelling, we have also used prior meta-information such as relative importance of the sensors, and initial estimate of the sufficient resolution. Although the definition of privacy constraint in its general form is a high-level one, we have demonstrated possibility of computationally efficient ways of checking it with satisfactory precision. We have also checked practically that the saving for privacy budget can be made by as far as the mutual information between the measurement is high, and this is reflected in prior knowledge.

In future work, this scheme may be also to the system updates that means adding or removing the sensors. Earlier we have mentioned re-binning of old features as an alternative to over-binning of new features. This may be actual when a new sensor is added to the systems, and its weight is high enough, so that allocation of the privacy budget to the old measurements has to be revised. Re-binning can also take into account that some features lose their actuality with time, in this case stronger binning can be applied to older features, which means that old allocations ($\varepsilon_j - \varepsilon_{j-1}$) are decreased for smaller j in favor of larger ones.

List of Notations

$\Leftarrow, \leftarrow$, partial orders on the measurements – Sect. 4
[], the rounding function – Sect. 3.2
$\mathbf{A}$, a privacy transformation – Sect. 3.2
B, the result of a privacy transformation $\mathbf{A}$ – Sect. 3.2
D, individual data record – Sect. 4
$b_j, \overline{b}_j$, general/average result of rounding privacy transformation – Sect. 3.2
d_j, a specific measurement – Sect. 4

f_i, estimated frequency of measurements taken by a sensor i – Sect. 4.3
I_{ij}, a threshold for MI – Sect. 4.1
K, the number of Monte-Carlo simulations – Sect. 4.4
m, the number of measurements in a data record – Sect. 4
P_0, P_1, P_2, examples of density functions – Sect. 4.6
P, the knowledge in the form of a class of density functions – Sect. 4.1
$\mathbf{P}_A$, $\mathbf{P}_D$, $\mathbf{P}_I$, example of **P** – Sect. 4.6
Q, Q_0, Q_1, Q_2, auxiliary functions for calculations – Sect. 4.6.1
q, the number of sensors – Sect. 4.4
R, precision of scanning the values of the resolution levels – Sect. 4.4
R_j, resolution level r_j as a function of previous measurements – Sect. 3.2
r_j, resolution level for a measurement – Sect. 3.2
$\hat{r}_i$, desirable resolution level for a sensor's measurement – Sect. 4.3
s_j, the sensor taking a specific measurement j – Sect. 4
t_j, the time moment of a specific measurement j – Sect. 4
w_i, weight of a sensor – Sect. 4.3
ε, overall privacy budget – Sect. 3.1
ε_j, part of the privacy budget – Sect. 3.1
ε^r :, remaining funds of the privacy budget – Sect. 4.4
ρ_{ij}, correlation coefficients – Sect. 4.6
σ_1, σ_2, examples of variance – Sect. 4.6.1

Acknowledgements This work was supported by Technology Integrated Health Management (TIHM) project awarded to the School of Mathematics and Information Security at Royal Holloway as part of an initiative by NHS England supported by InnovateUK. It was also supported by European Union grant 671555 ("ExCAPE") and AstraZeneca grant "Machine Learning for Chemical Synthesis" (R10911).

References

1. Darwish S, Nuretdinov I, Wolthusen S (2017) Towards composable threat assessment for medical IoT (MIoT). In: The Fourth International Workshop on Privacy and Security in Healthcare (PSCare17), vol 113, pp 627–632
2. Samaratiy P, Sweeney L (1998) Pierangela: protecting privacy when disclosing information: k-anonymity and its enforcement through generalization and suppression. Tech. Rep., SRI International
3. Dwork C, Roth A (2014) The algorithmic foundations of differential privacy. Found Trends Theor Comput Sci 9(3/4):211–407. https://doi.org/10.1561/0400000042
4. Darwish S, Nuretdinov I, Wolthusen S (2018) A dynamic distributed architecture for preserving privacy of medical IoT monitoring measurements. In: 16th International Conference on Smart Homes and Health Telematics (ICOST'2018), pp 106–115
5. Vovk V (2006) Well-calibrated predictions from on-line compression models. Theor Comput Sci 364(1):10–26. https://doi.org/10.1016/j.tcs.2006.07.038
6. Duncan G (2006) Statistical confidentiality: is synthetic data the answer?. https://web.archive.org/web/20060905115447/http://www.idre.ucla.edu/events/PPT/2006_02_13_duncan_Synthetic_Data.ppt

7. Dalenius T, Reiss SP (1982) Data-swapping: a technique for disclosure control. J Stat Plann Inference 6(1):73–85
8. Reiter J (2004) Simultaneous use of multiple imputation for missing data and disclosure limitation. Surv Methodol 30:235–242
9. HESA (2017) Rounding and suppression to anonymise statistics. https://www.hesa.ac.uk/about/regulation/data-protection/rounding-and-suppression-anonymise-statistics
10. Lin Z, Hewett M, Altman RB (2002) Using binning to maintain confidentiality of medical data. In: Proceedings of AMIA Symposium, pp 454–8
11. Lee Y, Band H, Kim DJ (2016) How to establish clinical prediction models. Endocrinol Metab 31:38–44
12. Kattan MW (ed) (2009) Encyclopedia of medical decision making. https://doi.org/10.4135/9781412971980
13. Zagorecki A, Orzechowski P, Holownia K (2013) A system for automated general medical diagnosis using bayesian networks. In: MEDINFO 2013 – Proceedings of the 14th World Congress on Medical and Health Informatics, 20–13 Aug 2013, Copenhagen, pp 461–465. https://doi.org/10.3233/978-1-61499-289-9-461
14. Franklin SS, Khan SA, Wong ND, Larson MG, Levy D (1999) Is pulse pressure useful in predicting risk for coronary heart disease? Circulation 100(4):354–360. https://doi.org/10.1161/01.CIR.100.4.354. http://circ.ahajournals.org/content/100/4/354
15. Li N, Li T, Venkatasubramanian S (2007) t-closeness: privacy beyond k-anonymity and l-diversity. In: 2007 IEEE 23rd International Conference on Data Engineering, pp 106–115. https://doi.org/10.1109/ICDE.2007.367856
16. Hadian M, Liang X, Altuwaiyan T, Mahmoud MMEA (2016) Privacy-preserving mHealth data release with pattern consistency. In: IEEE Global Communications Conference, pp 1–6
17. Liu C, Chakraborty S, Mittal P (2016) Dependence makes you vulnerable: differential privacy under dependent tuples. In: Network and Distributed System Security Symposium
18. Malekzadeh M, Clegg RG, Haddadi H (2017) Replacement autoencoder: a privacy-preserving algorithm for sensory data analysis. CoRR abs/1710.06564. http://arxiv.org/abs/1710.06564
19. Liese F, Vajda I (2006) On divergences and informations in statistics and information theory. IEEE Trans Inf Theory 52(10):4394–4412. https://doi.org/10.1109/TIT.2006.881731
20. Cuff P, Yu L (2016) Differential privacy as a mutual information constraint. In: Proceedings of the 2016 ACM SIGSAC Conference on Computer and Communications Security (CCS'16), pp 42–54

Security and Privacy Techniques for the Industrial Internet of Things

Yuexin Zhang and Xinyi Huang

Abstract The wide employment of Internet of Things (IoT) across industrial sectors creates the Industrial Internet of Things (IIoT). In practical applications, however, the IIoT has many attack surfaces. As a result, the IIoT is vulnerable to kinds of attacks, including physical attacks (such as the invasive hardware attacks, side-channel attacks and reverse-engineering attacks), malicious code (such as Trojans, viruses and runtime attacks), and other attacks (such as phishing and sabotage). To ensure the security and privacy of the IIoT, many countermeasures have been proposed, a non-exhaustive list includes authentication techniques, secure routing techniques, intrusion detection techniques, signature techniques, and key establishment techniques. As a fundamental countermeasure, key establishment has been extensively and intensively studied. In this chapter, we will present a survey and taxonomy of the key establishment protocols. Specifically, we will review the conventional key establishment protocols which are designed at higher layers and the physical layer. By reviewing the conventional key establishment protocols, we aim to illustrate the necessity of designing cross-layer key establishment protocols for the IIoT. Then, we will provide the detailed review of cross-layer key establishment protocols. The review illustrates that, the cross-layer design enables the IIoT devices to establish communication keys without the trusted entity and the secret sharing assumption. At the end of this chapter, we will provide a conclusion and point out some future research trends of the IIoT.

Y. Zhang
State Key Laboratory of Integrated Service Networks, Xidian University, Xi'an, China

School of Software and Electrical Engineering, Swinburne University of Technology, Hawthorn, VIC, Australia
e-mail: zhangyuexin@xidian.edu.cn

X. Huang (✉)
Fujian Provincial Key Laboratory of Network Security and Cryptology, College of Mathematics and Informatics, Fujian Normal University, Fuzhou, China
e-mail: xyhuang@fjnu.edu.cn

C. Alcaraz (ed.), *Security and Privacy Trends in the Industrial Internet of Things*, Advanced Sciences and Technologies for Security Applications,
https://doi.org/10.1007/978-3-030-12330-7_12

Keywords Industrial Internet of Things (IIoT) · Security and privacy · Key establishment · Cross-layer design

1 Introduction

The cross-employment of the Internet of Things (IoT) among several industrial sectors, such as the manufacturing, logistics, transportation and other sectors, creates the so-called Industrial Internet of Things (IIoT). Typically, in the industrial control systems, the notion of security refers to the safety, i.e., the protection of humans, environment and machines against consequences of system failures. However, protection against cyberattacks becomes the major design goal of the IIoT when it is integrated with information technology [36].

In practical applications, the IIoT systems have many attack surfaces. Take the smart factories as examples. In smart factories, there are several Cyber-Physical Production Systems (CPPS), including electronics and monitors. Specifically, the electronics are driven by certain software such as the embedded operating systems and applications. It interacts with human beings and other CPPS via various network connections. In practice, electronics are vulnerable to physical attacks, such as the invasive hardware attacks, side-channel attacks and reverse-engineering attacks. Additionally, the software also can be compromised by malicious code, including Trojans, viruses and runtime attacks. Even our human operating CPPS may subject to attacks such as phishing and social engineering [36].

To combat these security and privacy threats, a holistic cybersecurity countermeasure should be designed for the IIoT. Specifically, different aspects should be considered, including secure engineering, security and privacy management, identity management, industrial rights management, platform security, and communication security and privacy.

Security and privacy goals of the IIoT Availability is the most important goal of the IIoT, and it requires the prevention of any unnecessary delay in production. Specifically, the delay in the IIoT may result in significant loss of productivity and loss of revenues. Thus, in order to prevent the unnecessary delay, the security and privacy techniques, which are proposed to against the attacks such as the Denial of Service (DoS) attacks, should be carefully designed for the IIoT. Additionally, preventing of any system failure is another important goal of the IIoT. It is due to the reason that in practice, any failure of the IIoT system may lead to physical damage or harm to humans. To prevent the system failure, the integrity should be achieved. Typically, the integrity requires the protection against sabotage. Moreover, any unintended use of counterfeit components should be thwarted. Besides the above two goals, in the IIoT, other security and privacy countermeasures should be provided in order to protect the IIoT against the industrial espionage and protect the privacy of customers and employees.

Until now, many techniques have been proposed, and they can be employed in order to ensure the security and privacy of the IIoT, a non-exhaustive list includes authentication techniques [49, 50, 55, 56], secure routing techniques [38, 42, 54], privacy-preserving techniques [29, 41, 57], intrusion detection techniques [16, 30, 40], signature techniques [10, 19, 20, 26, 51, 72], and key establishment techniques. Due to the length limitation, we will concentrate on reviewing the key establishment protocols in the following sections.

As a fundamental technique, key establishment has been extensively and intensively studied. In this chapter, we will present a survey and taxonomy of the key establishment protocols. Specifically, by reviewing the conventional key establishment protocols in Sect. 2, we aim to illustrate the necessity of designing cross-layer key establishment protocols for the IIoT. Then, Sect. 3 provides the detailed introductions of the cross-layer key establishment protocols (including the protocols which are designed in asymmetric and symmetric key settings). In Sect. 4, we will conclude this chapter and discuss future research trends in the IIoT.

2 Conventional Key Establishment Protocols

Numerous key establishment protocols have been proposed, and they can be employed for devices in order to ensure the security and privacy of the IIoT. Typically, conventional key establishment protocols can be classified into two categories (which are designed at higher layers), including asymmetric key establishment protocols and symmetric key establishment protocols. In the past two decades, some key extraction protocols were proposed at the physical layer. Specifically, these key extraction protocols are designed using the physical layer key extraction techniques or using the keyless cryptography techniques. Figure 1 overviews the category of the conventional key establishment protocols. In this section, we will briefly review the conventional key establishment protocols which are designed at higher layers or at the physical layer.

Key establishment protocols which are designed at higher layers At higher layers, the key establishment protocols are designed in symmetric key setting or in asymmetric key setting.

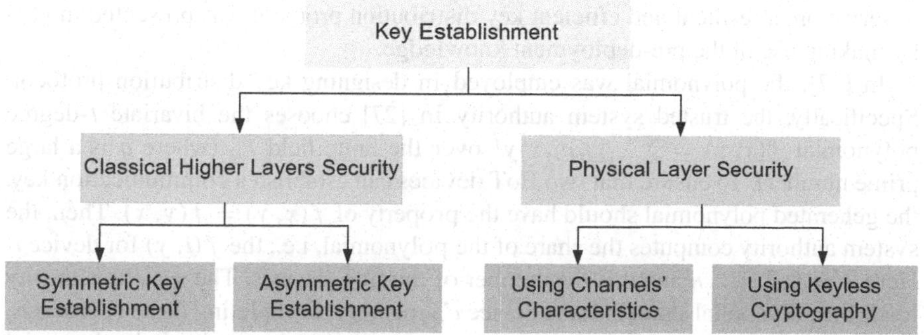

Fig. 1 An overview of the conventional key establishment protocols [69]

Typically, *symmetric key establishment protocols* are lightweight. Namely, the IIoT devices can establish the communication keys with less energy consumption (compared with that of the asymmetric key establishment protocols). Kinds of symmetric key establishment protocols have been proposed, such as entity-based protocols, probabilistic-based protocols, polynomial-based protocols, matrix-based protocols, and tree-based protocols.

In entity-based key establishment protocols, a trusted entity is involved in establishing communication keys. Take the protocol [25] as an example, the trusted entity distributes a common master key for each IIoT device. Then, any two devices can establish a communication key by making use of the master key and random chosen numbers. the protocol presented in [25] achieves the properties include low storage costs and high scalability. However, it achieves low level of security. Namely, the security and privacy of the IIoT devices will be exposed as long as the master key is compromised by the adversary. To improve the security of [25], in [71], the IIoT devices are required to erase the master key after they establish the communication keys. Furthermore, the base station was involved in the key establishment protocol of [32]. Specifically, the base station in [32] serves as the trusted entity. Additionally, it is assumed that each IIoT device shares a unique master key with the base station. Then, two devices can establish a communication key with the assistance of the trusted base station.

A probabilistic key distribution protocol was proposed in [17], then, it was improved by [12]. These probabilistic key distribution protocols [12, 17] consist of three phases, including the key pre-distribution phase, the shared key discovery phase, and the path key establishment phase. In the key pre-distribution phase, a trusted system authority chooses t keys randomly from the key pool P for each IIoT device. Using these t chosen keys, a key ring can be generated. Then, the system authority loads the key ring and the keys' identifiers into the device's memory. These operations need to be completed before the deployment of the devices. In the shared key discovery phase, the IIoT devices broadcast the loaded key identifiers such that they can find the keys which are shared with neighbor devices. Then, neighbor devices can establish a communication key using the shared keys. In the path key establishment phase, two IIoT devices can establish a path key with the help of these devices which located at two different links between them. In some applications, the pre-deployment knowledge of the IIoT devices is available. Motivated by this observation, a resilient and efficient key distribution protocol was presented in [14] by making use of the pre-deployment knowledge.

In [27], the polynomial was employed in designing key distribution protocol. Specifically, the trusted system authority in [27] chooses the bivariate t-degree polynomial $f(x, y) = \sum_{i,j=0}^{t} a_{ij} x^i y^j$ over the finite field F_q (where q is a large prime number). To ensure that two IIoT devices can establish a communication key, the generated polynomial should have the property of $f(x, y) = f(y, x)$. Then, the system authority computes the share of the polynomial, i.e., the $f(i, y)$ for device i. Here, $i = 1, 2, \ldots, n$ and n is the number of the IIoT devices. The system authority loads the polynomial share into the device i's memory. Completing these operations, any two devices can establish a communication key using the pre-loaded polynomial

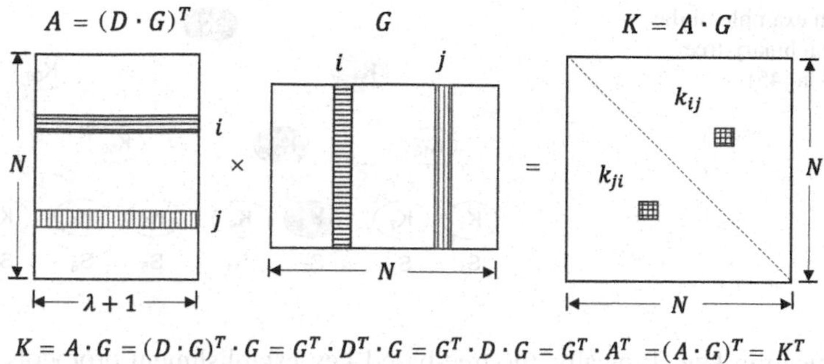

Fig. 2 The main idea of Blom's matrix-based key distribution protocol [63]

shares. Moreover, employing the bloom filter technique, the improved polynomial-based key distribution protocol was presented in [60].

In 1984, a matrix-based key distribution protocol was presented in [7]. Specifically, the protocol ensures that any two IIoT devices can directly establish a communication key as long as they pre-load secrets from the same matrix space. Figure 2 provides the basic idea of [7]. In [7], the system authority chooses the $(\lambda + 1) \times N$ public matrix G and the $(\lambda + 1) \times (\lambda + 1)$ secret symmetric matrix D over the finite field F_q, where N is the number of the IIoT devices. Using matrices G and D, the system authority computes the $N \times (\lambda + 1)$ matrix $A = (D \cdot G)^T$. Then, the system authority loads the ith row of matrix A and ith column of matrix G for the ith device. These operations ensure that two IIoT devices can establish a communication key by exchanging the column of the public matrix G. the protocol [7] was improved in [13] by providing a new design of matrix G. Additionally, it was optimised in [15] by taking advantage of the deployment knowledge.

To reduce the communication and storage costs of energy-constraint IIoT devices, a new matrix-based key distribution protocol was presented in [53]. Different from the aforementioned matrix-based key distribution protocols, in [53], the matrix A was used as the public matrix and the matrix G was used as the secret matrix. Furthermore, the pre and/or post deployment knowledge was employed in [61–63, 65] in order to improve the performance of [53].

In [45], the hierarchical binary tree was introduced to design the group key establishment protocol. The group controller in [45] constructs the hierarchical binary tree using keys. Specifically, each node in the tree denotes a key encryption key. Additionally, each IIoT device S_i in the group represents a leaf of the tree. To establish a group communication key, each device needs to load the nodes' key encryption keys from the leaf to the root. Figure 3 provides an example of the hierarchical binary tree employed in [45]. For a balanced tree, each IIoT device needs to load $log_2 n$ keys. We denote by n the number of group members and denote by $log_2 n$ the height of the tree. For instance, the IIoT device S_3 in Fig. 3 needs to load $log_2 8 = 3$ keys, i.e., K_3, K_{34} and K_{14}, in order to establish a group

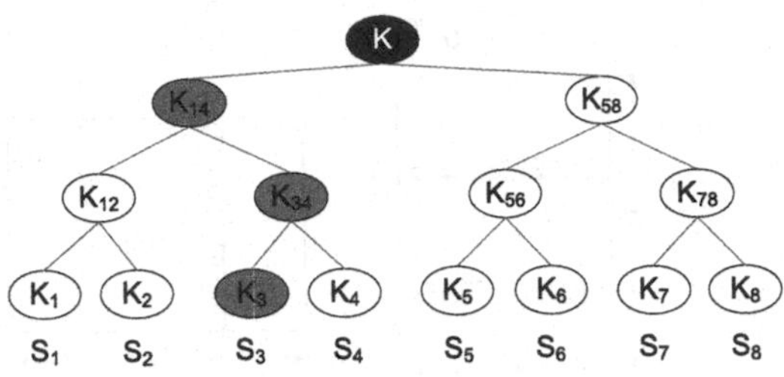

Fig. 3 An example of the hierarchical binary tree employed in [45]

communication key. Typically, the tree-based key establishment protocols support dynamic update, and the re-key message size is less than $O(2log_2n)$.

Combinatorial design theory also can be used in designing the key establishment protocols, examples in this field can be found in [9, 35, 37, 39]. For example, to decide the number of pre-loaded secret keys, the combinatorial design theory was employed in [9]. Specifically, the Balanced Incomplete Block Design (BIBD) and Generalized Quadrangles (GQ) were mapped in [9] in order to obtain efficient key distribution. To against the node capture attacks, a new key establishment protocol, which is called the Strong Steiner Trade, was presented in [35] using the BIBD design. Moreover, in [39], a block design-based key establishment protocol was presented using the Symmetric Balanced Incomplete Block Design (SBIBD). The analysis in [39] shows that it can be flexibly extended according to the structure of the SBIBD design.

Asymmetric key establishment protocols have been widely studied. Take the classical RSA and the Diffie-Hellman key exchange protocols as examples, they have been widely employed in our modern commercial systems and the IIoT systems, such as the web servers, banking systems and e-mail systems, in order to provide security and privacy services.

Two-Party Password Authenticated Key Exchange (2PAKE) protocols facilitate two IIoT devices to establish a communication key by utilizing a pre-shared short password. For instance, some 2PAKE protocols [5, 6] have been designed and proved secure in the ideal-cipher model. To avoid employing the random-oracle model or the ideal-cipher model, researchers try to design 2PAKE protocols in the standard model. However, only a few 2PAKE protocols are proved secure in the standard model [47, 48]. The initial practical 2PAKE protocol was proposed in [24]. Specifically, it is assumed in [24] that some public parameters, i.e., the so-called common reference string, are generated by the system authority, and all IIoT devices can access these parameters. Then, two devices can establish a communication key by running the improved Cramer-Shoup encryption algorithm. Moreover, the protocol of [24] was proved secure in the standard model under the DDH assumption. Inspired by [24], a few 2PAKE protocols [1, 18, 22, 23] were proposed. For example, making use of the smooth projective hash functions, Katz and Vaikuntanathan presented a round optimal 2PAKE protocol [23]. Jiang and Gong designed a lightweight protocol in [22]. Specifically, two IIoT devices in [22] only need to exchange 5 group elements within three flows in order to establish a communication key.

Extending the idea of 2PAKE protocols, a few Password Authenticated Group Key Exchange (GPAKE) protocols were proposed. The seminal work in this area was proposed in [8]. Specifically, it has constant rounds of communications. In the initialization phase of [8], the trusted system authority generates a cyclic group $\mathbb{G}$ with prime order q. Then, n group IIoT devices form a logical ring. The n devices can establish a group communication key by performing the following operations: (1). In the first round, each device D_i chooses an integer x_i randomly from $\mathbb{Z}_q^*$, computes $z_i = g^{x_i}$ (where g is the generater of $\mathbb{G}$), and broadcasts z_i. Additionally, device D_i records received messages z_js which are broadcasted by other devices, here $j = 1, 2, \ldots, n$ and $j \neq i$; (2). In the second round, each device D_i computes $Z_i = z_{i-1}^{x_i}$ and $Z_{i+1} = z_{i+1}^{x_i}$, and broadcasts $X_i = Z_{i+1}/Z_i$. Completing the above operations, the IIoT device D_i can directly compute and obtain the group communication key $K_i = Z_i^n X_i^{n-1} X_{i+1}^{n-2} \cdots X_{i+n-2} = g^{x_1x_2+x_2x_3+\cdots+x_nx_1}$.

Examples of GPAKE protocols, which are designed and proved secure in the ideal-cipher model, can be found in [2, 46]. Furthermore, several GPAKE protocols, such as [1, 3, 4, 44], are designed and proved secure in the standard model. For example, Abdalla and Pointcheval presented a scalable GPAKE protocol in [1]. Additionally, a GPAKE protocol compiler was proposed in [3]. Specifically, the protocol can transform any provably secure 2PAKE protocol into a provably secure GPAKE protocol with only two extra communication rounds. In [3], the IIoT device D_i needs to compute and broadcast the message $X_i = K_i^l \oplus K_i^r$. Where K_i^l and K_i^r are the output values when device D_i calls an existing 2PAKE protocol. Then, device D_i sets $K_i = K_i^l$ and $K_{i-j} = K_i^l \oplus X_{i-1} \oplus \cdots \oplus X_{i-j}$, for $j = 1, \ldots, n-1$. Completing the above operations, each of the n IIoT devices in [3] can establish a group communication key by making use of the $K_1, \ldots, K_n$.

Key establishment protocols which are designed at the physical layer In the past two decades, many physical layer key extraction protocols were proposed by making use of *the characteristics of wireless fading channels*. In practical multipath environments, the wireless channel between two IIoT devices, say Alice and Bob, undergoes the time-varying and stochastic fading between the exchanged wireless signals. Specifically, the fading is unique, location-specific and reciprocal. More precisely, the fading has the property that it is invariant within the channel coherence time no matter the signals are transmitted from Alice to Bob or vice-versa. In wireless communication theory, the channel coherence time is a statistical measurement of time duration over which the channel impulse response is essentially invariant. Additionally, the fading decorrelates over the distances of the order of half a wavelength, i.e., $\lambda/2$. For instance, the signals transmitted between Alice and Bob and the signals transmitted between Alice and the adversary will experience independent fading if the adversary locates more than $\lambda/2$ away from Alice and Bob. Namely, the adversary cannot extract any useful information when it is more than $\lambda/2$ away from the genuine participants. Moreover, it can be evaluated that, $\lambda/2 \approx 17.28$ cm when the frequency band is 868 MHz; $\lambda/2 \approx 16.39$ cm when the frequency band is 915 MHz; and $\lambda/2 \approx 6.25$ cm when the frequency band is 2400 MHz.

The Received Signal Strength (RSS) or the Channel Impulse Response (CIR) was employed to extract secret bits in existing physical layer key extraction protocols. Specifically, existing physical layer key extraction protocols consist of three phases, i.e., *Quantization* phase, *Reconciliation* phase and *Privacy Amplification* phase. In the Quantization phase, the IIoT devices sample the transmitted signals at a pre-defined frequency. Then, the sampled signals are quantized using pre-defined thresholds. At the end of this phase, two IIoT devices obtain two initial binary bit sequences. In practical applications, the two bit sequences may not be exactly the same due to the effect of imperfect reciprocity and noise. In the Reconciliation phase and Privacy Amplification phase, two devices can remove the mismatch bits and make the bit sequences has sufficient entropy by employing the techniques such as the error correcting code and secure sketch techniques.

Many physical layer key extraction protocols have been proposed, such as [21, 28, 31, 34, 43, 52, 58, 59, 73]. For instance, two IIoT devices in [28] can extract and obtain a communication key by evaluating the envelopes of signals and quantifying the evaluations. Additionally, the experimental analysis shows that the key extraction rate of [28] is around 1 bit/sec when two devices are placed in the indoor environment. In [21], Jana et al. investigated the practicality of the physical layer key extraction protocol in different environments and settings. In [59], the key extraction protocol was implemented using the off-the-shelf 802.11n multiple-antenna devices. Specifically, it is analyzed in [59] that the key extraction rate is increased by 4 times than that of devices equipped with the single-antenna. The S-box was generated in [52] in order to accelerate the key extraction rate. Additionally, the analysis in [52] shows that the key extraction rate can achieve an order of magnitude faster than other protocols.

A few key establishment protocols [11, 33, 69] were designed at the physical layer by making use of the *keyless cryptography techniques*. In these protocols, the characteristics of anonymous channels were utilized. Specifically, the wireless channel can be used as an anonymous channel as long as the channel achieves source indistinguishability. In practice, the source indistinguishability requires that the adversary cannot identify the source of the transmitted signals correctly. For example, in [11], two IIoT devices generate some random bits and convent the bits into wireless signals according to certain pre-defined rules. Then, the two devices can obtain secret bits by verifying the correctness or incorrectness of the transmitted messages. In order to ensure the protocol [11] achieves source indistinguishability, the devices need to be shook during the implementation of the protocol. To minimize human intervention, the protocol of [11] was improved by [33, 69]. Specifically, the wireless channel in [69] was transformed into the anonymous channel by introducing artificial randomness to the transmitted signals. The analysis shows that the energy consumption of [69] is around 176 times cheaper than that of the classical Diffie-Hellman key exchange protocol, and it takes about 159.04 ms to establish a 112-bit communication key (the analysis was conducted using the standard IEEE 802.15.4).

3 Cross-Layer Key Establishment Protocols

In Sect. 2, we reviewed some types of conventional key establishment protocols. In certain IIoT applications, however, these protocols cannot be directly implemented. For instance, in these key establishment protocols which are designed at higher layers, it is assumed that the IIoT devices are pre-loaded with certain secrets. In some applications of the IIoT, the IIoT devices are produced by different factories, and the secret sharing assumption cannot be met. As a result, the conventional protocols cannot be directly implemented by the IIoT when the assumption cannot be fulfilled. In the physical layer key extraction protocols, the IIoT devices do not need to pre-load secrets. However, the key extraction rate of these protocols is slow. To alleviate these problems and ensure the security and privacy of the IIoT, a few cross-layer key establishment protocols have been proposed by cooperatively utilizing the characteristics of higher layers and the physical layer. In this section, we will review these cross-layer key establishment protocols which are designed in asymmetric/symmetric key setting.

3.1 Cross-Layer Key Establishment in Asymmetric Key Setting

Preliminary Before reviewing the cross-layer key establishment protocols [66, 70], here we introduce the Decisional Diffie-Hellman (DDH) assumption in order to facilitate understanding. Let $\mathcal{G}$ be an efficient algorithm. Giving the security parameter k, it outputs the description of a cyclic (multiplicative) group $\mathbb{G}$ with prime order q. As in [70], the DDH problem is defined via the following definition.

Giving the algorithm $\mathcal{G}$, and for any algorithm D, define

$$DH_{D,\mathcal{G}}(k) \stackrel{def}{=} \Pr[\mathbb{G} \leftarrow \mathcal{G}(1^k); g \leftarrow \bar{\mathbb{G}}; a \leftarrow \mathbb{Z}_q^*; b \leftarrow \mathbb{Z}_q^* \setminus \{a\} : D(\mathbb{G}, g, g^a, g^b, g^{ab}) = 1]$$

and

$$Rand_{D,\mathcal{G}}(k) \stackrel{def}{=} \Pr[\mathbb{G} \leftarrow \mathcal{G}(1^k); g \leftarrow \bar{\mathbb{G}}; h \leftarrow \mathbb{G}; a \leftarrow \mathbb{Z}_q^*; b \leftarrow \mathbb{Z}_q^* \setminus \{a\} : D(\mathbb{G}, g, g^a, g^b, h) = 1].$$

The DDH problem is hard in $\mathcal{G}$ if $|DH_{D,\mathcal{G}}(k) - Rand_{D,\mathcal{G}}(k)|$ is a negligible probability for all PPT algorithms D. The DDH assumption assumes that there exists an algorithm $\mathcal{G}$ such that the DDH problem is hard in $\mathcal{G}$. Informally, we say the DDH problem is hard in $\mathbb{G}$ (recall that group $\mathbb{G}$ is output by $\mathcal{G}$).

In [70], a variant of password authenticated key exchange protocol was proposed. Specifically, it is a cross-layer design, i.e., the IIoT devices in [70] extract short secrets at the physical layer in order to shorten the key extraction time. Then, the extracted short secrets are used as "passwords". Taking advantage of the passwords, two IIoT devices can establish a communication key at higher layers. The proposed protocol [70] consists of four phases, including the **Initialization** phase, the **Passwords Extraction** phase, the **Key Exchange** phase, and the **Key Establishment** phase.

In the **Initialization** phase, the trusted system authority generates system parameters. For example, giving a security parameter k, the system authority computes the cyclic group $\mathbb{G}$. Specifically, the group $\mathbb{G}$ has prime order q, and the length of q is k bits, namely, $|q| = k$. It is assumed in [70] that the DDH assumption is hard in $\mathbb{G}$. Completing these operations, the system authority chooses a hash function $H(x)$ from the collision-resistant hash family $\mathbb{H}$. Then, the system authority chooses group generators $g_1, g_2, h \in \bar{\mathbb{G}}$. At the end of this phase, the system authority publishes public parameters $< H(x); \mathbb{G}; g_1, g_2, h \in \bar{\mathbb{G}} >$. These operations presented in the **Initialization** phase can be completed when the trusted system authority is off-line.

In the **Passwords Extraction** phase, two IIoT devices extract a short password at the physical layer by running the physical layer key extraction algorithm. In [70], the key extraction algorithm has $|\mathbb{D}|/\log_2(q)$ rounds.[1] For each round, there are two time slots (ST_1 and ST_2). For example, in the time slot ST_1, the IIoT device U chooses initial phase ϕ_1 randomly from $[0, 2\pi]$, generates the sinusoidal signal $x(t) = A\sin(w_c t + \phi_1)$, and sends the signal to device V. The signals are modulated by the wireless multipath fading channels. The steady-state portion of the signal received by the IIoT device V is $y_{UV}(t) = A_{U\to V}\sin(w_c t + \phi_1 + \phi_{UV}) + n_{UV}(t)$. Thus, at the end of the time slot ST_1, device V has phase offset $\phi_{U\to V} = \phi_1 + \phi_{UV}$.[2]

In the time slot ST_2, the IIoT device V executes the same operations. Namely, device V chooses initial phase ϕ_2 randomly from $[0, 2\pi]$, generates the sinusoidal signal $x'(t) = A\sin(w_c t + \phi_2)$, and sends the signal to device U. Similarly, the steady-state portion of the signal received at the IIoT device U is $y_{VU}(t) = A_{V\to U}\sin(w_c t + \phi_2 + \phi_{VU}) + n_{VU}(t)$. Thus, at the end of the time slot ST_2, device U obtains phase offset $\phi_{V\to U} = \phi_2 + \phi_{VU}$.

At the end of the first round, the IIoT devices U and V obtain the phase components Φ_1

$$\textbf{device } U: \quad \Phi_1 = \underline{\phi_{V\to U}} + \phi_1 = \underline{\phi_2 + \phi_{VU}} + \phi_1 \textbf{ mod } 2\pi,$$

$$\textbf{device } V: \quad \Phi_1 = \underline{\phi_{U\to V}} + \phi_2 = \underline{\phi_1 + \phi_{UV}} + \phi_2 \textbf{ mod } 2\pi.$$

[1] Here $|\mathbb{D}|$ is the size of the password dictionary $\mathbb{D}$. Recall that physical layer key extraction algorithms extract secrets bits using the channel's randomness, and the algorithms are designed without assuming the computationally-bounded adversary. Namely, the physical layer key extraction algorithms achieve information-theoretical secrecy. Thus, in [70], the extracted passwords are independently and uniformly distributed in the password dictionary $\mathbb{D}$. Furthermore, $|\mathbb{D}| \ll q$, i.e., $\mathbb{D} \subset \mathbb{Z}_q^*$.

[2] In practice, at the end of the time slot ST_1, device V has phase offset $\phi_{U\to V} = \phi_1 + \phi_{UV}$ and amplitude deviation $A_{U\to V} = A + A_{UV}$. Both $\phi_{U\to V}$ and $A_{U\to V}$ can be used to extract secrets. In [70], only the phase offsets are used to extract secrets in order to simplify the descriptions.

According to the communication theory, the phase offsets ϕ_{VU} and ϕ_{UV} are highly relevant (i.e., $\phi_{VU} \approx \phi_{UV}$) when the signals $x(t)$ and $x'(t)$ are exchanged during the channel coherence time. Then, the IIoT devices U and V quantize Φ_1 using the formula:

$$Q_x = k \quad if \;\; x \in [\frac{2\pi(k-1)}{q}, \frac{2\pi k}{q}),$$

where $k = 1, 2, \ldots, q$. For each round, the above quantization generates $\log_2(q)$-bit secrets. Thus, devices U and V can extract the "password" with length of $|\mathbb{D}|$-bit by repeating the above operations $|\mathbb{D}|/\log_2(q)$ rounds. In practice, other techniques, such as the *Secret Key Reconciliation* and *Privacy Amplification*, need to be employed in order to reconcile the differences (i.e., the mismatch bits) in the extracted passwords. The differences are introduced due to the factors, such as the presence of noise and interference, manufacturing variations, half-duplex mode of communication, and estimation errors. At the end of the **Passwords Extraction** phase, the IIoT devices U and V obtain a short password p.

In the **Key Exchange** phase, the IIoT devices U and V execute the following operations (Fig. 4):

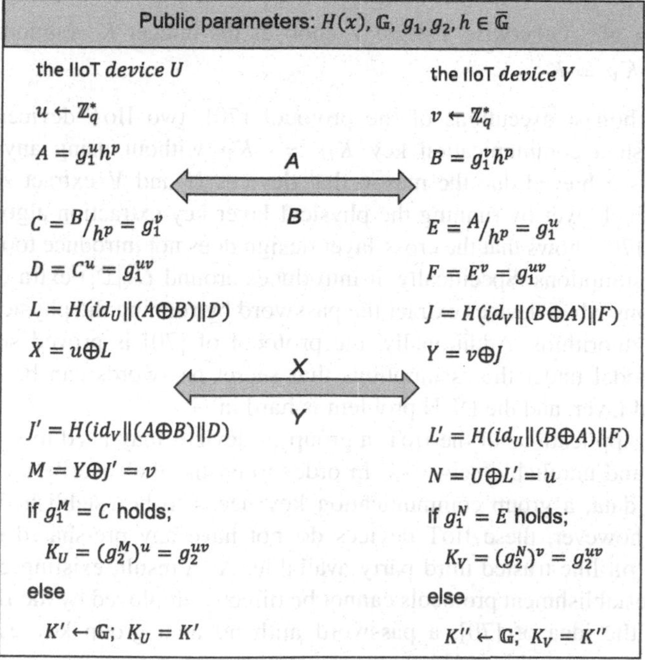

Fig. 4 The main operations in the **Key Exchange** and **Key Establishment** phases of [70]

1. Device U chooses the integer u randomly from $\mathbb{Z}_q^*$, and computes $A = g_1^u h^p$. Similarly, device V chooses the integer v from $\mathbb{Z}_q^*$, and computes $B = g_1^v h^p$. Recall that the p is the password extracted in the **Passwords Extraction** phase.
2. Device U and device V exchange messages $< id_U, A >$ and $< id_V, B >$, where id_U and id_V are the identifiers of device U and device V.
3. Receiving the message $< id_V, B >$, device U computes $C = B/h^p = g_1^v$, $D = C^u = g_1^{uv}$, $L = H(id_U \| (A \oplus B) \| D)$, and $X = u \oplus L$. Device V executes the same operations. Namely, receiving the message $< id_U, A >$, device V computes $E = A/h^p = g_1^u$, $F = E^v = g_1^{uv}$, $J = H(id_V \| (B \oplus A) \| F)$, and $Y = v \oplus J$. We denote by $\oplus$ the XOR operation, and denote by "$\|$" the concatenation operation.
4. The IIoT devices U and V exchange messages $< id_U, X >$ and $< id_V, Y >$.

In the **Key Establishment** phase, the IIoT devices U and V compute the communication keys by executing the following operations:

1. Device U computes $J' = H(id_V \| (A \oplus B) \| D)$, $M = Y \oplus J'$, and g_1^M. If $g_1^M = C$, device U computes $K_U = (g_2^M)^u = g_2^{uv}$. Otherwise, device U chooses the integer K' randomly from $\mathbb{G}$, and lets $K_U = K'$.
2. Device V executes the same operations. Namely, device V computes $L' = H(id_U \|$
$(B \oplus A) \| F)$, $N = X \oplus L'$, and g_1^N. If $g_1^N = E$, device V computes $K_V = (g_2^N)^v = g_2^{uv}$. Otherwise, Device V chooses the integer K'' randomly from $\mathbb{G}$, and lets $K_V = K''$.

For the honest executions of the protocol [70], two IIoT devices U and V can establish a communication key $K_U = K_V$ without using any pre-shared secrets. It is achieved due the reason that devices U and V extract short secrets at the physical layer by running the physical layer key extraction algorithms. The analysis in [70] shows that the cross-layer design does not introduce too much extra energy consumptions. Specifically, it introduces around $O(|\mathbb{D}|)$ extra computation consumptions when devices extract the password by running the physical layer key extraction algorithms. Additionally, the protocol of [70] is proved secure in the standard model under the assumptions that secret passwords can be extracted at the physical layer, and the DDH problem is hard in $\mathbb{G}$.

In some applications of the IIoT, a group of devices may need to share data via the public and unreliable networks. In order to ensure the security and privacy of the shared data, a group communication key needs to be established. In certain scenarios, however, these IIoT devices do not have any pre-shared secrets, and there is no on-line trusted third party available. As a result, existing conventional group key establishment protocols cannot be directly employed by the IIoT devices. Extending the idea of [70], a password authenticated group key establishment protocol was proposed in [66].

The password authenticated group key establishment protocol [66] is a cross-layer design, and it has four phases, i.e., the **Initialization** phase, the **Passwords Extraction** phase, the **Group Key Exchange** phase, and the **Group Key Establishment** phase.

In the **Initialization** phase, the public parameters are generated by the trusted system authority. Specifically, giving the security parameter k, the system authority generates a cyclic group $\mathbb{G}$ with prime order q, where $|q| = k$. Additionally, the system authority chooses a hash function $H(x)$ from the collision-resistant hash family $\mathbb{H}$. The hash function $H(x)$ is used to map $\{0, 1\}^*$ to members of $\mathbb{G}$. Note that operations presented in the **Initialization** phase can be completed when the trusted system authority is off-line.

In the **Passwords Extraction** phase, the IIoT devices extract passwords at the physical layer by running the physical layer key extraction algorithms. We let $\mathbb{U}$ denote the set of devices in a group. Additionally, it is assumed in [66] that each legitimate device $U \in \mathbb{U}$. Before extracting passwords, the n devices need to form a logical ring, which are indexed as $U_1, U_2, \ldots, U_n$. We have $U_{n+1} = U_1$ and $U_{1-1} = U_n$ (due to the reason that the devices are indexed in a cycle). Then, neighbor IIoT devices in the ring, i.e., $< U_i, U_{i+1} >$ where $i = 1, 2, \ldots, n$, move into proximity (i.e., within $\lambda/2$ distance) and extract short secrets by running the physical layer key extraction protocols.

To facilitate understanding, we provide an example. In the example, we assume that there are three participants, i.e., IIoT devices U_1, U_2 and the eavesdropper Eve. Additionally, we assume that there is a public radio frequency (RF) source. When the public RF source S broadcasts the sinusoidal signal $x(t) = A\sin(w_c t + \varphi_0)$, the signal is modulated by the multipath wireless fading channels, mobile environments and noise, the signal received at IIoT devices U_1, U_2 and the eavesdropper Eve can be represented as follows:

$$\begin{aligned}
y_{S\to U_1}(t) &= A_{S\to U_1}\sin(w_c t + \varphi_0 + \varphi_{S\to U_1}) + n_{S\to U_1}(t),\\
y_{S\to U_2}(t) &= A_{S\to U_2}\sin(w_c t + \varphi_0 + \varphi_{S\to U_2}) + n_{S\to U_2}(t),\\
y_{S\to E}(t) &= A_{S\to E}\sin(w_c t + \varphi_0 + \varphi_{S\to E}) + n_{S\to E}(t).
\end{aligned}$$

In communication theory, the deviated phases $\varphi_{S\to U_1} \approx \varphi_{S\to U_2}$ when device U_1 and device U_2 are physically close to each other. Typically, the distance between them should be no more than $\lambda/2$ in order to ensure $\varphi_{S\to U_1} \approx \varphi_{S\to U_2}$. Here, the λ is the wavelength of the ambient wireless signal $x(t)$. Additionally, the deviated phases $\varphi_{S\to E}$ and $\varphi_{S\to U_1}$, $\varphi_{S\to E}$ and $\varphi_{S\to U_2}$ are independent statistically when the eavesdropper is more than $\lambda/2$ away from device U_1 and device U_2. Namely, the eavesdropper Eve cannot obtain any useful secrets by utilizing the eavesdropped signals $y_{S\to E}(t)$ as long as the Eve is far away from the IIoT devices U_1 and U_2. Then, devices U_1 and U_2 can quantize $\varphi_{S\to U_1}$, $\varphi_{S\to U_2}$ and obtain a short password $pw_{1,2}$. Execute the same operations, each IIoT device U_i extracts two passwords $pw_{i,i-1}$ and $pw_{i,i+1}$ (which are shared with neighbor devices U_{i-1} and U_{i+1} in the ring) using the ambient wireless signals, where $i = 1, 2, \ldots, n$.

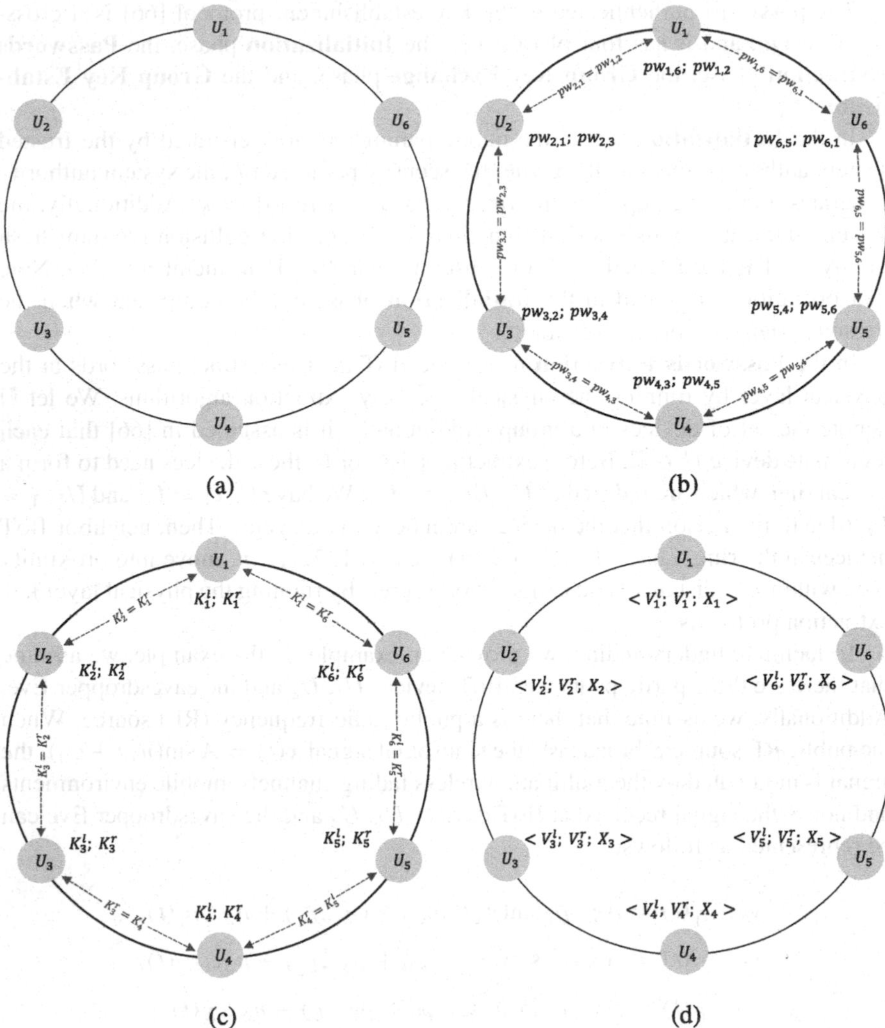

Fig. 5 Password authenticated group key establishment protocol in [66]. Specifically, there are $n = 6$ IIoT devices in this example, i.e., $\mathbb{U} = \{U_1, U_2, U_3, U_4, U_5, U_6\}$. (**a**) Form a logical ring. (**b**) Extract passwords. (**c**) Call *2PAKE*(ů,ů) round. (**d**) Broadcast messages round

In [66], an example is provided. Specifically, in the example (please refer to Fig. 5a, b), it is assumed that there are six IIoT devices (i.e., $n = 6$) in the group. To establish a group communication key, these devices form a logical ring, and they are indexed as U_1, U_2, U_3, U_4, U_5, and U_6. To extract and obtain passwords, neighbor devices in the ring run the physical layer key extraction algorithms. For instance, executing the operations provided in the **Key Extraction** phase (as reviewed above), the IIoT device U_6 extracts and obtains two short passwords $pw_{6,1}$ and $pw_{6,5}$, which are shared with neighbor IIoT devices U_1 and U_5, respectively.

In the **Group Key Exchange** phase, There are two rounds, i.e., *Call* $2PAKE(\cdot,\cdot)$ *round* and *Broadcast messages round.*

Call $2PAKE(\cdot,\cdot)$ *round.* Using the extracted passwords $pw_{i,i-1}$, $pw_{i,i+1}$ and calling $2PAKE(U_i, U_{i-1})$, $2PAKE(U_i, U_{i+1})$, the IIoT device U_i obtains two secret values K_i^l and K_i^r. Here, the $2PAKE(\cdot,\cdot)$ denote a black-box of the employed two-party password authenticated key exchange (2PAKE) protocol. It is showed in [66] that, the $2PAKE(\cdot,\cdot)$ returns the output of the employed 2PAKE protocol when inputting $U_i, U_j \in \mathbb{U}$. For instance, device U_6 obtains secret value K_6^l (shared with device U_5) by calling $2PAKE(U_6, U_5)$. Similarly, device U_6 obtains secret value K_6^r (shared with device U_1) by calling $2PAKE(U_6, U_1)$ (as shown in Fig. 5c).

Broadcast messages round. In this round, the IIoT device U_i computes $V_i^l = K_i^l \cdot H(K_i^r)$, $V_i^r = K_i^r \cdot H(K_i^l)$, and $X_i = H(K_i^l) \oplus H(K_i^r)$. At the end of this round, device U_i broadcasts $m_i = < U_i, U_{i-1}, V_i^l; U_i, U_{i+1}, V_i^r; X_i >$, and records the exchanged messages m_js, where $j = 1, 2, \ldots, n$. Take Fig. 5d as an example, device U_6 obtains two secret values K_6^l and K_6^r when it complete the *Call* $2PAKE(\cdot,\cdot)$ *round.* In this round, device U_6 computes $V_6^l = K_6^l \cdot H(K_6^r)$, $V_6^r = K_6^r \cdot H(K_6^l)$, and $X_6 = H(K_6^l) \oplus H(K_6^r)$. At the end of this round, device U_6 broadcasts $m_6 = < U_6, U_5, V_6^l; U_6, U_1, V_6^r; X_6 >$, and records the exchanged messages m_js, where $j = 1, 2, \ldots, 6$.

In the **Key Establishment** phase, each IIoT device U_i $(i = 1, 2, \ldots, n)$:

1. Authenticates its neighbor IIoT devices U_{i-1} and U_{i+1} by verifying the received messages m_{i-1} and m_{i+1}. For example, in order to authenticate the neighbor device U_{i-1}, device U_i computes $E_i^l = V_{i-1}^r / K_i^l$, and verifies if $X_{i-1} = H(K_i^l) \oplus E_i^l$. Similarly, to authenticate the neighbor device U_{i+1}, device U_i computes $E_i^r = V_{i+1}^l / K_i^r$, and verifies if $X_{i+1} = H(K_i^r) \oplus E_i^r$.
2. Checks if $X_1 \oplus X_2 \oplus \cdots \oplus X_n = 0$. The IIoT device U_i terminates the executions of the protocol immediately when any one of these verifications fails. Otherwise, device U_i lets $K_i = H(K_i^l)$ and computes other $n-1$ values $K_{i-j} = H(K_i^l) \oplus X_{i-1} \oplus \cdots \oplus X_{i-j}$, where $(j = 1, \ldots, n-1)$.
3. Computes the group communication key $K_g = K_1 \cdot K_2 \cdots K_n$.

The proposed protocol of [66] is proved secure in the standard model by assuming that: (1) Secret passwords can be extracted by running the physical layer key extraction algorithms; and (2) The $2PAKE(\cdot,\cdot)$ is a secure 2PAKE protocol (in the standard model).

In this section, we reviewed the cross-layer key establishment protocols which are designed in asymmetric key setting. In the next section, we will review these cross-layer key establishment protocols which are designed in symmetric key setting.

3.2 Cross-Layer Key Establishment in Symmetric Key Setting

In [68], a matrix-based cross-layer key establishment protocol was proposed. Specifically, the protocol was designed for one of the applications of the IIoT, i.e., smart homes. In smart homes scenarios, home IIoT devices sense the environment and collect the sensed data to the home gateway. Analysing the collected data, the home gateway distributes commands to the smart home devices for the purpose of optimizing residents' comfort. In practice, the data and commands are transmitted wirelessly, and this contributes to the smart homes networks' vulnerabilities. To ensure the security and privacy of smart home networks, communication keys need to be established. In practice, however, the home IIoT devices are manufactured by different factories. In this scenario, it is not practical to assume that the IIoT devices are pre-loaded with certain secrets when they leave factories. As a result, conventional key establishment protocols cannot be directly employed. Implementing the matrix-based cross-layer key establishment protocol [68], any two IIoT devices can directly establish a communication key (without using any pre-shared secrets). The protocol of [68] has four phases, i.e., the **Initialization** phase, the **Master Key Extraction** phase, the **Key Seed Distribution** phase and the **Session Key Establishment** phase.

In the **Initialization** phase, the home gateway, serves as the trusted system authority, generates system parameters. It is assumed that the M IIoT devices (i.e., the home appliances) are connected to the home gateway wirelessly. The home gateway performs the following operations:

- Giving the security parameter 1^k, the trusted home gateway generates N independent key seeds $s_1, s_2, \ldots, s_N$ from the finite field $GF(q)$, where q has the length of k-bit. To ensure the scalability, in practice, we let $N > M$.
- The home gateway chooses system parameter λ, and generates the $(\lambda + 1) \times N$ secret matrix G

$$G = \begin{pmatrix} s_1 & s_2 & \cdots & s_N \\ (s_1)^2 & (s_2)^2 & \cdots & (s_N)^2 \\ \vdots & \vdots & \ddots & \vdots \\ (s_1)^{\lambda+1} & (s_2)^{\lambda+1} & \cdots & (s_N)^{\lambda+1} \end{pmatrix}. \tag{1}$$

- The home gateway generates ω secret symmetric $(\lambda + 1) \times (\lambda + 1)$ matrices $D_1, D_2, \ldots, D_\omega$ in $GF(q)$, and computes the public matrices $A_1 = (D_1 \cdot G)^T, A_2 = (D_2 \cdot G)^T, \ldots, A_\omega = (D_\omega \cdot G)^T$. Here, we denote by "·" the matrix dot product, and denote by "T" the matrix transpose.
- The home gateway chooses the hash function $H(x)$ from the collision-resistant hash family $\mathbb{H}$. The hash function $H(x)$ is employed to map $\{0, 1\}^*$ to members of the field $GF(q)$. At the end, the home gateway publishes the hash function $H(x)$.

Note that the operations presented in the **Initialization** phase can be completed when the home gateway is off-line.

In the **Master Key Extraction** phase, master keys are extracted at the physical layer by running the physical layer key extraction algorithms. Specifically, the master keys in [68] are shared between home IIoT devices and the home gateway. Let $\mathbb{P}_{\mathbb{M}}$ be the set of M home devices. In this phase, the ith IIoT device P_i ($P_i \in \mathbb{P}_{\mathbb{M}}$ and $i = 1, 2, \ldots, M$) runs the physical layer key extraction algorithms (as reviewed in Sect. 3.1) and obtains a master key k_i which is shared with the home gateway. Then, the home gateway sends the identifer id_i for the home IIoT device P_i. Note that the identifer id_i is the public information, and the public distribution of it does not introduce security and privacy issues. Completing these operations, each home IIoT device obtains an identifer and a master key (which is shared with the home gateway).

In the **Key Seed Distribution** phase, the trusted home gateway distributes a secret key seed for each home IIoT device. To facilitate understanding, in Fig. 6, we take the ith home IIoT device P_i as an example. To ensure the security and privacy of the key seed s_i, the home gateway and the device P_i perform the following operations:

- The ith home IIoT device P_i sends the request to the home gateway: {req key seed distribution: id_i, id_{HG}}.
- Receiving the request from device P_i, the home gateway randomly chooses a number r_x from the field $GF(q)$, computes $c_1 = H(k_i) \oplus s_i$ and $c_1^* = H(k_i - 1) \oplus r_x$. Completing these operations, the home gateway sends the message $V_1 =< id_{HG}, id_i, c_1, c_1^* >$ to the IIoT device P_i.
- Receiving the message V_1, the home IIoT device P_i computes $H(k_i)$ and $H(k_i - 1)$ using the extracted master key k_i. Then, the P_i recovers and obtains the key seed s_i and random number r_x by computing $H(k_i) \oplus c_1$ and $H(k_i - 1) \oplus c_1^*$. Completing these operations, the home IIoT device P_i chooses a number r_y randomly from the field $GF(q)$, computes $c_2 = H(r_x) \oplus r_y$ and $c_2^* = H(r_x - 1) \oplus H(r_y)$. Then, the home IIoT device P_i sends the message $V_2 =< id_i, id_{HG}, c_2, c_2^* >$ to the home gateway.
- Receiving the message V_2, the home gateway computes $r_y = H(r_x) \oplus c_2$ and $r_y^* = H(r_x - 1) \oplus c_2^*$, and verifies $r_y^* = H(r_y)$. If the verification passes successfully, the home gateway computes $c_3 = H(r_y - 1)$, and sends $V_3 =< id_{HG}, id_i, c_3 >$ to the home IIoT device P_i; Otherwise, the home gateway chooses a number r_z randomly from the field $GF(q)$, sets $c_3 = r_z$, and sends $V_3 =< id_{HG}, id_i, c_3 >$ to the home IIoT device P_i.
- Receiving the message V_3, the home IIoT device P_i verifies the received c_3 by computing $H(r_y - 1)$. If the verification passes successfully, the home IIoT device P_i accepts the key seed s_i; Otherwise, it directly outputs the undefined symbol "$\perp$" and immediately terminates the communications.

At the end of **Key Seed Distribution** phase, each home IIoT device obtains a secret key seed.

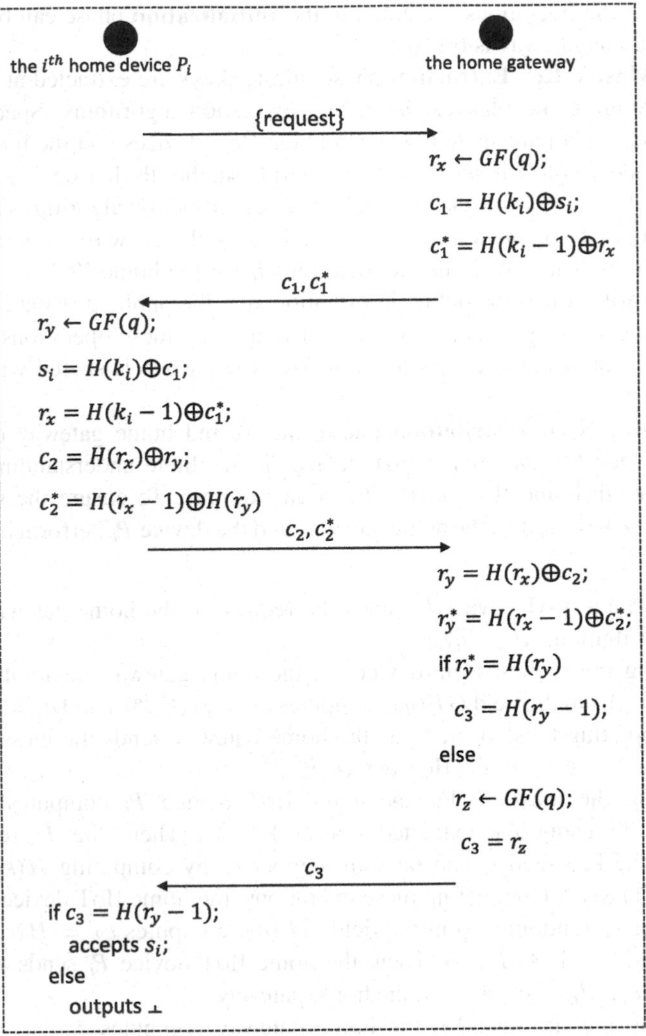

Fig. 6 Main operations in the **Key Seed Distribution** phase of [68]

In the **Session Key Establishment** phase, the ith and jth home IIoT devices can establish a communication key by sending the request {req: id_i, id_j} to the home gateway. Receiving the request from home IIoT devices, the home gateway chooses a matrix, e.g., A_c, from $A_1, A_2, \ldots, A_\omega$. Then, the home gateway sends the ith and jth IIoT devices with jth and ith rows of matrix A_c. Receiving the messages from the home gateway, the ith and jth home IIoT devices can directly establish a communication key k_{ji} (k_{ji}) by performing the following calculations:

The ith home IIoT device P_i:

- Computes the ith column of matrix G using its key seed s_i: $(s_i, s_i^2, \ldots, s_i^{\lambda+1})^T$;
- Let $(a_{j1}, a_{j2}, \ldots, a_{j(\lambda+1)})$ denote the jth row of matrix A_c, the home IIoT device P_i computes the communication key k_{ji} as

$$k_{ji} = (a_{j1}, a_{j2}, \ldots, a_{j(\lambda+1)}) \cdot (s_i, s_i^2, \ldots, s_i^{\lambda+1})^T = \sum_{r=1}^{\lambda+1} a_{jr} \cdot (s_i)^r.$$

Similarly, the jth home IIoT device P_j:

- Computes the jth column of matrix G using its key seed s_j: $(s_j, s_j^2, \ldots, s_j^{\lambda+1})^T$;
- Let $(a_{i1}, a_{i2}, \ldots, a_{i(\lambda+1)})$ be the ith row of matrix A_c, the home IIoT device P_j computes the communication key k_{ij} as

$$k_{ij} = (a_{i1}, a_{i2}, \ldots, a_{i(\lambda+1)}) \cdot (s_j, s_j^2, \ldots, s_j^{\lambda+1})^T = \sum_{r=1}^{\lambda+1} a_{ir} \cdot (s_j)^r.$$

Note that the matrix $K = A_c \cdot G$ is a symmetric matrix, i.e.,

$$K = A_c \cdot G = (D_c \cdot G)^T \cdot G = G^T \cdot D_c \cdot G = (A_c \cdot G)^T = K^T. \tag{2}$$

Namely, $k_{ji} = k_{ij}$. Thus, the ith and jth home IIoT devices can establish a communication key by performing the aforementioned operations.

Furthermore, a cross-layer key establishment model was designed in [67] by extending the idea of [68]. The model can convent existing key pre-distribution protocols into cross-layer key establishment protocols such that the IIoT devices can establish communication keys without the secret sharing assumption. Additionally, in [64], a hybrid cross-layer key establishment protocol was presented. The protocol is designed for the scenarios when two remote IIoT devices, who are out of each other's communication range, to establish a communication key. Specifically, two IIoT devices in [64] extract a partial key by employing the physical layer key extraction algorithms. Additionally, the two IIoT devices obtain another partial key at higher layers using the XOR coding technique. The communication key is computed by making use of these two partial keys. The security analysis shows that the security of [64] is guaranteed by two-fold. Namely, a partial key computed by employing the XOR coding at higher layers and the other partial key extracted at the physical layer by taking advantage of the wireless fading channels' characteristics.

In this section, we reviewed the cross-layer key establishment protocols which are designed in symmetric key setting. In the next section, we will conclude this chapter and discuss some future research trends in terms of the security and privacy of the IIoT.

4 Conclusions and Future Research Trends

In this chapter, we briefly introduced the IIoT and its characteristics. In practical applications, however, the IIoT has many attack surfaces. Thus, we reviewed the security and privacy vulnerabilities of the IIoT. Then, we highlighted the security an privacy goals of the IIoT (including the availability, integrity, etc). To ensure the security and privacy of the IIoT, many techniques, such as authentication techniques, intrusion detection techniques, secure routing techniques, privacy-preserving techniques, signature techniques, have been designed. Due to the length limitation, in this chapter, we only reviewed the key establishment protocols.

In Fig. 1, we showed the category of conventional key establishment protocols for the IIoT. Then, we reviewed these protocols. Specifically, for the symmetric key establishment protocols, we reviewed the entity-based protocols, probabilistic-based protocols, polynomial-based protocols, matrix-based protocols, tree-based protocols, and combinatorial design-based protocols. For the asymmetric key establishment protocols, we reviewed the password authenticated (group) key establishment protocols. Then, we reviewed the physical layer key extraction protocols which are designed by making use of the characteristic of wireless fading channels or using the keyless cryptograph technique. By reviewing these conventional protocols, we aim to illustrate the necessity of designing cross-layer key establishment protocols for the IIoT. Thus, in Sect. 3, we reviewed the cross-layer key establishment protocols. Specifically, the cross-layer design enables the IIoT devices to establish communication keys without the secret sharing assumption or the trusted entity.

In terms of designing key establishment protocols for the IIoT, there are several future research trends:

- As we reviewed in Sect. 3, cross-layer designs of key establishment protocols can alleviate the problems exist in conventional key establishment protocols (which constraint them to directly implement in the IIoT). Until now, only a few cross-layer key establishment protocols have been proposed. It is a promising research trend to design new types of cross-layer key establishment protocols for the IIoT.
- It has been investigated that wireless signals, lights, and ambient noise can be utilized to extract secret bits. In certain applications of the IIoT, various physical quantities are sensed and gathered by IIoT devices. Thus, it is a new research direction to extract secret keys using the different physical quantities.
- In the IIoT, different physical quantities can be used to extract secret bits. In certain applications, some physical quantities might be more efficient to extract secret bits than other physical quantities. In this scenarios, we are interested in designing smart key establishment protocols such that the IIoT devices can automatically choose several quantities (according to the unique features of their surrounding environments) to extract communication keys.
- In some applications of the IIoT, the devices might be the energy-constrained devices. Thus, lightweight key establishment protocols should be designed for the IIoT.

References

1. Abdalla M, Pointcheval D (2006) A scalable password-based group key exchange protocol in the standard model. In: Advances in Cryptology – ASIACRYPT 2006, 12th International Conference on the Theory and Application of Cryptology and Information Security, Shanghai, 3–7 Dec 2006, Proceedings, pp 332–347
2. Abdalla M, Bresson E, Chevassut O, Pointcheval D (2006) Password-based group key exchange in a constant number of rounds. In: Public Key Cryptography – PKC 2006, 9th International Conference on Theory and Practice of Public-Key Cryptography, New York, 24–26 Apr 2006, Proceedings, pp 427–442
3. Abdalla M, Bohli J, Vasco MIG, Steinwandt R (2007) (Password) authenticated key establishment: from 2-party to group. In: Theory of Cryptography, 4th Theory of Cryptography Conference, TCC 2007, Amsterdam, 21–24 Feb 2007, Proceedings, pp 499–514
4. Abdalla M, Chevalier C, Granboulan L, Pointcheval D (2011) Contributory password-authenticated group key exchange with join capability. In: Topics in Cryptology – CT-RSA 2011 – The Cryptographers' Track at the RSA Conference 2011, San Francisco, 14–18 Feb 2011. Proceedings, pp 142–160
5. Abdalla M, Benhamouda F, MacKenzie P (2015) Security of the J-PAKE password-authenticated key exchange protocol. In: 2015 IEEE Symposium on Security and Privacy (SP), pp 571–587
6. Bellare M, Pointcheval D, Rogaway P (2000) Authenticated key exchange secure against dictionary attacks. In: Advances in Cryptology – EUROCRYPT 2000, International Conference on the Theory and Application of Cryptographic Techniques, Bruges, 14–18 May 2000, Proceeding, pp 139–155
7. Blom R (1984) An optimal class of symmetric key generation systems. In: Advances in Cryptology: Proceedings of EUROCRYPT 84, A Workshop on the Theory and Application of Cryptographic Techniques, Paris, 9–11 Apr 1984, Proceedings, pp 335–338
8. Burmester M, Desmedt Y (1994) A secure and efficient conference key distribution system (extended abstract). In: Advances in Cryptology – EUROCRYPT'94, Workshop on the Theory and Application of Cryptographic Techniques, Perugia, 9–12 May 1994, Proceedings, pp 275–286
9. Çamtepe SA, Yener B (2007) Combinatorial design of key distribution mechanisms for wireless sensor networks. IEEE/ACM Trans Netw 15(2):346–358
10. Cao X, Xu L, Zhang Y, Wu W (2012) Identity-based proxy signature for cloud service in saas. In: 2012 Fourth International Conference on Intelligent Networking and Collaborative Systems, INCoS 2012, Bucharest, 19–21 Sept 2012, pp 594–599
11. Castelluccia C, Mutaf P (2005) Shake them up! a movement-based pairing protocol for CPU-constrained devices. In: Proceedings of the 3rd International Conference on Mobile Systems, Applications, and Services, MobiSys 2005, Seattle, 6–8 June 2005, pp 51–64
12. Chan H, Perrig A, Song DX (2003) Random key predistribution schemes for sensor networks. In: 2003 IEEE Symposium on Security and Privacy (S&P 2003), 11–14 May 2003, Berkeley, p 197
13. Du W, Deng J, Han YS, Varshney PK (2003) A pairwise key pre-distribution scheme for wireless sensor networks. In: Proceedings of the 10th ACM Conference on Computer and Communications Security, CCS 2003, Washington, 27–30 Oct 2003, pp 42–51
14. Du W, Deng J, Han YS, Chen S, Varshney PK (2004) A key management scheme for wireless sensor networks using deployment knowledge. In: Proceedings IEEE INFOCOM 2004, The 23rd Annual Joint Conference of the IEEE Computer and Communications Societies, Hong Kong, 7–11 Mar 2004
15. Du W, Deng J, Han YS, Varshney PK (2006) A key predistribution scheme for sensor networks using deployment knowledge. IEEE Trans Dependable Secure Comput 3(1):62–77

16. Du H, Li J, Zhang Y, Li T, Zhang Y (2012) Certificate-based key-insulated signature. In: Data and Knowledge Engineering – Third International Conference, ICDKE 2012, Wuyishan, 21–23 Nov 2012. Proceedings, pp 206–220
17. Eschenauer L, Gligor VD (2002) A key-management scheme for distributed sensor networks. In: Proceedings of the 9th ACM Conference on Computer and Communications Security, CCS 2002, Washington, 18–22 Nov 2002, pp 41–47
18. Groce A, Katz J (2010) A new framework for efficient password-based authenticated key exchange. In: Proceedings of the 17th ACM Conference on Computer and Communications Security, CCS 2010, Chicago, 4–8 Oct 2010, pp 516–525
19. Huang X, Susilo W, Mu Y, Zhang F (2005) On the security of certificateless signature schemes from asiacrypt 2003. In: Cryptology and Network Security, 4th International Conference, CANS 2005, Xiamen, 14–16 Dec 2005, Proceedings, pp 13–25
20. Huang X, Mu Y, Susilo W, Wong DS, Wu W (2007) Certificateless signature revisited. In: Information Security and Privacy, 12th Australasian Conference, ACISP 2007, Townsville, 2–4 July 2007, Proceedings, pp 308–322
21. Jana S, Premnath SN, Clark M, Kasera SK, Patwari N, Krishnamurthy SV (2009) On the effectiveness of secret key extraction from wireless signal strength in real environments. In: Proceedings of the 15th Annual International Conference on Mobile Computing and Networking, MOBICOM 2009, Beijing, 20–25 Sept 2009, pp 321–332
22. Jiang S, Gong G (2004) Password based key exchange with mutual authentication. In: Selected Areas in Cryptography, 11th International Workshop, SAC 2004, Waterloo, 9–10 Aug 2004, Revised Selected Papers, pp 267–279
23. Katz J, Vaikuntanathan V (2013) Round-optimal password-based authenticated key exchange. J Cryptol 26(4):714–743
24. Katz J, Ostrovsky R, Yung M (2001) Efficient password-authenticated key exchange using human-memorable passwords. In: Advances in Cryptology – EUROCRYPT 2001, International Conference on the Theory and Application of Cryptographic Techniques, Innsbruck, 6–10 May 2001, Proceeding, pp 475–494
25. Lai B, Kim S, Verbauwhede I (2002) Scalable session key construction protocol for wireless sensor networks. In: IEEE Workshop on Large Scale RealTime and Embedded Systems (LARTES), p 7
26. Li J, Du H, Zhang Y, Li T, Zhang Y (2014) Provably secure certificate-based key-insulated signature scheme. Concurr Comput Pract Exp 26(8):1546–1560
27. Liu D, Ning P (2003) Establishing pairwise keys in distributed sensor networks. In: Proceedings of the 10th ACM Conference on Computer and Communications Security, CCS 2003, Washington, 27–30 Oct 2003, pp 52–61
28. Mathur S, Trappe W, Mandayam NB, Ye C, Reznik A (2008) Radio-telepathy: extracting a secret key from an unauthenticated wireless channel. In: Proceedings of the 14th Annual International Conference on Mobile Computing and Networking, MOBICOM 2008, San Francisco, 14–19 Sept 2008, pp 128–139
29. Peng K, Zhang Y (2012) A secure mix network with an efficient validity verification mechanism. In: Internet and Distributed Computing Systems – 5th International Conference, IDCS 2012, Wuyishan, Fujian, 21–23 Nov 2012. Proceedings, pp 85–96
30. Peng J, Choo KR, Ashman H (2016) User profiling in intrusion detection: a review. J Netw Comput Appl 72:14–27
31. Peng Y, Wang P, Xiang W, Li Y (2017) Secret key generation based on estimated channel state information for TDD-OFDM systems over fading channels. IEEE Trans Wirel Commun 16(8):5176–5186
32. Perrig A, Szewczyk R, Wen V, Culler DE, Tygar JD (2001) SPINS: security protocols for sensor netowrks. In: MOBICOM 2001, Proceedings of the Seventh Annual International Conference on Mobile Computing and Networking, Rome, 16–21 July 2001, pp 189–199
33. Pietro RD, Oligeri G (2013) COKE crypto-less over-the-air key establishment. IEEE IEEE Trans Inf Forensics Secur 8(1):163–173

34. Premnath SN, Jana S, Croft J, Gowda PL, Clark M, Kasera SK, Patwari N, Krishnamurthy SV (2013) Secret key extraction from wireless signal strength in real environments. IEEE Trans Mob Comput 12(5):917–930
35. Ruj S, Nayak A, Stojmenovic I (2013) Pairwise and triple key distribution in wireless sensor networks with applications. IEEE Trans Comput 62(11):2224–2237
36. Sadeghi A, Wachsmann C, Waidner M (2015) Security and privacy challenges in industrial Internet of Things. In: Proceedings of the 52nd Annual Design Automation Conference, San Francisco, 7–11 June 2015, pp 54:1–54:6
37. Shen J, Moh S, Chung I (2012) Identity-based key agreement protocol employing a symmetric balanced incomplete block design. J Commun Netw 14(6):682–691
38. Shen J, Wang A, Wang C, Hung PCK, Lai C (2017) An efficient centroid-based routing protocol for energy management in WSN-assisted IoT. IEEE Access 5:18469–18479
39. Shen J, Zhou T, He D, Zhang Y, Sun X, Xiang Y (2017, to be appear) Block design-based key agreement for group data sharing in cloud computing. IEEE Trans Dependable Secure Comput
40. Shen J, Zhou T, Lai CF, Li J, Li X (2017) Hierarchical trust level evaluation for pervasive social networking. IEEE Access 5:1178–1187
41. Shen J, Zhou T, Wei F, Sun X, Xiang Y (2018) Privacy-preserving and lightweight key agreement protocol for v2g in the social internet of things. IEEE Internet Things J 5(4):2526–2536
42. Shen J, Zhou T, Chen X, Li J, Susilo W (2018) Anonymous and traceable group data sharing in cloud computing. IEEE Trans Inf Forensics Secur 13(4):912–925
43. Shimizu T, Iwai H, Sasaoka H (2011) Physical-layer secret key agreement in two-way wireless relaying systems. IEEE Trans Inf Forensics Secur 6(3–1):650–660
44. Tang Q, Choo KR (2006) Secure password-based authenticated group key agreement for data-sharing peer-to-peer networks. In: Applied Cryptography and Network Security, 4th International Conference, ACNS 2006, Singapore, 6–9 June 2006, Proceedings, pp 162–177
45. Wallner D, Harder E, Agee R (1999) Key management for multicast: issues and architectures. No. RFC 2627
46. Wan Z, Deng RH, Bao F, Preneel B (2007) *n*PAKE+: a hierarchical group password-authenticated key exchange protocol using different passwords. In: Information and Communications Security, 9th International Conference, ICICS 2007, Zhengzhou, 12–15 Dec 2007, Proceedings, pp 31–43
47. Wang D, Zhang Z, Wang P, Yan J, Huang X (2016) Targeted online password guessing: an underestimated threat. In: Proceedings of the 2016 ACM SIGSAC Conference on Computer and Communications Security, Vienna, 24–28 Oct 2016, pp 1242–1254
48. Wang D, Cheng H, Wang P, Huang X, Jian G (2017) Zipf's law in passwords. IEEE Trans Inf Forensics Secur 12(11):2776–2791
49. Wang D, Cheng H, He D, Wang P (2018) On the challenges in designing identity-based privacy-preserving authentication schemes for mobile devices. IEEE Syst J 12(1):916–925
50. Wang D, Li W, Wang P (2018) Measuring two-factor authentication schemes for real-time data access in industrial wireless sensor networks. IEEE Trans Ind Inf 14(9):4081–4092
51. Wang M, Zhang Y, Ma J, Wu W (2018, to appear) A universal designated multi verifiers content extraction signature scheme. Int J Comput Sci Eng
52. Xi W, Qian C, Han J, Zhao K, Zhong S, Li X, Zhao J (2016) Instant and robust authentication and key agreement among mobile devices. In: Proceedings of the 2016 ACM SIGSAC Conference on Computer and Communications Security, Vienna, 24–28 Oct 2016, pp 616–627
53. Xu L, Zhang Y (2014) Matrix-based pairwise key establishment for wireless mesh networks. Futur Gener Comput Syst 30:140–145
54. Xu S, Mu Y, Susilo W, Chen X, Huang X, Zhang F (2006) Online/offline signatures and multisignatures for AODV and DSR routing security. IACR Cryptol ePrint Archive 2006, 236
55. Xu L, Cao X, Zhang Y, Wu W (2013) Software service signature (S3) for authentication in cloud computing. Clust Comput 16(4):905–914

56. Yang X, Zhang Y, Liu JK, Zeng Y (2016) A trust and privacy preserving handover authentication protocol for wireless networks. In: 2016 IEEE Trustcom/BigDataSE/ISPA, Tianjin, 23–26 Aug 2016, pp 138–143
57. Ye A, Zheng Y, Xu L, Zhang Y (2017) A road-network based privacy-preserving approach in trajectory publishing. J Internet Technol 18(4):867–876
58. Zan B, Gruteser M, Hu F (2013) Key agreement algorithms for vehicular communication networks based on reciprocity and diversity theorems. IEEE Trans Veh Technol 62(8):4020–4027
59. Zeng K, Wu D, Chan AJ, Mohapatra P (2010) Exploiting multiple-antenna diversity for shared secret key generation in wireless networks. In: INFOCOM 2010. 29th IEEE International Conference on Computer Communications, Joint Conference of the IEEE Computer and Communications Societies, 15–19 Mar 2010, San Diego, pp 1837–1845
60. Zhang Y, Xu L, Huang X (2012) Polynomial based key predistribution scheme in wireless mesh networks. J Comput Inf Syst 8(6):2539–2549
61. Zhang Y, Xu L, Huang X, Li J (2013) Matrix-based pairwise key establishment with pre and post deployment knowledge for wireless mesh networks. In: Seventh International Conference on Innovative Mobile and Internet Services in Ubiquitous Computing, IMIS 2013, Taichung, 3–5 July 2013, pp 153–158
62. Zhang Y, Xu L, Xiang Y, Huang X (2013) Matrix-based pairwise key establishment in wireless mesh networks using deployment knowledge. In: Proceedings of IEEE International Conference on Communications, ICC 2013, Budapest, 9–13 June 2013, pp 1604–1608
63. Zhang Y, Xu L, Xiang Y, Huang X (2013) A matrix-based pairwise key establishment scheme for wireless mesh networks using pre deployment knowledge. IEEE Trans Emerg Top Comput 1(2):331–340
64. Zhang Y, Xiang Y, Huang X, Xu L (2014) A cross-layer key establishment scheme in wireless mesh networks. In: Computer Security – ESORICS 2014 – 19th European Symposium on Research in Computer Security, Wroclaw, 7–11 Sept 2014. Proceedings, Part I, pp 526–541
65. Zhang Y, Xu L, Huang X, Li J (2015) Matrix-based key pre-distribution schemes in WMNS using pre and post deployment knowledge. Int J Ad Hoc Ubiquitous Comput 20(4):262–273
66. Zhang Y, Xiang Y, Huang X (2016) Password-authenticated group key exchange: a cross-layer design. ACM Trans Internet Technol 16(4):24:1–24:20
67. Zhang Y, Xiang Y, Huang X (2017) A cross-layer key establishment model for wireless devices in cyber-physical systems. In: Proceedings of the 3rd ACM Workshop on Cyber-Physical System Security, CPSS@AsiaCCS 2017, Abu Dhabi, 2 Apr 2017, pp 43–53
68. Zhang Y, Xiang Y, Huang X, Chen X, Alelaiwi A (2018) A matrix-based cross-layer key establishment protocol for smart homes. Inf Sci 429:390–405
69. Zhang Y, Xiang Y, Wang T, Wu W, Shen J (2018) An over-the-air key establishment protocol using keyless cryptography. Futur Gener Comput Syst 79:284–294
70. Zhang Y, Xiang Y, Wu W, Alelaiwi A (2018) A variant of password authenticated key exchange protocol. Futur Gener Comput Syst 78:699–711
71. Zhu S, Setia S, Jajodia S (2003) LEAP: efficient security mechanisms for large-scale distributed sensor networks. In: Proceedings of the 10th ACM Conference on Computer and Communications Security, CCS 2003, Washington, DC, 27–30 Oct 2003, pp 62–72
72. Zhu F, Zhang Y, Lin C, Wu W, Meng R (2017) A universal designated multi-verifier transitive signature scheme. In: Information Security and Cryptology – 13th International Conference, Inscrypt 2017, Xi'an, 3–5 Nov 2017, Revised Selected Papers, pp 180–195
73. Zhu X, Xu F, Novak E, Tan CC, Li Q, Chen G (2017) Using wireless link dynamics to extract a secret key in vehicular scenarios. IEEE Trans Mob Comput 16(7):2065–2078

Part V
Application Scenarios

IIoT in the Hospital Scenario: Hospital 4.0, Blockchain and Robust Data Management

Luca Faramondi, Gabriele Oliva, Roberto Setola, and Luca Vollero

Abstract The Industrial Internet of Things (IIoT) consists of the pervasive application of the IoT paradigm in conjunction with analytics and artificial intelligence (AI) in industrial scenarios. Industry 4.0 (I4.0) extends further the IIoT model with the inclusion of robotics and automation, whereas Hospital 4.0 (H4.0) is the application of the I4.0 paradigm to the healthcare sector. H4.0 relies on cyber-physical systems managing several devices and software components, and the exchange of a huge amount of sensible data that includes medical records. Medical records can be much more valuable to criminals than financial data, indeed the control of medical data allows criminals to plan and realize different frauds, that the victims may identify only too late. Furthermore, the complexity of a typical H4.0 cyber-physical system makes healthcare records particularly vulnerable. Blockchain is today an emerging technology for the management of data that may avoid or mitigate the impact of threats related to data storage and management, in general, and to the administration, in particular, of healthcare records. The blockchain technology relies on cryptography and distributed consensus to guarantee data integrity, accountability and security. The exploitation of such technology is considered in this chapter, showing the advantages when used in a H4.0 scenario.

Keywords IIoT · Blockchain · Hospital 4.0 · Data management · Data security

L. Faramondi · G. Oliva · R. Setola
Complex Systems & Security Lab, University Campus Bio-Medico of Rome, Roma, Italy
e-mail: l.faramondi@unicampus.it; g.oliva@unicampus.it; r.setola@unicampus.it

L. Vollero (✉)
Neurophysiology and Neuroengineering of Human-Technology Interaction Lab and Computational Systems and Bioinformatics Lab, University Campus Bio-Medico of Rome, Roma, Italy
e-mail: l.vollero@unicampus.it

C. Alcaraz (ed.), *Security and Privacy Trends in the Industrial Internet of Things*, Advanced Sciences and Technologies for Security Applications,
https://doi.org/10.1007/978-3-030-12330-7_13

1 Introduction

The Industrial Internet of Things (IIoT) consists of the pervasive application of the IoT paradigm with analytics and artificial intelligence (AI) in industrial scenarios. IIoT aims at facilitating the optimization of processes and producing new value and services [1]. Industry 4.0 (I4.0) extends further the IIoT model with the inclusion of robotics and automation. The vision of I4.0 is that of industrial businesses built on top of a cyber-physical system composed of machinery, factories, and warehousing facilities [2]. Eventually, Hospital 4.0 (H4.0) is the application of the I4.0 model to the healthcare sector, transforming a pure physical environment, the classical hospital, into a cyber-physical ensemble of systems that may extend the provisioning of care services even to the patients houses. H4.0 is a revolution characterized by the exploitation and fusion of technologies such as artificial intelligence, robotics, the IoT (Internet of Things), autonomous vehicles, 3-D printing, biotechnology, materials science, nanotechnology and energy storage [3], with the aim of improving healthcare processes. H4.0 represents an abrupt change in the organization and functioning of the systems, control of assets, management of personnel, assistance of patients, administration, analysis and protection of data.

The term medical record (MR) refers to the ensemble of documents on the status of the patient and his medical history and care across time. MRs can be much more valuable to criminals than financial data. Indeed, due the rich amount of information they store, MRs can be more valuable than credit card numbers on the deep web [4]. Hospital MRs include, for instance, social security numbers, clinical information, medical diagnoses and demographic data. Fraudsters can use this information to buy medical drugs or file fictional claims with insurers, just to cite few examples of threats. MRs are housed as paper and as electronic health records (EHRs). As shown in recent studies [5], paper and film-based records comprised the vast majority of hospital data breaches (about 65%), which represent one third of all healthcare breaches. Even if network servers are the least common location of breached data, their breaches affect the greatest number of patients. Furthermore, despite some impediment from physicians that delayed the adoption of EHRs [6], the amount of data electronically stored in hospitals increases continuously, due the advantages that electronic records have in terms of management, history inspection and data analysis, and the different governative actions stimulating their adoption, e.g. the HITECH (Health Information Technology for Economic and Clinical Health) Act of 2009 [7]. This is making the protection of EHRs increasingly important over time.

Another element that makes data integrity and privacy protection even more important is the introduction of robotics and automation in hospitals, i.e. the increasing adoption in the care environment of systems that could autonomously intervene on patients and whose behavior relies on the dependability of provided data. Data driven processes in the industry, in general, and in hospitals, in particular, require special care in terms of privacy and security. Indeed, especially in the healthcare sector, a data vulnerability problem may abruptly escalate in a safety problem, primarily for the patient and secondarily for the hospital.

Blockchain has emerged today as a polyvalent technology for the management of data and, due to its distributed consensus approach which translates into a high Byzantine fault tolerance, it is particularly suitable for the recording of medical records [8, 9], for identity management [10], transaction processing [11, 12] and documents traceability [13, 14]. Born in the 1991 as a mechanism for the protection of documents timestamps [15], the modern blockchain technology was defined by Satoshi Nakamoto in 2008, when he proposed the Bitcoin and used his proposed blockchain to manage its ledger [11]. Since then, other proposals followed and, among them, those generalizing from the management of cryptocurrencies to the management of other digital assets, such as the smart contracts [12]. The digital nature of blockchain technologies with the ability to manage almost all kind of digital information in a fault-tolerant and tampering-resistant way, is making the blockchain a serious candidate in different applications related to the management of sensible information including healthcare data.

In this chapter we discuss the application of the blockchain technology in the scenario of the Hospital 4.0. After a brief discussion on the relations between IIoT, I4.0 and H4.0 and the Hospital 4.0 model, respectively, in Sects. 2 and 3, and on the role of the EHR in different healthcare applications, in Sect. 4, we discuss the blockchain technology in general, in Sect. 5, and, then, its application to the considered scenario, in Sect. 6. Section 7 ends the chapter with conclusion and future perspectives.

2 IIoT, Industry 4.0 and Hospital 4.0

The Industrial Internet of Things (IIoT) extends and adapts the IoT paradigm to industrial scenarios and processes. Under the IIoT paradigm industries incorporate networks of connected devices, each network forming a system for the generation and exchange of data. Systems for data analysis collect this data and generate valuable insights that enable industrial companies to promptly make smarter decisions. Predictive maintenance is just an example of the IIoT paradigm and involves the analysis of data generated from IIoT systems to predict defects and avoid factory or line shutdown. Other examples are assets tracking, customers and facilities monitoring which enable the optimization of resources and the improvement of customer satisfaction and employees productivity.

Industry 4.0 (I4.0) incorporates the IIoT paradigm, as well as robotics and automation, in the context of the current industrial revolution. As shown in Fig. 1, humanity has seen the advent of three globally recognized industrial revolutions, and is now facing the fourth one. In the first revolution, the production of assets was mechanized by the introduction of mechanical systems powered by steam energy. The second industrial revolution was characterized by the introduction of electricity and assembly lines, whereas in the third electronics and information technologies were brought in to automate production. The fourth industrial revolution is building on top of the third [3] and over three pillars: IIoT, automation and robotics. I4.0 is

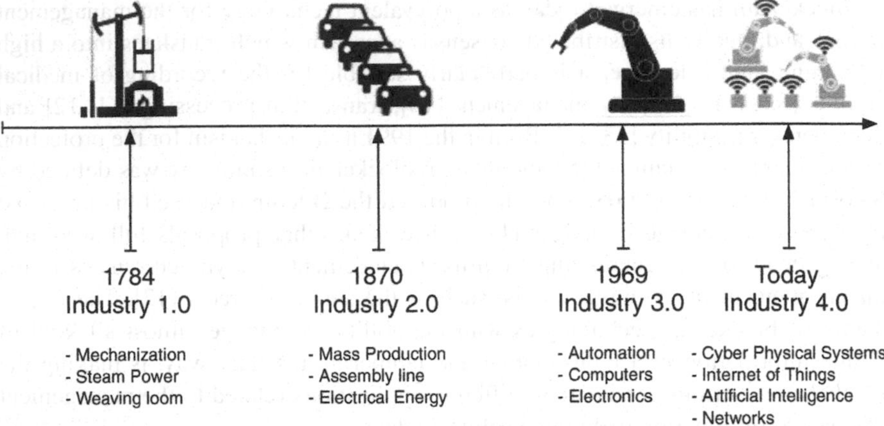

Fig. 1 Industry revolutions timeline

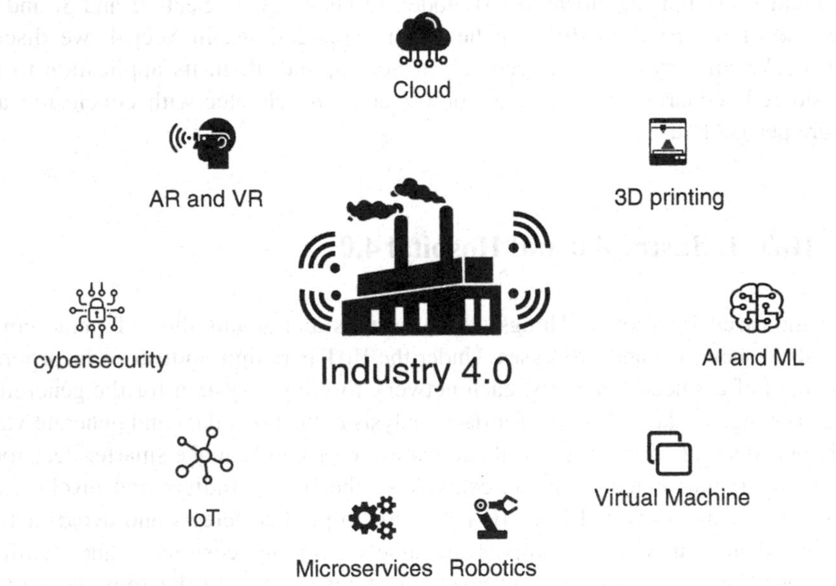

Fig. 2 The I4.0 ecosystem

characterized by its disruptiveness in terms of velocity, scope and impact. The speed of the current revolution has no historical precedent; indeed, it is evolving at an exponential pace, with a scope that is disrupting almost every industry worldwide, and impacting the entire systems of production, management and governance. As shown in Fig. 2, I4.0 benefits from several technologies that populate its ecosystems, each showing exponential transformation and providing highly impacting solution to several industrial problems. Artificial Intelligence and Machine Learning (AI and

ML), 3D printing, cloud computing, IoT and robotics are fields that are completely changing the production under I4.0.

The concept of Hospital 4.0 is an extension of the concept of I4.0 and it assumes the communication and cooperation between cyber-physical systems, Internet of Things devices, Internet services and humans in providing and consuming healthcare services [16]. As in the case of I4.0, the requirements for a true adoption of the H4.0 paradigm can be summarized in three main items:

1. widespread diffusion of (cheap) sensors, able to continuously measure the patient and the environment;
2. large communication capabilities (e.g., IoT paradigm), in order to collect such sensorial information and
3. capability to store, retrieve and elaborate large amounts of data (i.e., Big Data).

The presence of such factors, both in the infrastructure/production sectors and in the healthcare, is shifting the perspective of control systems, from passive (e.g., traditional mechanisms to monitor the hospital and the condition of the patients) to pro-active. In other terms, systems will be able to predict in advance the displacement between desirable and near-future situation and will operate in order to avoid abnormality and maintain the system under safe conditions. In this view, the ability to predict anomalies, leveraging also on the huge amount of data collected and on techniques such as artificial intelligence, will represent a valuable tool.

H4.0 represents a new hospital model, as shown Fig. 3, distributed over the territory, exploiting all emerging technologies (high efficiency miniaturized sensors, wearable devices, advanced computational tools for modeling complex systems, for the analysis of big data and for artificial intelligence, internet of things, cloud computing, automation, advanced robotics and additive manufacturing, etc.) to improve the effectiveness, efficiency, safety and sustainability of personal care services in acute and chronic conditions, but also for prevention towards a healthy life and active aging. New technologies require the design and development of

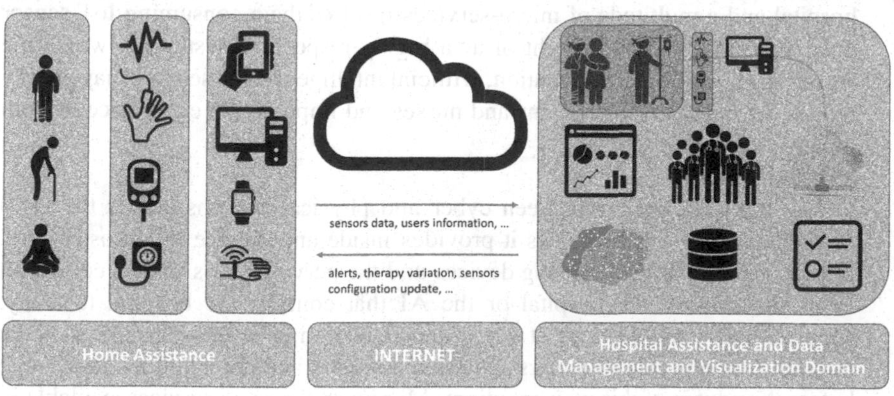

Fig. 3 Hospital 4.0 simple model

new forms of Machine-to-Machine (M2M) interactions and, above all, interactions with humans (Human-Machine Interaction – HMI). These innovative technologies will deeply influence the design and development of new processes involving healthcare professionals and patients, allowing an adequate control of the transfer of digital information and instructions to users and the real world. In this respect, the integration of these processes with hospital information systems (HIS) is essential. At the same time, it should be guaranteed that social value of personal care will be preserved and the human acts characterizing these processes will not be altered in their ethical and moral nature by the introduction of new technologies.

3 Hospital 4.0 Principles

Hospital 4.0 exploits several principles: (i) the interoperability between cyber and physical systems, (ii) the virtualization of services in the context, for instance, of eHealth and mHealth applications, (iii) a thorough and pervasive maintenance of systems by means of distributed sensors and monitoring services and (iv) the ability to spread the intelligence across the infrastructure, increasing the hospital dependability.

- **Interoperability**
  The number of services and devices currently running in a hospital infrastructure is huge and it is constantly increasing. From one side, standard services, for instance related to the management of patients and diagnosis, are already highly interoperable, whereas IoT devices, whose diffusion is more recent, lack of a real path of integration with the hospital system. Among the other reasons, this is due, in particular, to well motivated security concerns related to the complexity of an integrated infrastructure and the effort required to secure and maintain safe such an infrastructure. However, in a fully functional H4.0, we should expect a complete integration between the traditional services of the hospital and a multitude of micro-services most of them consuming IoT sensor data. The H4.0 can be thought of as a highly responsive system, in which the seamless integration of automation, artificial intelligence and sensing may greatly reduce the burden of physicians and nurses and improve the experience of both inpatients and outpatients.
- **Virtualization**
  The tight integration between cyber and physical systems allows the H4.0 to be adaptable in the services it provides inside and outside its infrastructure. Indeed, the diffusion of sensing devices and the pervasiveness of data collection allow the staff of the hospital or the AI that controls the hospital to study, simulate and adapt services at any scale. Based on a virtualized replica of the hospital it is possible to understand behaviors and attempt solutions unfeasible before the advent of this new paradigm. Moreover, the same services available in the hospital can be extended outside the infrastructure towards other federated

infrastructures or even towards the houses of the patients in the context of innovative telemedicine approaches based, for instance, on telepresence and virtual/augmented reality. Eventually, virtualization of services is a key element to improve dependability and robustness against hardware/software faults or cyber attacks.

- **Maintenance**
  The availability of pervasive sensing enables new approaches in checking the systems availability and in aiding people to maintain them. Proactive and reactive approaches could coexist in order to reduce the occurrence of service interruption and to promptly react in case of an unpredicted fault. Proactive maintenance is today a reality that allows industries, and will allow in the future hospitals, to reduce costs related to device unavailability and avoid patients trouble and distress. Virtual and augmented reality could be useful tools also in this context allowing for virtual inspection and remote control of repairing and medical robots.
- **Distributed dependability**
  The H4.0 is characterized by distributed sensing and actuation, but it can exploit its pervasiveness further, allowing its composing cyber systems to perform their tasks as autonomously as possible. The distribution of intelligence brings several advantages in terms of reliability and computational burden reduction. Even in the absence of a centralized control the infrastructure preserves a minimum level of responsiveness, whereas attacks, faults or errors have limited propagation chances. Furthermore, computation, data management and decision can be locally delegated reducing centralized requirements.

It is clear that data is the key factor for the existence and the proper provisioning of services under H4.0. Several sources of data can be identified inside a H4.0 system, including system diagnostic data, system control data, administrative data, users data. Users data, i.e. the data of the patients, is the most problematic to manage. Indeed, the management of patients data impacts on security and safety of both the patient and the hospital. The management of patients data requires the solution of a very though trade-off between privacy, controlled-access, accountability, non-repudiation and integrity.

4 EHR and Data Management Requirements

Hospital Data Management is the process of storing, protecting, and analyzing data collected within the Hospital and originating from patients, personnel and devices. The managed data may be enriched from external sources in case of direct connection with other infrastructures, such as other hospitals, regional and national health and security departments. In usual scenarios, data is manually entered by human actors or it is generated by the vast variety of devices that the Hospital hosts for the monitoring of activities and patients. Once in the system, available data is a useful historical archive of the activities in the hospital and an asset exploitable in the creation of holistic views of personnel and patients, in the definition of personalized procedures and treatments, and in enhancing procedures in order to improve office efficiency and health outcomes. Eventually, the collected data include personal information managed under privacy laws, requiring special treatment by the hospital: indeed, even the partial disclosure of such data without a proper owner authorization or without a proper control may open serious security and legal problems.

The electronic health record (EHR) is the encoded collection of patient health data in a digital format [17]. As shown in Fig. 4, EHRs may include patient's personal information like age and weight, billing information, demographics and medical history, medication and allergies, laboratory test results and radiology images. EHR can be stored, replicated and shared between hospitals and with the patient. Due to their information content, the EHR systems have to take into account [18] (i) the confidentiality, (ii) the control and monitoring, (iii) the integrity and non-repudiation and (iv) the legal value of the managed information.

- **Confidentiality**
  EHR systems require a strict control on the data in order to protect it against unwanted disclosures. Indeed patient information is private and confidential, and its access requires owner consent, as stated by laws [19]. This requirement is currently provided through encryption and well defined policies on authentication and accounting for record access. Unluckily, these approaches are prone to different attacks, from password spoofing to virus and malware attacks, to password cracking and ransomware.
- **Control and monitoring**
  EHR systems should control and track the access to the stored data. This requirement simplifies the identification of confidentiality breaches and their causes. It is a mandatory effort for the Hospital in the sense of a cyclic improvement model of their internal security and as an helping measure in case of legal actions. EHR systems implement logging functionalities that may help in this direction, however also these systems are not immune from the attacks

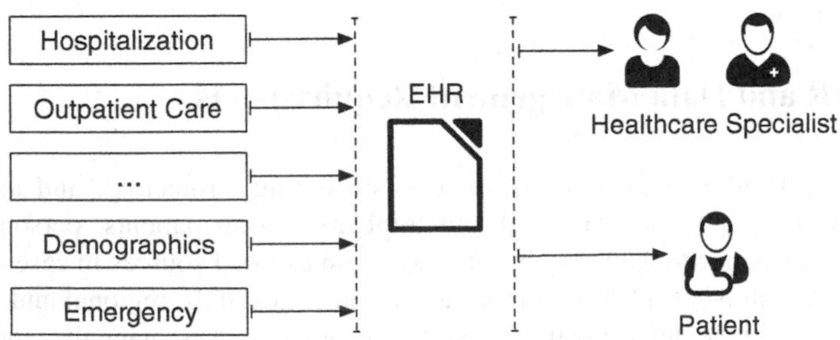

Fig. 4 EHR sources of information and consumers in pre-Hospital 4.0 scenario

cited before. Moreover, they may be also prone to DoS and DDoS attacks if the system is not properly configured.

- **Integrity and non-repudiation**
  Data integrity and non-repudiation are important aspects of EHR data management. Data integrity can be fundamental for the safety of the patient. Indeed, in case of integrity violation the status of the patient or, worse, the therapy could be altered, and this can translate in delayed or wrong treatments that, in the worst case, may bring patient to death. Non repudiation is fundamental in case of legal controversies. Integrity can be enforced through checksum, whereas non repudiation is enforced today by using conventional signing procedures, outside the reign of EHR.
- **Legal value of managed data**
  EHRs store patients data and the medical actions of the physicians. All the stored data should have legal value, i.e. in case of legal challenges, the physician must be able to use EHRs to justify his decisions and actions. This is an important aspect that an hospital has to take care of.

Current technologies employed in the implementation of EHR allows to achieve almost all the above requirements. Unluckily, most solutions are adaptation of existing technologies/products, with intrinsic and extrinsic vulnerabilities, the former coming from the specific product (relational database, authentication/authorization model, ...) and the latter due to the integration of such products. This poses the problem of experimenting and testing new data management systems and models in which the above requirements are integrated directly in the design of the considered solution.

As a final remark, most of the requirements of EHRs are shared in other data management scenarios of the I4.0 and the IIoT. For instance, products and assets traceability is a field in which requirements such as data integrity and non repudiation are fundamental, whereas sensor data may require strict confidentiality in mission critical applications. The legal value requirement is important in almost all the scenarios in which valuable digital assets are considered or the managed assets may have a direct or indirect impact that may cause legal troubles.

5 Blockchain and Data Management

The Blockchain is the technology at the basis of cryptocurrencies. The idea behind the Blockchain is that of a distributed ledger, i.e. a distributed database storing a record of transactions. The validation (non-repudiation) and the immutability (integrity) of transactions are based on asymmetric encryption and a distributed consensus algorithm. The absence of a centralized server and intermediary services ensures trust, traceability and security in systems that exchange data or assets.

As an example, let us consider the Bitcoin's mechanisms. Bitcoins are registered to addresses, also known as wallets. Bitcoins ownership is expressed in relationship to a wallet and the owner of that address is allowed to create transactions. In practice, the owner of a wallet is allowed to transfer a given amount of bitcoins, i.e. a digital asset, belonging to that wallet. A transaction is a signed assertion that a given amount of bitcoins is being transferred from the address of the owner to another address. A bitcoin transaction is checked for validity by verifying the signatures of all the transactions in the history of the transferred bitcoins. Asymmetric cryptography ensures that all above steps are secure for the owner of the wallet. To prevent double spending and guarantee full accountability, Bitcoin relies on a P2P protocol that manages a distributed timestamp service providing a fully-serialized log of all the transactions. The transactions are placed into a block and all the blocks are linked together forming the Bitcoin Blockchain as shown in Fig. 5. The Bitcoin Blockchain requires that any block of the chain includes a hash of the block that precedes it in the chain. Hence, any change in a node breaks the chain leaving behind a clear evidence, making it impossible to alter recorded transactions.

The Bitcoin Blockchain is unique and its shared representation is obtained through a distributed consensus algorithm [20]. Special nodes, the miners, collect all the transactions and compete in adding up new nodes in the chain. Node integrity is managed by means of an optimized transactions hashing based on a Merkle Tree structure [21]. The competition is based on a proof-of-work puzzle, a computation difficult to perform but whose result is easy to verify (the nonce field of Fig. 5 is the result of this puzzle). The incentive to the mining process is the possibility for the

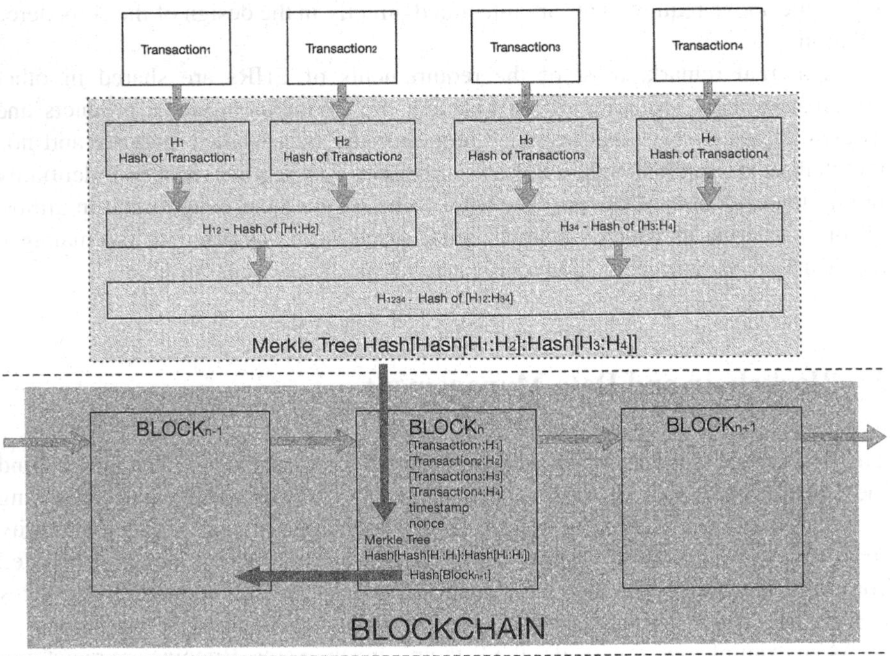

Fig. 5 The Bitcoin Blockchain model

miner to add to the new block a special transaction with a number of new Bitcoins rewarded to himself. Different miners can add blocks concurrently and a distributed consensus algorithm is implemented to solve this problem: the chain may include different temporary branches, and the one growing faster survives.

The above described Blockchain model guarantees transactions integrity and accountability, and it does not support privacy. However, several solutions include privacy also in the model, such as [22, 23]. For instance, privacy can be added with introduction of proper designed obfuscation techniques based on data encryption. The characteristics of a blockchain as Bitcoin translate in protection by design against several typologies of cyber and physical attacks. By design, a blockchain technology provides data integrity and non repudiation, moreover the public nature of the ledger allows for monitoring and control of data. Eventually, encryption and data obfuscation may be introduced to guarantee privacy and confidentiality. Furthermore, the legal value can be justified by the complexity to alter data inside a blockchain-based data system. As a final remark, moving from transactions to other typology of data is only a matter of terminology and blockchain data management design. This makes the Blockchain technology a perfect candidate to the requirements listed in Sects. 2 and 4.

6 EHR Blockchain-Based Data Management in Hospitals

It is worth mentioning that, as of 2008, less than 10% of medical records were digitally stored, while today almost all data is in the form of EHRs [24]. Yet, most of such data are non-portable and the different ways information is encoded do not guarantee interoperability. A major reason for this phenomenon lies on the commercial side; in other words, the institutions collecting EHR data are hardly interested in making such data accessible or portable, fearing that interoperability would translate in a loss of patients/clients [24]. Indeed, healthcare institution tend to perceive EHR data as their property; if this is certainly true in a legal sense [25], it generates costs and inconveniences for patients in a need to move their data to another healthcare provider [24].

On top of that, as noted in [24], several institutions, not only in the US, are timid in the adoption of more advanced data management options due to the need to fulfill the *Health Insurance Portability and Accountability Act* (HIPAA) requirements [26]. However, efforts to move in the opposite direction are rapidly gaining momentum. In particular, with the introduction of the *Medicare Access and CHIP Reauthorization Act* (MACRA) in 2015, the *Medicare EHR Incentive Program*, commonly referred to as *Meaningful use*, is requiring providers to give patients the capability to view, download and move their records to other locations [24, 27].

In this view, the blockchain technology appears a viable way to foster the diffusion and the interoperability of medical information. Yet, in the literature only a handful of solutions or frameworks have been proposed, and only a small fraction has been tested in a real world scenario. Before discussing such applications, it is worth mentioning that the Abid Hospital in Islamabad, Pakistan is the first Asian hospital to accept a cryptocurrency called PakCoin [28, 29].

Let us now survey the existing approaches in the literature for EHR sharing via blockchain. In [30], several examples are discussed. For instance, the *Gem Health Network*, a framework based on the Ethereum Blockchain technology has been proposed [31]. Another relevant example is the system implemented in Estonia in 2011 with the collaboration of Guardtime, a company that provides a blockchain-based healthcare platform [32]. As of today, several Estonian actors (e.g., citizens, healthcare providers, health insurance firms, etc.) are able to obtain information on medical treatments in Estonia using such a system [32]. In [33, 34] it is argued that, in order to avoid exchanging too large amount of information, it seems reasonable to feature only metadata about health and medical events on the blockchain, while the actual records are stored in a separate universal health cloud.

However, in the literature there is no general consent on the requirements a blockchain-based medical data management system should have. A first attempt in this sense is done in [35], where the authors identify seven criteria the system should satisfy:

1. it must be HIPAA-compliant (in particular, data should be secure and encrypted);
2. it should support Turing-completeness, i.e., it should contain programming features capable of solving any computation problem;
3. it should support user identifiability and authentication;
4. it should support structural interoperability (i.e., data can be exchanged with other systems, subject to interpretation), if not semantic interoperability (i.e., data can be exchanged and automatically interpreted by other systems);
5. scalability across large populations of healthcare participants;
6. cost-effectiveness compared to existing approaches
7. support of patient-centered care model.

However, to date, no existing solution is able to satisfy all the above requirements.

Let us now discuss existing implementations. In [36–38], the authors introduce MedRec, a blockchain-based architecture for managing EHRs; such a framework is not proprietary but consists of public APIs. A first implementation of the approach has been done at the Beth Israel Deaconess Medical Center (Harvard Medical School Teaching Hospital). One of the most interesting features of the MedRec framework is that medical stakeholders (researchers, public health authorities, etc.) that play the role of the "miners", i.e., that validate the transactions, are rewared by gaining access to aggregate or anonymized data. In [39] an approach focused on sharing EMR data for cancer patient care is provided and implemented at the Stony Brook University Hospital. In particular, in [39] two approaches are suggested: institution-based and case-specific. In the institution-based setting, the network is composed of trusted peers (e.g., medical practitioners) and collects data from several patients. In the case-specific setting, e.g., for hospitalized oncological patients, a network connecting doctors, nurses, but also family is envisaged, with the aim to minimize human-made mistakes and maximize the effectiveness of treatment. In [40] the authors develop their approach, namely BlocHIE; specifically, the authors employ two loosely-coupled Blockchains to handle different kinds of healthcare data. In [41] a system that combines blockchain and group-based secret sharing is provided. In [42] a blockchain approach is provided with a focus on privacy-preserving machine learning over medical data.

7 Conclusion

In this document we discuss the characteristics of Hospital 4.0 and Blockchain, and the application of the Blockchain technology in Hospital 4.0 for the robust management of EHRs. The characteristics of blockchain, especially the impossibility to change data without being tracked back, allow the robust management of data, guaranteeing integrity against attacks and faults. Furthermore, the distributed nature of a blockchain protects against other attacks such as ransomware. Eventually, an adoption at a national/regional level could allow portability of EHRs, facilitating and fostering safe communications among different Hospital reality.

References

1. Daugherty P, Banerjee P, Negm W, Alter AE (2015) Driving unconventional growth through the industrial internet of things. Accenture Technology, New York
2. Gilchrist A (2016) Industry 4.0: the industrial internet of things, Apress
3. Schwab K (2017) The fourth industrial revolution. Crown Business, London
4. NHS Cyber Attack (2018) Why stolen medical information is so much more valuable than financial data, June 2018
5. Gabriel MH, Noblin A, Rutherford A, Walden A, Cortelyou-Ward K (2018) Data breach locations, types, and associated characteristics among us hospitals. Am J Manage Care 24(2):78–84
6. MacIver K, Ngafeeson MN (2018) Electronic health record (EHR) diffusion and an examination of physician resistance. In: Khosrowpour M (ed) Encyclopedia of information science and technology, 4th edn. IGI Global, Hershey, pp 3678–3688
7. One Hundred Eleventh Congress of the United States of America (2009) American recovery and reinvestment act of 2009
8. Yuan B, Lin W, McDonnell C (2016) Blockchains and electronic health records. Mcdonnell. mit. edu
9. Ekblaw A, Azaria A (2016) MedRec: medical data management on the Blockchain. Viral Communications. Available at https://viral.media.mit.edu/pub/medrec
10. Ebrahimi A (2017) Identity management service using a Blockchain providing certifying transactions between devices, 1 Aug 2017. US Patent 9,722,790
11. Nakamoto S (2008) Bitcoin: a peer-to-peer electronic cash system. Available at http://bitcoin.org/bitcoin.pdf
12. Wood G (2014) Ethereum: a secure decentralised generalised transaction ledger. Ethereum Proj Yellow Paper 151:1–32

13. Benchoufi M, Porcher R, Ravaud P (2017) Blockchain protocols in clinical trials: transparency and traceability of consent. F1000Research 6:66
14. Jeppsson A, Olsson O (2017) Blockchains as a solution for traceability and transparency. Student Paper
15. Haber S, Stornetta WS (1990) How to time-stamp a digital document. In: Conference on the Theory and Application of Cryptography, pp 437–455. Springer
16. Thuemmler C, Bai C (2017) Health 4.0: application of industry 4.0 design principles in future asthma management. In: Health 4.0: how virtualization and big data are revolutionizing healthcare. Springer, Cham, pp 23–37
17. Gunter TD, Terry NP (2005) The emergence of national electronic health record architectures in the united states and Australia: models, costs, and questions. J Med Internet Res 7(1):e3
18. Wainer J, Campos CJR, Salinas MDU, Sigulem D (2008) Security requirements for a lifelong electronic health record system: an opinion. Open Med Inform J 2:160
19. Carey P (2018) Data protection: a practical guide to UK and EU law. Oxford University Press, Inc., Oxford
20. Kroll JA, Davey IC, Felten EW (2013) The economics of Bitcoin mining, or Bitcoin in the presence of adversaries. In: Proceedings of WEIS, vol 2013, p 11
21. Becker G (2008) Merkle signature schemes, Merkle trees and their cryptanalysis. Ruhr-University Bochum, Technical Report
22. Zyskind G, Nathan OZ et al (2015) Decentralizing privacy: using Blockchain to protect personal data. In: 2015 IEEE Security and Privacy Workshops (SPW), pp 180–184. IEEE
23. Kosba A, Miller A, Shi E, Wen Z, Papamanthou C (2016) Hawk: the Blockchain model of cryptography and privacy-preserving smart contracts. In: 2016 IEEE Symposium on Security and Privacy (SP), pp 839–858. IEEE
24. Ivan D (2016) Moving toward a Blockchain-based method for the secure storage of patient records. In: ONC/NIST Use of Blockchain for Healthcare and Research Workshop, Gaithersburg. ONC/NIST
25. Who owns your healthcare information? https://andersonhunterlaw.com/blog/healthcare-information
26. Hipaa's Use as Code of Silence Often Misinterprets the Law: http://www.nytimes.com/2015/07/21/health/hipaas-use-as-code-of-silence-often-misinterprets-the-law.html?_r=0
27. Meaningful Use and the Shift to the Merit-based Incentive Payment System: https://www.healthit.gov/topic/federal-incentive-programs/meaningful-use
28. Pilkington M (2017) Can Blockchain technology help promote new tourism destinations? The example of medical tourism in moldova. Available at SSRN: https://ssrn.com/abstract=2984479
29. Abid Hospital: https://www.pakcoin.io/client/abid-hospital/
30. Mettler M (2016) Blockchain technology in healthcare: the revolution starts here. In: 2016 IEEE 18th International Conference on e-Health Networking, Applications and Services (Healthcom), pp 1–3. IEEE
31. The Blockchain for Healthcare: Gem Launches Gem Health Network With Philips Blockchain Lab – https://bitcoinmagazine.com/articles/the-blockchain-for-heathcare-gem-launches-gem-health-network-with-philips-blockchain-lab-1461674938
32. Estonia is using the technology behind bitcoin to secure 1 million health records: http://www.businessinsider.com/guardtime-estonian-health-records-industrial-blockchain-bitcoin-2016-3?r=uk&ir=t
33. Linn LA, Koo MB (2016) Blockchain for health data and its potential use in health it and health care related research. In: ONC/NIST Use of Blockchain for Healthcare and Research Workshop, Gaithersburg. ONC/NIST
34. Adopting Blockchain Technology for Electronic Health Record Interoperability: https://dokumen.tips/documents/adopting-blockchain-technology-for-electronic-health-record-.html
35. Zhang P, Walker MA, White J, Schmidt DC, Lenz G (2017) Metrics for assessing Blockchain-based healthcare decentralized apps. In: 2017 IEEE 19th International Conference on e-Health Networking, Applications and Services (Healthcom), pp 1–4. IEEE

36. Ekblaw A, Azaria A, Halamka JD, Lippman A (2016) A case study for Blockchain in healthcare:"medrec" prototype for electronic health records and medical research data. In: Proceedings of IEEE Open & Big Data Conference, vol 13, p 13
37. Azaria A, Ekblaw A, Vieira T, Lippman A (2016) Medrec: using Blockchain for medical data access and permission management. In: International Conference on Open and Big Data (OBD), pp 25–30. IEEE
38. MedRec. https://medrec.media.mit.edu/technical/
39. Dubovitskaya A, Xu Z, Ryu S, Schumacher M, Wang F (2017) Secure and trustable electronic medical records sharing using blockchain. arXiv preprint arXiv: 1709.06528
40. Jiang S, Cao J, Wu H, Yang Y, Ma M, He J (2018) BlocHIE: a BLOCkchain-based platform for healthcare information exchange. In: IEEE International Conference on Smart Computing (SMARTCOMP 2018). IEEE
41. Cyran MA (2018) Blockchain as a foundation for sharing healthcare data. Blockchain in Healthcare Today. Available at https://blockchainhealthcaretoday.com/index.php/journal/article/view/13
42. Kuo T-T, Ohno-Machado L (2018) ModelChain: decentralized privacy-preserving healthcare predictive modeling framework on private Blockchain networks. arXiv preprint arXiv: 1802.01746

Design and Realization of Testbeds for Security Research in the Industrial Internet of Things

Nils Ole Tippenhauer

Abstract Research on the (cyber) security of industrial control systems requires holistic understanding of practical systems in the field. In particular, important differences to IT security scenarios are related to industrial networking protocols and programming languages such as ladder logic. Arguably, access to realistic testbeds with physical process and related controls would enable researchers to understand the scenarios better, to develop attacks, and test countermeasures. While the implementation of such testbeds presents significant investments and efforts, the implementation process itself is often not discussed in literature. In this chapter, we discuss the design and realization of such industrial control system testbeds for security research. In particular, we discuss a process in which testbeds are designed by security researchers to resemble existing (and future) plants, and are then implemented by commercial system integrators using industry's best practises. As use cases, we provide details on design decisions, cost, and outcomes for three testbeds established at the Singapore University of Technology and Design.

Keywords Testbed design · ICS security · IIoT security · Applied security

1 Introduction

The discovery of Stuxnet demonstrated to the public that advanced attacks on Industrial Control Systems (ICS) exist in practice [3, 43], and motivated academic researchers to investigate security aspects of ICS and the future Industrial Internet of Things (IIoT). While reliability and safety issues are well understood by the engineering community, the introduction of cyber-security threats and solutions in ICS and the IIoT is still challenging for several reasons. Traditional (proprietary) industrial protocols feature no security mechanisms to guarantee confidentiality,

N. O. Tippenhauer (✉)
CISPA Helmholtz Center for Information Security, Saarbrücken, Germany
e-mail: tippenhauer@cispa.saarland

C. Alcaraz (ed.), *Security and Privacy Trends in the Industrial Internet of Things*, Advanced Sciences and Technologies for Security Applications,
https://doi.org/10.1007/978-3-030-12330-7_14

authenticity, and freshness of exchanged traffic [12, 30]. Access controls for functionality of the protocols are usually non-existent, and the assumption is that anyone on the local network is trusted. The wide-spread use of legacy devices in existing ICS prevents fast introduction of protocols with modern security concepts. Legacy devices cannot be patched or upgraded without major investments, and risks to breaking working production systems, leading to major costs due to interruptions.

For that reason, academic research often focuses on passive attack detection measures to be introduced to complement existing systems [17, 26]. In that setting, the diversity of (proprietary) industrial protocols is challenging for the introduction of an Intrusion Detection System (IDS), as only limited documentation of many protocols is openly available. If active countermeasures are designed, their effects on operations in the IIoT need to be tested in a holistic manner, in real-world systems.

In addition, the actual physical process presents novel challenges and opportunities for research on attacks that leverage physical interconnections, and aim to achieve physical damage as impact [45]. For example, attacks in which sensor values are spoofed can have serious consequences on the control stability of a physical process [28], but investigations into that topic require detailed physical process models or real test systems.

The outlined challenges motivate the construction of realistic IIoT testbeds. The testbeds would contain both (scaled-down) physical processes, industrial control, and potentially the Supervisory Control And Data Acquisition Systems (SCADA) part of IIoT. Among other things, such testbeds would allow academics to (i) understand state-of-the-art industrial protocols and devices, (ii) experiment on attacks and countermeasures leveraging physical processes, and (iii) develop holistic detection and defense systems, and to test them with actual attacks.

Although such testbeds require considerable resources to set up and maintain, strong interest into IIoT security by funding agencies and the industry has resulted in construction of several by now (see a comprehensive survey in [22]). Example research results at top (computer science) security venues resulting such testbeds are [9, 14, 41]. While the implementation of such testbeds presents significant investments and efforts, the implementation process itself is often not discussed in literature.

In this chapter, we discuss the design and implementation process for IIoT security testbeds, based on experiences gained through the construction of three such testbeds at the Singapore University of Technology and Design (SUTD).[1] We provide a general discussion of our approach, together with details on the testbeds as use cases, lessons learned, and specific recommendations for future projects.

This chapter is structured as follows. In Sect. 2, we summarize the motivation to set up a testbed, and provide an introduction to components of IIoT. We present a general approach for design, specification and implementation of testbeds in Sect. 3. In Sect. 4, we discuss three testbeds at SUTD as use cases. Application of testbeds

[1]The author was employed at SUTD during most work for this chapter, and involved in design and use of all three testbeds.

for research, advantages and disadvantages, and lessons learned are discussed in Sect. 5, and related work is summarized in Sect. 6. The chapter is concluded in Sect. 7.

2 Industrial Control System Testbeds

In this section, we discuss motivations for the implementation of IIoT testbeds, design alternatives and trade-offs, and general architectures.

2.1 Motivation for Testbeds

There are three main motivations to construct a testbed for security research on Industrial Control Systems:

- **Discovery**: Discover knowledge on the subject that cannot be learned from simulations (without investing more in simulations than the real system).
- **Demonstration**: Demonstrate applicability of research findings to make them more convincing for researchers, funding agencies, and end users.
- **Education**: Leverage the testbed to educate students, researchers, stakeholders.

Components of an IIoT testbed could be real, emulated, or simulated (and a hybrid system mixing those types of components). For example, a testbed could contain only real components: Industrial devices, a real physical process, together with the SCADA system required to operate everything. Instead of a real process, high fidelity process emulators could also be used, for example Real Time Digital power Systems (RTS) in the domain of power systems [34]. Such emulation systems do real-time simulations of complex electromagnetic transients in distributed power systems, and are able to represent the simulated system state through analog or digital interfaces to other real industrial devices. If the overall system is mostly emulated (or simulated) and includes only few real devices, the setup is usually referred to as *hardware-in-the-loop*. If less precision is required, process or system stages can be simulated using domain specific software, or standard tools such as Python or Matlab. Simulations can be offline, or connected to other simulated systems in real-time. Typically, simulated systems would not exchange real network traffic with industrial protocols any longer, but use other (less overhead) messaging mechanisms between the systems components (direct network sockets, Message Queuing Telemetry Transport/MQTT).

The decision which components are required to be real, which should be emulated, and what can be simulated is certainly one of the most important decisions when designing a testbed for security research. A number of trade-offs have to be considered depending on the individual project scope, domain, and intended research:

- **Adaptability**: Cyber-Physical testbeds that contain real devices and real processes will require more effort to reconfigure and adapt to different settings. Fully simulated processes can likely be changed more quickly by updating the process topology. Real Programmable Logic Controllers (PLCs) will have to be reprogrammed with new control logic, while simulated systems might adapt automatically.
- **Fidelity**: Testbeds with real (although scaled down) processes and real devices allow to take observations that are expected to be closest to operational systems. Systems that are planned and implemented by industrial vendors are expected to resemble commercial systems better.
- **Observability**: Experimental work will likely require data collection, and potentially manipulations of physical processes and control logic. Data collection from real systems can leverage the SCADA and historian systems (if available), but introduction of additional sensors will cost money and potentially disrupt the process. The design of the system should have observability of the process in mind, and potentially allow for later addition of sensors if required. While data collection from simulations and emulations is usually easier, the data itself if expected to carry less details and surprising results.
- **Safety**: Simulated and emulated testbeds have less or no risk of damaging components, and endangering human lives. On the other hand, attack that threaten devices (or even humans) cannot be faithfully replicated in such systems.

We note that in [36], related requirements are stated (in a non-security context): Fidelity, repeatability, measurement accuracy, safe execution of tests. As noted in [22], repeatability and accuracy are likely less important in a security context (as it can be assumed that attacks/defenses need to work in a range of operating conditions). The observability we discussed is related to accuracy, but extends the concept to cover additional sensors and insights from the running system. The authors of [22] also note that a minority of existing testbeds provides arguments why they faithfully represent setups from industry.

For our testbeds at SUTD, we decided to design and implement fully functional testbeds without emulation of simulation. We made this choice to ensure high fidelity, to enable use of testbeds for other educational and training use, and in order to develop solutions that could directly be translated to industry. In terms of research goals, we planned to leverage the fully functional physical process to explore process-based attack detection and novel attacks, including physical interactions between process stages. To achieve diversity in physical processes, protocols, and vendors, we decided to build several testbeds, that could be interconnected. For the remainder of this chapter, we discuss similar setups.

2.2 Abstract IIoT Model

We start with an informal definition of IIoT testbeds. In general, cyber-physical systems contain a "cyber" part (i.e., communications, controllers, digital components), and a "physical" part (i.e., the physical process to be controlled, the actuators and sensors). An IIoT testbed can cover one or both of those areas. In the following, we refer to testbeds just covering the physical process as *physical testbeds*, testbeds that are covering only cyber components as *cyber testbeds*, and testbeds that cover both as *cyber-physical testbeds*. In this chapter, we focus on IIoT testbeds as sub-class of general CPS testbeds, but many aspects should also be generally applicable for other types of CPS testbeds.

2.3 General Structure of IIoT Testbeds

A general structure of IIoT testbeds is as follows (see Fig. 1): One (or more) physical process segments (which can be separate, or connected physically), the corresponding sensing and control devices, together with the required networking. To simplify discussion, we only refer to major components in the figure. A historian (a database to keep historical values of sensor measurements and actuator states in the IIoT) and SCADA (for supervisory control, monitoring by operators, and manual control) are responsible for overall control and monitoring. In individual process segments, local Human-Machine-Interfaces (HMI) can be found (that allow operators to directly interact with local PLCs and other components), together with the Programmable Logic Controller (PLC) that performs the distributed local real-time control. The PLCs are connected to the sensors and actuators through analog connections or field-buses.

Industrial Networking As shown in Fig. 1, the networking is often segmented into areas to accommodate logical and physical proximity, aggregated through *higher level* networks to forward the data. In the industry, the hierarchy of network segments (from local field-buses to SCADA and office networks) is often referred to as *Purdue Architecture* [44]. The lower levels carry sensor readings and direct commands to actuators using protocols such as Modbus/TCP, Ethernet/IP, and GOOSE. Physically, the protocols were traditionally spoken over bus architectures, with a trend in recent years to run legacy protocols over Ethernet and IP, e.g., with minimal changes as payload of a TCP connection (as in the case of Modbus/TCP). An in-depth introduction of related protocols can be found in [12].

The choice of industrial protocol spoken on the lower levels of the industrial control system often determines which vendors can be considered for individual components such as PLCs, as not all products support all protocols. As largest common denominator, Modbus/TCP is often chosen, as the protocol is so simple and common that it is supported by most devices. For future IIoT applications, publish/subscribe protocols such as Message Queuing Telemetry Transport/MQTT

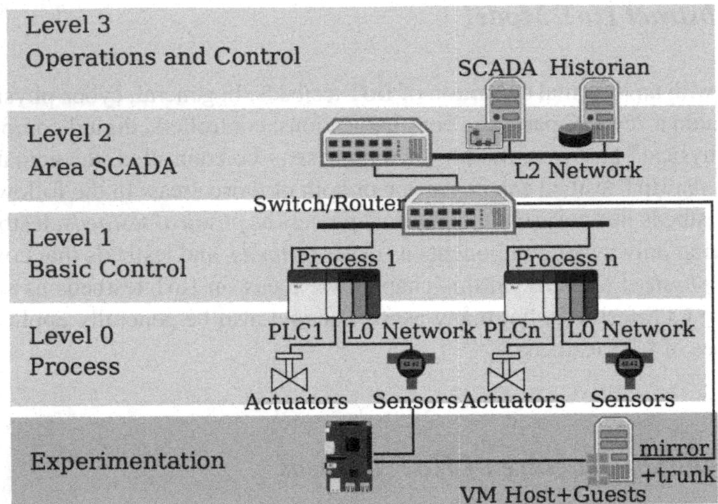

Fig. 1 General networking structure of our Industrial Control System testbeds

are also often discussed. Protocols used in the testbed will also depend on the application domain, as the type of industry and physical process (e.g., electric power) determines the standardization organization (for legacy reasons). As result, testbeds usually only use few different protocols, and cannot cover a large range. To mitigate that issue, testbeds could implement discrete sections with processes from different industries [20].

Physical Process Physical processes are diverse, and different domains such as electric power, water and oil processing, water and gas distribution, transportation and manufacturing have vastly different characteristics. In general, the control system interacts with the physical process through sensors (that observe the current state of the process), and actuators (that manipulate the physical state of the process). Actuation of the process will result in change of the state of the process, and that change will be observable through the sensors. Unlike digital systems, in which the spread of information can be precisely measured and contained, in physical processes different process segments and components can interact in a multitude of ways, from subtle changes (e.g., due to environmental temperature change) to more direct ways (e.g., exchange of product material).

Experimentation Platform While the testbed aims to replicate practical scenarios, additional components will be required to instrument the testbed (i.e., additional sensors, traffic taps), and to conduct active experiments (traffic manipulators, process manipulators). We refer to this second system as *experimentation platform* (in [20], dedicated networking for experiments is called a *management network*). The experimentation platform can be used to collect additional sensor measurement (that are not directly required by or used in the control process), obtain ground truth measurements in case of practical attacks that manipulate traffic, and to deploy countermeasures that are intended to be unreachable by an attacker. As such, the experimentation platform itself is usually assumed to be non-existent from the perspective of the attacker. In addition, the experimentation platform itself should never actively interfere or influence the normal operations of the system. We provide more details of such an experimentation platform when discussing the testbed in Sect. 4.1.

3 Design and Realization of Industrial Control Systems

In this section, we discuss the overall process to design and build an IIoT testbed, based on experience gained by constructing three large testbeds at SUTD (explained in more details in Sect. 4). In the following, we assume that the testbed is planned to be sufficiently large to contain (parts of) closed physical processes, and estimated cost will be more than 100,000 USD.

3.1 Preparations

Acquisition of Funding As the design and implementation of any testbed will require significant effort in manpower, it is likely that funding sources will have to be sought for before starting detailed planning. In particular, the overall time-frame of the testbed construction should be considered at this stage. As we will discuss later, the *time to operate* (i.e., time from starting to write a tender specification, to opening the testbed) was at least 12 months for the testbeds at SUTD, and additional time will be required to establish appropriate tool-chains to use the testbeds productively in research. As such, grants and deliverable time-lines should allow for at least 18 months of delay until results can be obtained. Ideally, construction of testbeds themselves will be seen as deliverables and milestones in the related grants. From experience, it pays to reserve larger contingency sums for costs related to the testbeds, to ensure that the overall system can be finished in face of unplanned costs.

At SUTD, funding for all three testbeds was sought before the procurement and implementation process was started, based on initial cost approximations. After funding was secured, the testbeds were then designed in more detail to work out specification documents. Using those specification documents, a public tender was called for each testbed, asking related system integrators and manufacturers to submit bids on construction of the whole system (or parts of the system).

Prerequisites In addition to funding sources and related projects, we briefly want to highlight other requirements that might be overlooked. In general, a testbed will present potential hazards to humans and assets, so such risks need to be addressed and mitigated. Sufficient permanent space has to be found for the testbed, together with infrastructure to enable continuous operations (depending on the testbed, for example power supplies, water supplies, waste water disposal, sprinkler systems, and even network connectivity). If industrial processes are replicated, it can be expected that components with heavy weight are required. In that case, it has to be ensured that the floor's load capacity will not be violated by larger components, and that leakages can be contained. Noise emanations from the testbed should not interfere with other users of nearby spaces. Heat produced in the testbed should be appropriately addressed through ventilation and air conditioning.

We also found that testbeds should be designed with sufficient free space in mind in order to allow tours for visitors, and ideally have demonstrations in mind when designing setup of HMIs, monitors, and similar. Researchers naturally want to work close to the testbed when running experiments, so ideally seating in a nearby room (isolated from heat and noise generated by the testbed) should be available. In our testbeds, we also found that large windows connecting the researcher seating and testbeds were quite helpful for visual inspections of the current state of process (e.g., during attacks, when data reported was unreliable).

3.2 Testbed Specification for Tender Process

We recommend to draft the testbed design including physical process and cyber components by a committee formed involving faculty members (who are going to be the main users) and admin staff (who will support the procurement process). Based on those initial ideas, we then suggest to hire a consultant and involve stakeholders from the industry to work out further details. In particular, we found that the physical process specification has to be quite detailed to allow for a smooth tender process. For example, a fair evaluation of the tender bids requires that all bidders will propose roughly the same system, so ambiguities in the specifications should be avoided. The tender specification should also mention specific standards and protocols to use in the industrial control, if such requirements exist from the research side. We found that the physical process specification and the specification of the main control system were relatively straight forward to discuss with the vendors. More attention had to be paid to specify requirements related to the experimentation platform, and instrumentation related to the research nature of the testbed. As an example, we requested the vendors to add wireless networking options in our testbeds, and in general segment networks more than strictly required to "simply make it work". In the end, additional work by us was required to really integrate the networking setup with our general back-end, and enable seamless manipulation and monitoring of traffic at the different levels of the network. On average, the technical specification for the tenders at SUTD were around 45 pages long, and significant effort was spent on those documents.

3.3 Tender and Construction

Tender process We assume that your local policies mandate a public tender process to ensure appropriate spending of funding (as it was the case in Singapore). As part of the tender process, the specifications are published, and bidders have a time window to submit bids (usually, several weeks). In addition, tender briefings and on-site visits will be held. After bids are submitted by interested companies, they will be reviewed by a committee, and requests for clarifications (or a second tender stage) will be asked if required. For our testbeds at SUTD, time from publishing of the tender spec to awarding the tender was 90 days (on average).

The tender specification documents outline operational requirements such target flow rate of the process, total power to be generated, safety requirements, requirements on the documentation, and similar. In addition, evaluation criteria were outlined, together with and overview of the evaluation process. For our testbeds, the technical tender specification had an average length of 45 pages.

Construction process The average time to build the testbeds (from day of awarding, to official opening after the user acceptance test) was around 9 months. During that time, manpower was required on our side to liaise with the vendors to answer questions during the implementation, and supervise the progress. At the end, a comprehensive user acceptance test was conducted in each case, in which the fulfillment of the requirements from the tender specification was verified and certified.

We also found it useful to require the vendor to provide training on the software and hardware used in the testbed for our researchers. In addition, control strategies used to program the PLCs had to be documented in a manual, together with details on the implementation such as comprehensive listing of all *tags* (i.e., variables) used in the system, comments on the protocols used, user-names and passwords, and similar. For future projects, we would recommend to add requirements on specifications of physical process parameters and relations between physical features measured (if possible). Ideally, suitable process simulation software would be available to complement the actual physical testbed, and to allow for easier simulation of the overall system.

4 Case Studies: SWaT, WADI, EPIC

In this section, we introduce the three testbeds at SUTD (see Fig. 2), and compare them on a number of dimensions relating to their design and construction, manufacturers and protocols, and outcomes. The testbeds are Secure Water Treatment (SWaT), Water Distribution (WADI), and Electric Power and Intelligent Control (EPIC). The testbeds all have fully functional (but scaled down) physical processes, controlled by state-of-the-art industrial controls (implemented by a commercial system integrator).

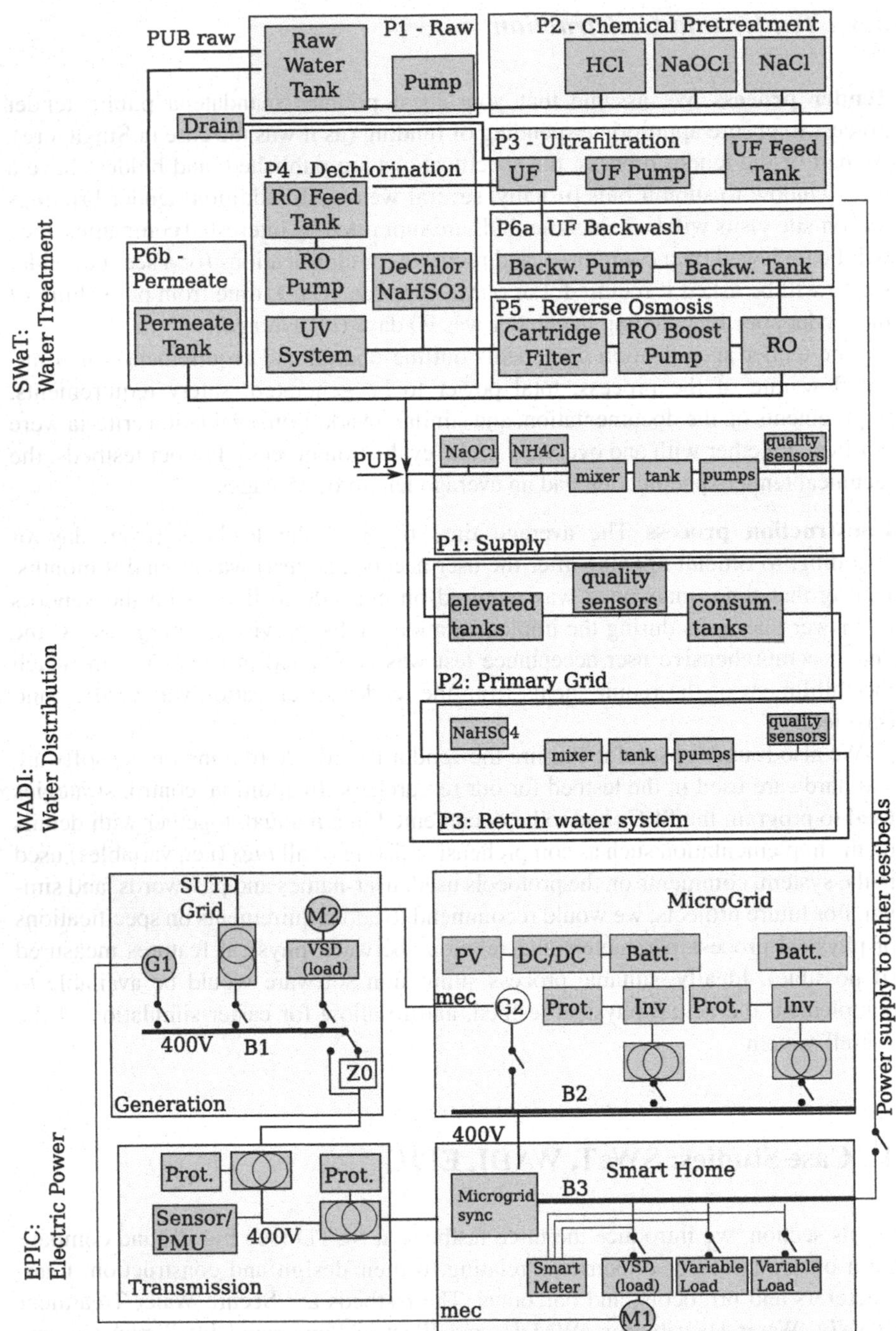

Fig. 2 Summary of physical process stages in the three use-case testbeds at SUTD. The water treatment testbed can be used to feed the water distribution testbed, and the power testbed is designed to be able to supply the other testbeds if needed for experiments on cascading effects

Table 1 Key information on design, procurement (costs approximate), and operation of SUTD testbeds

Name	Domain	Segments	Approx. cost	Opening	Time to open
SWaT	Water treatment	Chemical treatment, ultra-filtration, UV-dechlorination, reverse osmosis	750k USD	March 2015	15 months
WADI	Water distribution	Chemical treatment, gravity-fed distribution boosted distribution	750k USD	July 2016	12 months
EPIC	Electric power	Generation, transmission, micro-grid, smart home/consumption	750k USD	May 2017	21 months

Table 2 Summary of manufacturers and industrial protocols of main control devices used in the testbeds. RTU refers to Remote Terminal Unit, PLC refers to Programmable Logic Controller

Testbed	Manufacturer	Type	Model	Count	Protocols
SWaT	Allen Bradley	PLC	ControlLogix 1756	12	Ethernet/IP
WADI	National instruments	PLC	RIO-cRIO 9068	3	LOGOS (NI LabView)
WADI	Schneider electric	RTU	SCADAPack 334	2	Modbus/TCP
EPIC	Wago	PLC	750-8202	5	MMS (ISO 9506)
Skid 1	Allen Bradley	PLC	750-8202	3	Ethernet/IP
Skid 2	National instruments	PLC	cRIO 9063	1	LOGOS
Skid 2	Siemens	PLC	SIMACTIC S7-1500	1	S7
Skid 3	Schneider electric	PLC	CPU 651 50	1	Modbus/TCP

We start by summarizing their domain, the approximate cost to procure and construct, and the *time to open* (time from starting to write the specification, to officially opening the testbed) in Table 1. Outcomes from the testbeds are discussed in Sect. 5.

As each testbed covers a different physical process domain and was purchased separately, devices and their manufacturers differ between them. Table 2 summarizes main manufacturers, type, and models of the industrial devices used in the testbeds.

4.1 Secure Water Treatment Testbed

The SWaT Testbed was the first to be constructed at SUTD. Its construction was motivated by a collaboration with the Public Utility Board (PUB) in Singapore, who is operating (among other things) the water distribution and treatment processes for households and commercial customers in Singapore. The SWaT testbed was designed to resemble a modern water treatment process, which is leveraging chemical treatment and reverse osmosis membrane filters. More details on the SWaT testbed are available in [29].

Design and Procurement The physical process of SWaT is designed as distributed control of a interconnected plant, with six distinct process stages. The process stages include raw water, chemical pre-treatment, ultrafiltration, dechlorination, reverse osmosis, and backwash/permeate. As the names suggest, the physical process contains both hydraulic components (valves, pumps, tanks), and chemical treatment (using HCl, NaOCl, NaCl). The physical process design was lead by a consultant from the water treatment industry, together with few faculty members to ensure that ideal conditions for experimentation would be provided.

Preparation of the design specification for the tender was started around February 2014, with the main tender call published in August 2014. The overall technical design specification for the tender covered around 50 pages. The tender was awarded in October 2014 after several rounds of meetings with bidders and subsequent clarifications.

Experimentation platform We soon realized that practical attacks and attack detection in the fieldbus (level 0) communications requires devices in the network (in the case of SWaT, the device level ring between PLCs and the remote Input/Output Units (RIOs) and actuators). Such devices would require at least three networking interfaces: One that is connected to the *experimentation platform* (e.g., to control the devices, and aggregate data), and two interfaces to bridge the device level ring network without disturbing the normal communication flow. To minimize cost and space requirements, we are mostly using Raspberry Pi devices for that task, equipped with a wireless interface (to connect to the experimentation platform), and two wired Ethernet adapters (to bridge the device level ring). While establishment of a bridge for the ring network is straight forward with Linux networking (i.e., using *bridge-utils*), traffic manipulation is also possible, but requires additional work. As traffic manipulation usually requires parsing first, we implemented a framework using libnetfilter-queue and scapy with custom protocol support.[2]

In addition, we adapted the networking of the industrial system to allow for better monitoring of the traffic, and connection of additional devices. To achieve this, the central industrial switch (Moxa EDS-316) connecting most devices of level 1 and 2 in the network was replaced with a 24-port configurable switch with management interface (HP E3800). The switch allowed us to set up four monitoring ports, which were then directly connected to IDSs, and several servers that act as general Virtual Machine (VM) hosts. The VM hosts are 2U servers with processors that feature a larger number of logical cores, and sufficient RAM to operate around 10 VMs on each host. The VM hosts are connected through Virtual Local Area Network (VLAN) trunk connections to the switch (i.e., with attached 802.1Q headers), which

[2]See the swat-assault-crawler source at [40].

allows guests on the VM hosts to access any VLAN on the main switch. We then migrated the SCADA server, engineering workstation, and historian (all running on Windows OS) to be guests on the VM hosts. The VM host for Windows guests is also running Windows (to allow for easier management of licenses), while other research and experimentation guests are usually running on Linux, on an Ubuntu VM host. This setup allowed a very simply backup solution for important machines, ensuring that they could be restored after experimentation if required. In addition, sending the mirroring port traffic to VM hosts allowed to supply a larger number of VM guests with mirroring traffic in real-time, which would otherwise have been more challenging due to restrictions on the number of mirroring ports on the main switch.

Challenges As SWaT was our first testbed, we had to develop a range of tools to collect data from the testbed, and perform attacks. A main challenge in that context is the lack of open source tool support for the main protocol used in the testbed, Ethernet/IP, and the interface to the historian. For both, we had to implement our own software stack, before we were able to fully run experiments. In addition, reprogramming the PLCs with the industrial Studio 5000 software had to be learning through training sessions organized by the vendor. Related to this, our researchers had to be trained on the IEC 61131-3 compliant languages such as ladder logic and functional block diagrams.

Understanding and simulating the physical process was another challenge. Although we specified the physical process in detail in our tender specification, our work on attacks and detection mechanisms required detailed understanding of process dynamics. The related equations were not provided by the vendor and had to be derived by us. As the SWaT process contains physical and chemical aspects, finding an overall process model that covers all aspects was not possible. We currently model hydrological and chemical processes separately. Parameters for the simulations were derived from the process description in the manual by the vendor, and by matching models to observations.

4.2 *Water Distribution Testbed*

The WADI testbed focuses on a replication of a geographically distributed and centrally controlled water distribution network, for example of a city. The testbed features three main process stages: (i) Water storage and transmission grid, (ii) Consumers, and (iii) Water return (required in testbed to return "consumed" water to water storage). Each process stage is individually controlled by PLCs. The testbed is set up physically next to SWaT, in a room of approximately 80 m^2 size, and designed for a throughput of 10 US gallons of water per minute.

A major consideration in the design of WADI was the realistic simulation of gravity-based water pressure in the network. To achieve effects similar to the ones experienced in real systems, the water tanks in WADI are mounted at different heights, pipes with varying diameters are used, and booster pumps are available. In addition, a custom leak simulation setup was designed and implemented as part of WADI. The leak simulation allows to divert a defined percentage of water from the main distribution pipe that lowers pressure and volume of available water.

As WADI simulates geographically distributed centrally controlled system, it contains a number of remote terminal units and PLCs (NI compactRIO) that aggregate data from local sensors, and transmit that data to a central SCADA system. The transmission link for that communication can be switched between (simplistic) Ethernet communication, and 3G-based wireless communication. As such, a range of different attack and defense scenarios can be investigated. In WADI, the PLCs and RTU are connecting to most sensors and actuators directly. Where needed, Modbus/TCP is used for communications between RTUs and the SCADA, and the National Instruments proprietary Logos protocol between the PLCs and the SCADA. For further details on WADI, we refer to [2].

Procurement The WADI technical specification document was a bit shorter, due to a simpler physical process and control design. The tender was published in September 2015, and awarded in December 2015. The testbed was opened after 7 months in July 2016.

Challenges Similarly to SWaT, the PLCs (National Instrument CompactRio) and SCADA in WADI communicate with a proprietary protocol. In this case, it is what we assume to be the Logos Real-Time Protocol (related to LabVIEW), for which we did not find open source libraries that would support parsing or creation. We are currently working on a simulation model for the full physical process in the EPANET software, but face issues with matching the simulations with data from the testbed. In particular, several sensors in the actual system drift over time, which is challenging to account for when comparing data from real and simulation system.

4.3 Electric Power and Intelligent Control Testbed

The EPIC testbed consists of four process segment: Three-phase generation segment, transmission segment, micro-grid segment, and a smart home segment. Together, the generation, transmission, and smart home segments emulate a utility grid network. In addition, the micro-grid segment consists of photo-voltaic (PV) generation, transformers, battery-based storage, and protection devices. The testbed is briefly introduced in [37], with particular focus on cyber components.

Physical Process The generation part consists of three motorized generators (powered by the SUTD power grid). In particular, the SUTD grid is used to power the motors M2 and M3, which are mechanically linked to generators G2 and G3 in the micro-grid. Generator G1 is powered when M1 is powered through the smart home load bus.

In the transmission portion, a transformer is used to step down the voltage from the distribution levels to transmission levels. While in real systems, electricity is normally transmitted at high voltages in order to reduce losses, EPIC uses lower voltage for safety reasons. For slight changes in load, tap changes in the transformer adjust the voltages. The smart home segment consists of variable resistive, capacitive, and inductive loads. The loads are connected to the smart home section by a bus tie. The motor M1 is used as load. This motor can operate the generator G1, and provide power to the extra loads in the system.

Control System In general, the communication infrastructure and control system in EPIC is similar to that in WADI. In particular, it represents a spatially distributed but centrally controlled system. The individual process stages each have one PLC or RTU connected to the central SCADA system. The PLC in the generation process controls the different ways to generate power for the system, and collects data from the sensors and fault detection components. The PLC in the smart home section is used to control the variable loads from the SCADA. In addition, the smart home has several smart meters connected to the loads, which report the usage to the Advanced Metering Infrastructure (AMI).

Procurement The design specification for EPIC had approximately 45 pages, and was published as part of the tender call in September 2015. The winning bid of the tender was selected and awarded in January 2016. Construction of the testbed took about 16 months, in particular due to delays related to the more complex process (and safety considerations).

4.4 Training Skids

In addition to the three testbeds with full physical process, we also purchased several *training skids*. The training skids are usually employed in industry for staff training on PLC and RTU programming, and consist of 1–3 PLCs from different product lines of the same manufacturer, together with IO components, and dummy sensors and actuators. The dummy sensors are not connected to actual physical processes, but can be manipulated by users easily. For example, a thermometer is measuring environmental temperature, and its measured value can easily be changed by touching it. Other sensors are simple switches, that can be actuated by the user and provide binary signals to the attached PLCs. Similarly, the dummy actuators can be used to give feedback to the user, e.g., by lighting up lamps or activating a simple motor.

The training skids allow the researchers to interact with the devices without fearing to break the overall process, and to become familiar with the respective programming environments. In addition, we found that the skids provide good variety of the devices for traffic capture and vulnerability testing. Overall, while the price of a single training skid is significantly less than a full testbed (around 100k USD), we also found that their usefulness for most of our direct experimental research work is limited. As such, they are a good platform to familiarize new staff with the environment, but do not offer the functionality that more complete testbeds provide.

5 Discussion

We now discuss how the introduced testbeds were used for research, options for commercialization, data collection and sharing, remote access and collaboration, and other lessons learned.

5.1 Testbed Applications for Research

Since their opening, the testbeds have been leveraged in a significant number of publications authored by researchers from SUTD (to date, more than 50 conference and workshop papers, and around 5 journal publications). A number of quite productive research directions developed out of the testbeds, for example the experimental investigation of attacks, and collection of data samples during normal and attack scenarios, together with related attack detection schemes.

A semi-public dataset with traces of normal operations and several attacks in the SWaT testbed was collected in 2016, and made available on request online. To date, access to the dataset has been requested by (and granted to) more than 115 international researchers and teams. Details on the dataset can be found on the website[3] and in the companion publication [18].

For the other two testbeds (WADI and EPIC), similar datasets and companion papers are planned for the future. We feel that the existence of such common dataset will benefit the community and enable collective work towards better attack detection mechanisms with comparable performance evaluations. While similar datasets can be generated by use of physical process simulators (e.g., to host attack detector competitions [39]), datasets from real testbeds will have more realistic noise, artifacts, and the possibility to combine network and process data.

Attack Traces Research on security of IIoT faces the challenge that very little information on actual attacks are publicly available. In particular, no details network traffic captures or historian traces of actual attacks are published (to the best of our knowledge). While a number of attacks are prominently cited in academic works (e.g., Stuxnet [43], the Maroochi shire attack [38], the Aurora Attack [45]), those attacks are only described on a certain level of abstraction. As result, there is no common attack to help design or evaluate proposed countermeasures.

[3] Available online at [25].

IIoT testbeds can help with collection of traces of attacks to enable comparison of countermeasures, and drive the design process of protection mechanisms. In particular, there are four options to generate such traces: (i) Attacks on the physical process in which the attacker has control over HMI or SCADA can be conducted by simply overriding normal plant control manually on the HMI or SCADA, (ii) Attacks in which the attacker manipulates traffic in real-time can be conducted by using suitable devices that are placed as man-in-the-middle (e.g., as part of the experimentation platform discussed earlier), (iii) (trusted) Third parties could be asked to perform (unknown) attacks while traffic and historian data is recorded, and (iv) the testbed could be exposed to untrusted third parties (similar to a honeypot). Out of those options, we used (i)–(iii) to generate attack data from our testbeds, further details can be found in publications such as [1] (for (i)), [42] (for (ii)), and [5] (for (iii)). Option (iv) (attacks by untrusted third parties/honeypot) was not considered so far, as the risk of physically damaging the plant was considered too high, the likelihood of successful attacks by third parties was considered too low, and recovery of the system was estimated as too challenging.

Overall, we believe that generating such datasets is one of the key advantages of similar testbeds, and a promising way to provide contributions to the community. To the best of our knowledge, only few other such datasets have so far been made available. A collection of power system, gas and water system and energy management system data was made available by Oak Ridge National Laboratories (ORNL).[4]

Security Competitions The SWaT testbed was also used as centerpiece for two *SWaT Security Showdown* (S3) competitions held at SUTD in 2016 and 2017. In those competitions, 5–6 international teams of students, academic researchers, and industrial professionals were invited to develop and demonstrate attacks on the SWaT testbed. Teams were invited directly (in 2016), or had to qualify (in 2017). During an on-line stage teams had to demonstrate basic capabilities in IIoT security in a Capture-The-Flag (CTF)-like event. Afterwards, successful teams were invited to Singapore, and given around one day of time in the lab to prepare, and then had to demonstrate their attacks. Attacks were scored based on a formula that awarded impact and control over the physical process, while at the same time using weaker attacker models (i.e., only access over the network). More details on the event in 2016 can be found in [5], and related reports on the iTrust website [24].

Countermeasures We used the testbeds to test various countermeasures we developed, among them solutions integrated in PLCs, traffic monitoring on the fieldbus [42], and SCADA systems [16]. In addition, we were able to host a number of commercial platforms in the testbeds for extended periods of time, which (among other things) enabled potential government customers to become more familiar with the products. It also enabled the system vendors to demonstrate their capability in customizing the product to specific settings, and their support for proprietary

[4]Available online, at [31].

protocols. Most countermeasures (academic and commercial) were also evaluated as part of the S3 competitions (see above).

5.2 *Commercialization*

As discussed, testbeds represent significant investments in terms of money and effort. In addition, continued operations require funding for maintenance and replacement of consumables, and salaries for operational and admin staff. Testbeds can also be expected to operate over a longer time-frame, as the technology itself will likely still be relevant 10 years later (given that industrial technology in the field can have much longer lifetime).

In order to make such testbeds self-sustainable, either long-term funding through grants has to be acquired (ideally, 10+ years), multiple sequential grants have to be acquired (which is susceptible to unexpected delays in new grants, or failure to secure funding), or other funding sources have to be found. In particular, charging for access to the testbeds (for training or research purposes) for third parties promises to mitigate the funding problem. To the best of our knowledge, this practice is not yet widely established (a trial is currently running at SUTD). While charging will likely bring in additional funding, it will also require a professionalized management of resources and technical support, likely moreso than purely academic research environments usually provide.

5.3 *Data Collection and Sharing*

One of the advantages of real-world IIoT testbeds is the opportunity to collect traffic and process data over long time frames, under different operating conditions, and including artifacts such as the industrial protocol headers, temporal features, and noise. Collection of a dataset that unites traffic and historian process data, ideally with detailed labels and context information, would provide a very valuable source for applications such as machine-learning enabled attack detection, and verification of simulation and emulation environments.

Unfortunately, we found it challenging to collect such datasets in the SUTD testbeds for various reasons. Testbeds initially did not have appropriate means to capture relevant network traffic, extraction of process data from the historian required use of industrial software (e.g., manual queries for each individual tag, use of Windows-only development kits with libraries), and integration of both types of data together with labels requires suitable frameworks to be developed. While partial dataset have been semi-publicly release by SUTD (essentially on request), those datasets cover only historian data or network traffic, and not both. Labeling of the datasets is also only coarse, as periods are labeled as under attack, or normal traffic.

For future testbeds, we would recommend to leverage a central data collection and labeling platform, which combines data such as network traffic from various taps, historian data, information such as log-files produced, and enables to easily label the samples (perhaps semi-automatically based on process state data). This central platform could be used by all researchers to obtain and share data, and minimize engineering effort related to data acquisition.

5.4 Remote Access and Federation

While the IIoT network part of the network will likely not be directly connected to the Internet (not even indirectly through a NAT), other network segments will likely have Internet access, e.g., for researchers. In our testbeds, we use open source network applications such as pfSense [33] to provide NATing and firewalling. The central pfSense host is directly connected to our Internet uplink on a public IP.

Remote access to testbeds is often asked for by collaborators at SUTD. In general, remote access to the experimentation platform and IIoT network is not difficult to achieve. We use VPN server features of the pfSense application to allow remote users to connect with a VPN client, which will typically give them network-layer access to a special VLAN which also contains selected virtual machine guests. The VM guests are then used by our external collaborators to collect traffic or interact with selected other network segments. Based on virtual networking setup of the host the VM guests can have any number of virtual interfaces, in any of our VLANs. External collaborators can also connect to our HMI and SCADA system if allowed in our configuration, and can thus directly influence the physical process. Due to the related safety risks, we require on-site collaborators to be present for such experiments.

The setup would technically also allow to create link-layer bridges over VPN with remote sites, to integrate remote testbeds and SCADA with our testbeds. So far, the engineering effort, security and safety concerns have prevented us from setting up such systems in practice.

5.5 Lessons Learned

We conclude the discussion now with a quick summary of lessons learned from the process of building and operating our testbeds.

Network specification vs. implementation We found that it was relatively challenging to ensure that the final network design and implementation followed our specification closely enough to allow for easy experimentation and research. In all three testbeds, the original networking designs had to be revised several times during implementation. Most likely this is influenced by the fact that protocols used and data exchanged are influenced by control logic and devices used, and in normal industrial settings the vendor has more freedom to implement the network as deemed suitable. We recommend to at least stay in touch with the vendor on network design decisions to ensure that the resulting platform will be suitable for research.

Staff We found that permanent engineering staff is essential to support the testbeds. The staff does not only need to take care of maintenance of the process, but also acts as support for the researchers during their work. The engineer can oversee physical experiments to ensure no damage is done unintentionally, and support the development of tools to interact with the system. Of course, this job profile asks for a person with industrial engineering and programming experience, who is often hard to find. The networking setup and virtualization environments also benefit from a permanent staff member who is familiar with the setup. We have one permanent engineer position for each testbed. In addition, several admin staff members help with admin side of testbeds and collaborations.

Size and Complexity of Physical Process We found that the sheer size and complexity of the first testbed (SWaT)—while allowing for interesting experimental setups—made it hard to fully model the process in order to theoretically analyze it. Ideally, the testbed needs a mix of small closed control loops and process segments, and some larger and more complex ones. In practice, we found that in many of our publications, we used simple process stages such as the raw water tank (a tank with a pump, level sensor, several valves and flow meters). The simplicity of those process segments allowed us to more easily model the system, and spend less time explaining the setup in the publications. After more experience on systems and solutions was gained, larger process segments could be used.

6 Related Work

A number of general surveys on ICS security provide a good overview of challenges and current work [30, 46]. In [46], the authors review ICS intrusion detection and prevention systems. In [30], the authors review the general cybersecurity challenges and state of the art in defense in ICS networks. In particular, the authors also discuss advantages of hardware-in-the-loop testbeds. Work on ICS security is often focusing on power systems due to seminal works in that area [28]. An example for work in other domains is [8], in which is railway ICS security is discussed.

Testbeds A number of surveys provide an overview of ICS testbeds in different application domains and research fields [3, 17, 22]. While a full review of related testbeds is out of scope, we would like to mention a number of relevant ICS testbeds used for security research [6, 13, 20, 32]. In [32], the authors present a SCADA Security Laboratory and Power and Energy Research laboratory, which features several physical process segments with a RTU unit in each, and a central HMI to control the segments. The physical process segments are more diverse than the ones covered in the testbeds at SUTD, but are much smaller and less complex. In [21], the PowerCyber testbed at Iowa State University is introduced, which features RTS simulation of the physical process, and emulation of WAN communication. Full virtualization of testbeds is discussed in [35], with PLC and RTU components simulated in Python, and communicating over virtual network interfaces. The resulting testbed is also able to communicate with physical industrial devices. In [15], another framework is presented, which combines real-time simulation of physical processes (using Simulink), and emulation of cyber-components using Emulab. In [13], the authors discuss hybrid ICS testbeds for security research with special focus on emulation. In [10], an approach based on network emulation, process simulation, and SDN is presented in the context of power systems. A framework leveraging virtualized network emulation, device simulation using Linux guests, and an abstract physical process interface was proposed in [4]. In [19], the authors experimentally explore attacks on a vinyl acetate monomer plant by leveraging a fully simulated chemical process (the Tennessee Eastman challenge Process [11]). The process has also been made available as open source.[5]

Automotive ICS In this work, we mostly discussed testbeds related to utilities such as water and power, and plants such as they are used in manufacturing. A related research area is the one of automotive security. In that context, the threatened system is itself a commercially available product, which means that it is possible to just buy or rent the system *off the shelf*. Several high impact research works have done this, e.g., [7, 27]. To the best of our knowledge, rental of equipment for other ICS research has not been used otherwise, but could also be a promising avenue for systems that are relatively self-contained.

7 Conclusion

In this chapter, we provided an overview of motivations, design considerations, and alternatives for IIoT testbeds for security research. We focus on cyber-physical testbeds that combine a scaled-down physical process with real industrial devices such as PLCs, SCADA, sensors, and actuators. The design and implementation of such testbeds requires considerable resources and effort, and we provide and overview of the overall process, estimates on documentation and time required (based on our experiences in constructing three such testbeds).

Testbeds are long-term investments and will not only be used for single research projects. To show options for different usage scenarios, we summarize how our testbeds have been used so far, and what our plans for future use is. We also

[5]Available online at [23].

comment on general academic results so far, and point to promising directions to increase impact (e.g., through release of more extensively prepared datasets).

Overall, we conclude that the availability of cyber-physical testbeds provides unique opportunities for applied research, education, and outreach activities. Results and insights from such testbeds cannot be obtained from simulation and emulation alone. Given the required investments, suitable long-term funding and man-power is required to fully utilize those opportunities.

Acknowledgements Many people were involved at SUTD in the design, procurement, and operation of the testbeds. Most importantly, Aditya Mathur was responsible for the initial vision and overall leadership of the project. Ivan Lee, Mark Goh, and Angie Ng from iTrust additionally supported the process from the admin side. Writing of the SWaT process requirement documents was supported by Giedre Sabaliauskaite, while the WADI process was designed together with Stefano Galelli. Gareth Thomas acted as consultant on both water testbeds. The three testbeds were funded by grants awarded by the National Research Foundation of Singapore, the Singaporean Ministry of Defence, and the International Design Center at SUTD.

References

1. Adepu S, Mathur A (2016) Distributed detection of single-stage multipoint cyber attacks in a water treatment plant. In: Proceedings of the ACM ASIA Conference on Computer and Communications Security (ASIACCS). ACM, pp 449–460
2. Ahmed CM, Palleti VR, Mathur A (2017) WADI: a water distribution testbed for research in the design of secure cyber physical systems. In: Proceedings of the Workshop on Cyber-Physical Systems for Smart Water Networks (CySWATER), Apr 2017. ACM, pp 25–28
3. Antón SD, Fraunholz D, Lipps C, Pohl F, Zimmermann M, Schotten HD (2017) Two decades of scada exploitation: a brief history. In: Proceedings of the IEEE Conference on Application, Information and Network Security (AINS), Nov 2017, pp 98–104
4. Antonioli D, Tippenhauer NO (2015) MiniCPS: a toolkit for security research on CPS networks. In: Proceedings of Workshop on Cyber-Physical Systems Security & Privacy (SPC-CPS), co-located with CCS, Oct 2015
5. Antonioli D, Ghaeini HR, Adepu S, Ochoa M, Tippenhauer NO (2017) Gamifying ICS security training and research: design, implementation, and results of S3. In: Proceedings of Workshop on Cyber-Physical Systems Security & Privacy (SPC-CPS), co-located with CCS, Nov 2017
6. Candell R, Stouffer K, Anand D (2014) A cybersecurity testbed for industrial control systems. In: Process Control and Safety Symposium, International Society of Automation, Houston
7. Checkoway S, McCoy D, Kantor B, Anderson D, Shacham H, Savage S, Koscher K, Czeskis A, Roesner F, Kohno T (2011) Comprehensive experimental analyses of automotive attack surfaces. In: Proceedings of the USENIX Security Symposium
8. Chen B, Schmittner C, Ma Z, Temple WG, Dong X, Jones DL, Sanders WH (2015) Security analysis of urban railway systems: the need for a cyber-physical perspective. In: Koornneef F, van Gulijk C (eds) Computer safety, reliability, and security. Springer, Cham, pp 277–290
9. Chen Y, Poskitt CM, Sun J (2018) Learning from mutants: using code mutation to learn and monitor invariants of a cyber-physical system. arXiv preprint arXiv:1801.00903
10. Dong X, Lin H, Tan R, Iyer RK, Kalbarczyk Z (2015) Software-defined networking for smart grid resilience: opportunities and challenges. In: Proceedings of the 1st ACM Workshop on Cyber-Physical System Security. ACM, pp 61–68
11. Downs JJ, Vogel EF (1993) A plant-wide industrial process control problem. Comput Chem Eng 17(3):245–255

12. Galloway B, Hancke GP et al (2013) Introduction to industrial control networks. IEEE Commun Surv Tutor 15(2):860–880
13. Gao H, Peng Y, Jia K, Dai Z, Wang T (2013) The design of ICS testbed based on emulation, physical, and simulation (EPS-ICS testbed). In: Proceedings of the Conference on Intelligent Information Hiding and Multimedia Signal Processing, pp 420–423, Oct 2013
14. Garcia L, Brasser F, Cintuglu MH, Sadeghi A-R, Mohammed O, Zonouz SA (2017) Hey, my malware knows physics! attacking PLCs with physical model aware rootkit. In: Proceedings of the Annual Network & Distributed System Security Symposium (NDSS), Feb 2017
15. Genge B, Siaterlis C, Fovino IN, Masera M (2012) A cyber-physical experimentation environment for the security analysis of networked industrial control systems. Comput Electr Eng 38(5):1146–1161
16. Ghaeini H, Tippenhauer NO (2016) HAMIDS: hierarchical monitoring intrusion detection system for industrial control systems. In: Proceedings of Workshop on Cyber-Physical Systems Security & Privacy (SPC-CPS), Co-located with CCS, Oct 2016
17. Giraldo J, Sarkar E, Cardenas AA, Maniatakos M, Kantarcioglu M (2017) Security and privacy in cyber-physical systems: a survey of surveys. IEEE Design Test 34(4):7–17
18. Goh J, Adepu S, Junejo KN, Mathur A (2016) A dataset to support research in the design of secure water treatment systems. In: International Conference on Critical Information Infrastructures Security (CRITIS). Springer, pp 88–99
19. Gollmann D, Gurikov P, Isakov A, Krotofil M, Larsen J, Winnicki A (2015) Cyber-physical systems security: experimental analysis of a vinyl acetate monomer plant. In: Proceedings of the ACM Workshop on Cyber-Physical System Security. ACM, pp 1–12
20. Green B, Lee A, Antrobus R, Roedig U, Hutchison D, Rashid A (2017) Pains, gains and PLCs: ten lessons from building an industrial control systems testbed for security research. In: Proceedings of the USENIX Workshop on Cyber Security Experimentation and Test (CSET). USENIX Association
21. Hahn A, Ashok A, Sridhar S, Govindarasu M (2013) Cyber-physical security testbeds: architecture, application, and evaluation for smart grid. IEEE Trans Smart Grid 4(2):847–855
22. Holm H, Karresand M, Vidström A, Westring E (2015) A survey of industrial control system testbeds. In: Proceedings of Nordic Conference (NordSec), pp 11–26
23. Isakov A. Damn vulnerable chemical process – Tennessee eastman. https://github.com/satejnik/DVCP-TE
24. iTrust. Centre for research in cyber security. https://itrust.sutd.edu.sg/
25. iTrust. ICS testbed datasets. https://itrust.sutd.edu.sg/research/dataset/
26. Jardine W, Frey S, Green B, Rashid A (2016) SENAMI: selective non-invasive active monitoring for ICS intrusion detection. In: Proceedings of the ACM Workshop on Cyber-Physical Systems Security and Privacy (CPS-SPC), New York. ACM, pp 23–34
27. Koscher K, Czeskis A, Roesner F, Patel S, Kohno T, Checkoway S, McCoy D, Kantor B, Anderson D, Shacham H et al (2010) Experimental security analysis of a modern automobile. In: 2010 IEEE Symposium on Security and Privacy (SP). IEEE, pp 447–462
28. Liu Y, Ning P, Reiter MK (2011) False data injection attacks against state estimation in electric power grids. ACM Trans Inf Syst Secur (TISSEC) 14(1):13
29. Mathur A, Tippenhauer NO (2016) SWaT: a water treatment testbed for research and training on ICS security. In: Proceedings of Workshop on Cyber-Physical Systems for Smart Water Networks (CySWater), Apr 2016
30. McLaughlin S, Konstantinou C, Wang X, Davi L, Sadeghi AR, Maniatakos M, Karri R (2016) The cybersecurity landscape in industrial control systems. Proc IEEE 104(5):1039–1057
31. Morris TH. Industrial control system (ICS) cyber attack datasets. https://sites.google.com/a/uah.edu/tommy-morris-uah/ics-data-sets
32. Morris T, Srivastava A, Reaves B, Gao W, Pavurapu K, Reddi R (2011) A control system testbed to validate critical infrastructure protection concepts. Int J Crit Infrastruct Prot 4(2):88–103
33. pfSense. open source firewall. https://www.pfsense.org/
34. Real time digital simulator. www.rtds.com/index/index.html

35. Reaves B, Morris T (2012) An open virtual testbed for industrial control system security research. Int J Inf Secur 11(4):215–229
36. Siaterlis C, Garcia AP, Genge B (2013) On the use of emulab testbeds for scientifically rigorous experiments. IEEE Commun Surv Tutor 15(2):929–942
37. Siddiqi A, Tippenhauer NO, Mashima D, Chen B (2018) On practical threat scenario testing in an electric power ICS testbed. In: Proceedings of the Cyber-Physical System Security Workshop (CPSS), Co-located with ASIACCS, June 2018
38. Slay J, Miller M (2007) Lessons learned from the maroochy water breach. Springer, Boston
39. Taormina R, Galelli S, Tippenhauer NO, Salomons E, Ostfeld A, Eliades DG, Aghashahi M, Sundararajan R, Pourahmadi M, Banks MK, Brentan BM, Campbell E, Lima G, Manzi D, Ayala-Cabrera D, Herrera M, Montalvo I, Izquierdo J, Luvizotto E Jr, Chandy SE, Rasekh A, Barker ZA, Campbell B, Shafiee ME, Giacomoni M, Gatsis N, Taha A, Abokifa AA, Haddad K, Lo CS, Biswas P, Pasha M, Fayzul K, Kc B, Somasundaram SL, Housh M, Ohar Z (2018) The battle of the attack detection algorithms: disclosing cyber attacks on water distribution networks. J Water Res Plann Manag 144(8):04018048
40. Urbina D. The swat assault crawler. https://github.com/scy-phy/swat
41. Urbina D, Giraldo J, Cardenas AA, Tippenhauer NO, Valente J, Faisal M, Ruths J, Candell R, Sandberg H (2016) Limiting the impact of stealthy attacks on industrial control systems. In: Proceedings of the ACM Conference on Computer and Communications Security (CCS), Oct 2016
42. Urbina D, Giraldo J, Tippenhauer NO, Cárdenas A (2016) Attacking fieldbus communications in ICS: applications to the SWaT testbed. In: Proceedings of Singapore Cyber Security Conference (SG-CRC), Jan 2016
43. Weinberger S (2011) Computer security: is this the start of cyberwarfare? Nature 174:142–145
44. Williams TJ (1992) The Purdue enterprise reference architecture, a technical guide for CIM planning and implementation. Isa, Research Triangle Park
45. Zeller M (2011) Myth or reality—does the aurora vulnerability pose a risk to my generator? In: Proceedings of Conference for Protective Relay Engineers. IEEE, pp 130–136
46. Zhu B, Sastry S (2010) SCADA-specific intrusion detection/prevention systems: a survey and taxonomy. In: Proceedings of the Workshop on Secure Control Systems (SCS), vol 11

Printed in the United States
By Bookmasters